# Statistical Quality Control for the Food Industry

*Third Edition*

# Statistical Quality Control for the Food Industry

## Third Edition

## Merton R. Hubbard
*Consultant, Hillsborough, California*

**Kluwer Academic / Plenum Publishers**
New York, Boston, Dordrecht, London, Moscow

**Library of Congress Cataloging-in-Publication Data**

Hubbard, Merton R.
    Statistical quality control for the food industry/Merton R. Hubbard—3rd ed.
      p.   cm.
    Includes bibliographical references and index.
    ISBN 0-306-47728-9
     1.  Food industry and trade—Quality control.   I. Title.

   TP372.6.H83  2003
   664′.07—dc21

                                                2003047710

ISBN 0-306-47728-9

Copyright © 2003 by Kluwer Academic/Plenum Publishers, New York
233 Spring Street, New York, New York 10013

http://www.wkap.nl/

10  9  8  7  6  5  4  3  2  1

A C.I.P. record for this book is available from the Library of Congress.

Permissions for books published in Europe: *permissions@wkap.nl*
Permissions for books published in the United States of America: *permissions@wkap.com*

Printed in the United States of America

# Contents

*Preface to the third edition*                                                    ix

*Preface to the second edition*                                                   xi

*Preface to the first edition*                                                   xiii

*Acknowledgments*                                                                 xv

**1  Introduction**                                                                1

    Variability                                                 2
    Quality Control Programs                                     3
    Problems with Tool Selection                                 8
    Quality Control Tools                                        8

**2  Food Quality System**                                                        15

    The Formalized Quality System                               15
    Quality System Guidelines                                   16
    Malcolm Baldridge National Quality Award                    27
    Total Quality Management                                    28
    Team Quality Systems                                        30
    Computer Network Quality Systems                            30
    Summary                                                     30

**3  Control Charts**                                                             49

    The Importance of Charting                                  49
    Procedure for Constructing $X$-Bar and $R$ Charts           53
    Procedures for Constructing Attribute Charts                57

## 4  Fundamentals                                          71

Analysis of Data                                             71
Probability                                                  76
Binomial Distribution                                        78
The Normal Distribution                                      82
Distribution of Sample Means                                 84
Normal Approximation to the Binomial Distribution            90
t-Distribution                                               92
Confidence Limits for the Population Mean                    93
Statistical Hypotheses—Testing Hypotheses                    95
Distribution of the Difference Between Means                100
Paired Observations                                         103
F-Distribution                                              104
Analysis of Variance                                        105
Two Criteria of Classification                              111

## 5  Sampling                                             115

Sampling Plans                                              115
Why Sample?                                                 116
Samples from Different Distributions                        117
Sample Size                                                 118
How to Take Samples                                         123
Types of Samples                                            128
Sampling Plans                                              131
Types of Inspection                                         131
Classes of Defects                                          132
Sampling Risks                                              135
Selection of Population to be Sampled                       136
Selection of Sample Frequency and Location                 137
Hazard Analysis Critical Control Point                      138
Attribute Sampling Plans                                    149

## 6  Test Methods                                         151

General Analysis                                            153
Special Instrumentation                                     153
Microbiology                                                153
Sensory                                                     153

## 7  Product Specifications                               157

## 8  Product Capability                                   163

Capability Index                                            170
Benchmarking                                                173

## 9   Process Control                                        177

Chart Patterns                         179
Using the Control Chart as a Quality Management Tool     184

## 10   Sensory Testing                                 187

The Senses                           188
Sensory Testing Methods           189
Types of Panels             194
Selection and Training          197

## 11   Net Content Control                             201

Evaluation of Net Content Performance     205
Interpreting Net Content Control     205
Procedures for Setting Fill Targets     213

## 12   Design of Experiments                         219

Introduction           219
Elimination of Extraneous Variables     222
Handling many Factors Simultaneously     226
Full Factorial Designs     227
Fractional Factorial Designs     232
Response Surface Designs     236
Mixture Designs     239
Experimental Design Analysis by Control Chart     248

## 13   Vendor Quality Assurance                 253

Vendor–Vendee Relations     255
Specifications for Raw Materials, Ingredients, Supplies     257
Quality Assurance of Purchased Goods     259
Selecting and Nurturing a Supplier     263
Packaging Supplier Quality Assurance     266
Supplier Certification Programs     271

## 14   Implementing a Quality Control Program     275

Management Commitment     275
Getting Started     276
An In-House Program     277
Team Quality Systems     279
Stepwise Procedures for Team Problem Solving     282
Programs without Management Support     284
Training Quality Control Technicians     287
Summary     288

**15   The Computer and Process Control**                              **289**

Computer Integrated Management                                           289
Artificial Intelligence and Expert Systems                              291
Computer-controlled Processing                                          294
Summary                                                                 307

**16   Six-Sigma**                                                     **309**

Summary                                                                 313

*Appendix*                                                             *315*

*References*                                                           *335*

*Index*                                                               *339*

# Preface to the Third Edition

Since the second edition of *Statistical Quality Control for the Food Industry* was printed, the statistics involved in the quality control of food has not changed. Sigma is still sigma; the mean remains the mean. There have been some significant changes however in philosophies, particularly in the areas of quality management and food quality standards.

The Baldridge National Quality Program has moved another step away from the goal of product quality control by emphasizing business excellence as the major criteria for the Baldridge Award.

As the U.S. imports moved from one foreign country to another, the changing quality of imported manufactured goods in addition to the cost of foreign manufacture has substantially affected the U.S. national debt.

The major changes in ISO 9000 have resulted in two major concerns: (1) Do the currently certified processors have to be recertified? and (2) Since the ISO9000:2000 differs in so many areas, does that mean that the quality control procedures of the last five years were incorrect?

The success of many companies in meeting quality standards using the HACCP principles has been recognized by the FDA. As a direct result, the FDA is increasing the number of food products which must be produced using the HACCP principles. It should be noted that the FDA regulations are concerned with food safety, rather than food quality, and this is reflected in the new regulations. The need for statistical quality control principles are still required to meet a producer's needs for other critical food characteristics not included in the HACCP regulations (flavor, color, etc).

Considerable publicity for the six-sigma quality control system has suggested that conventional statistical quality control procedures are outmoded. This might be true in hardware manufacturing industries where warranties, returns, and repairs are part of the system, but certainly not in the food industries. However, there are some parts of the six-sigma approach which might be of value to the food industry as well.

The Net Content Control regulations have been modified somewhat, but the statistical approach to compliance remains essentially unchanged.

All of the above changes in the food industry quality control procedures are discussed in this third edition.

# Preface to the Second Edition

Within the six years since the first edition was published, ISO 9000, HACCP, Expert Systems, six-sigma, proprietary vendor certification programs, sophisticated team techniques, downsizing, new electronic and biochemical laboratory methods, benchmarking, computer-integrated management, and other techniques, standards, and procedures descended upon the quality control managers in the food industry with the impact of a series of tornados. Everything changed; it was time to rewrite *Statistical Quality Control for the Food Industry*.

Or so it seemed. But, as it turns out, everything has *not* changed. The concepts of variability, sampling, and probability are still the same. The seven basic tools of statistical quality control still work. Control charts still supply the information to control the process—although now the computer is doing most of the calculations and graph construction faster, and in color.

On close examination, even some of the major developments are not really all that new. For example, ISO 9000 closely resembles *Food Processing Industry Quality System Guidelines* published in 1986, and some other quality systems. The powerful Hazard Analysis Critical Control Point technique has also been around for some time, and many food companies have been using selected portions of it voluntarily. Now, however, it has become part of the food laws and has suddenly received widespread publicity.

There *have* been some real changes, however. The power of the computer has been applied to several phases of the food industry: Management has found that some computer applications can reduce the need for manpower. Other computers have been harnessed to processes to receive electrical information, analyze the input, and instantly send adjustment signals back—thus improving the process by reducing variability. Some have been used to instantly provide expert process advice to the operator. Still others have been used to extract data from process computers, and to analyze, calculate, and produce graphs, charts, and reports for product and process improvement studies for immediate use by all interested departments.

Considering the ability of food processing companies to consistently manufacture safe foods with uniform quality over the past 20 or 30 years *without* these new tools and new systems, one might expect that quality control improvements would be marginal. On the other hand, these changes have already provided substantial opportunities for process and product improvement. This second edition is intended to update the basic concepts and discuss some of the new ones.

# Preface to the First Edition

If an automobile tire leaks, or an electric light switch fails, or we are short-changed at a department store, or are erroneously billed for phone calls not made, or a plane departure is delayed due to a mechanical failure—these are fairly ordinary annoyances, which our culture has come to accept as normal occurrences.

Contrast this with a failure of a food product. If foreign matter is found in a food, if the product is discolored or crushed or causes illness or discomfort when eaten, the consumer reacts with anger, fear, and sometimes mass hysteria. The offending product is often returned to the seller, or a disgruntled letter is written to the manufacturer, or at worst, an expensive law suit may be filed against the company. The reaction is almost as severe if the failure is a difficult-to-open package or a leaking container. There is no tolerance for failure of food products.

Dozens of books on quality written for the hardware or service industries discuss failure rates, product reliability, serviceability, maintainability, warranty, and repairs. Manufacturers in the food industry do not use these measurements since food reliability must be 100%, failure rate must be 0%; serviceability, maintainability, warranty, and repairs are meaningless.

Consequently, this book on food quality does not concern itself with reliability and safety. It is assumed that manufacturers in the food industry recognize the intolerance of their customers and the rigid requirement of producing 100% safe and reliable product. Those few food processors who experienced off-flavor, foreign material, salmonella, botulism, or other serious defects in their products rarely survive.

What the book does cover are the various techniques which assure the safe production of uniform foods. All of the subjects covered are specifically tied to food industry applications. The chapter on fundamentals of statistics is made palatable by the use of examples taken directly from companies processing fruits, wine, nuts, and frozen foods. Many other food product examples are used to illustrate the procedures for generating control charts.

By now, most upper managers are aware that process control is a technique which long ago supplanted the "inspect and sort" concept of quality control. This book is intended to present upper managers with an understanding of what the technique includes. It is also targeted at the quality engineers, managers, and technicians who have been unable to find workable explanations for some of those quality techniques specifically used by the food industry. A new audience for this subject includes all of the departments in companies, embracing the concept of total quality control. Here is a collection of quality techniques that accounting, procurement, distribution, production, marketing, and purchasing can apply to their departments. Finally, the book is aimed at students hoping to enter the field of food quality control, and technicians who are aspiring to management positions in quality control.

Guidelines for overall quality control systems and suggestions for implementing a quality control program are discussed from a generic point of view. All of the other subjects are very specific "how to" discussions. For example, an entire chapter is devoted to a step-by-step procedure for controlling the net quantity of packaged foods. It explains how to obtain data, interpret government weight regulation, calculate both the legal and economic performance, and set target weights. For the most part, the calculations have been reduced to simple arithmetic.

Where possible, each chapter subject has been designed to stand alone. As an example, the chapter on process control explains how charts are interpreted and what actions should be taken. While reading this chapter on process control, it is not necessary to thumb through the pages to consult the Appendix tables or the chapter on methods for preparing control charts. Similarly, the design of experiments section uses some of the concepts introduced earlier, but does not require the reader to review the chapter on fundamentals. The subject of experimental design is complex, but the book reduces it to straightforward explanation and provides food processing examples, as well as a series of diagrams of the most useful designs.

The bibliography contains most of the common texts on statistical process control. In addition, the chapter on test methods provides a list of references, which have food industry applications. The Appendix tables include only those referred to in this book.

The author has attempted to avoid theories and generalities in order to make this book as practical and useful as possible. In the immense field of food processing, it is remarkable how little specific quality control information has been available. It is hoped that this book will fill that gap.

# Acknowledgments

The need for a book of this scope became apparent during the annual presentations of Statistical Process Control Courses for the Food Industry, sponsored by the University of California, Davis, California. I am primarily grateful to Robert C. Pearl, who spearheaded these quality control courses since the early 1960s, and to Jim Lapsley, his successor, for their continuing support during this book's development.

Many University staff and quality professionals have contributed to the preparation and instruction of these courses, and I must give special thanks to the following for permission to include portions of their unpublished notes in various sections of the book:

| | |
|---|---|
| Professor Edward Roessler | Wendell Kerr |
| Dr. Alan P. Fenech | Chip Kloos |
| Sidney Daniel | Ralph Leporiere |
| Ronnie L. De La Cruz | Jon Libby |
| Seth Goldsmith | Donald L. Paul |
| Randy Hamlin | Sidney Pearce |
| Gilbert F. Hilleary | Floyd E. Weymouth |
| Mary W. Kamm | Tom White |

Thanks to the Longman Group for permission to reprint Table XXXIII from the book *Statistical Tables for Biological, Agricultural and Medical Research*, 6th Edition, 1974, by Fisher and Yates.

Thanks are also due to my wife Elaine for her professional help as my editor, and for her encouragement and patience over the long haul.

# 1    Introduction

In 2002, the United States balance of trade with East Asia was negative $171,593,000. The prices were not necessarily lower than for merchandise produced in the United States, but the quality level and uniformity were excellent—a far cry from the shoddy reputation the Orient suffered throughout the first half of the 20th century. This has raised fears that the Orient would ultimately take over the production of all of our products, and that the United States has already turned into a service-industry nation. Statisticians have been known to generate analyses that are mathematically correct, but which occasionally are open to question if the data are presented out of context. The 171 billion dollars may fall into that category. There is no question that the 171 billion dollars represents a sizeable quantity of goods; but the yearly U.S. Gross National Product (GNP) in the early 2000s was 10 trillion dollars. This means that imports from East Asia accounted for only 1.7% of our GNP. Less than 2% does not seem to be a dangerously high level—certainly not high enough to suggest that we are in danger of having all of our products manufactured elsewhere.

Government statisticians have replaced GNP with Gross Domestic Product (GDP), but there is only a small difference between these figures. Perhaps the 1.7% figure might be overly pessimistic, because trade imbalances have built-in correcting devices. For example, during periods when domestic consumption slows, imports will slow as well. U.S. exports during the end of the century were actually growing at a 25% annual rate, and trade deficits with foreign countries had peaked. This improvement was masked by the fluctuating value of the dollar against foreign currencies. (Merrill Lynch Global Investment Strategy, 21 March 1995; also the Japan Business Information Center; Keizai Koho Center.)

The quality control and the quality level in the United States are not necessarily inferior, as implied by the cold numbers. Perhaps the impression that our manufacturing industry is about to be taken over by the Orient is due to their selection of some highly visible consumer products. They have done an excellent job of it in photography, optics, electronics and the auto industry. But even in these areas,

1

the United States still maintains a significant number of profitable operations with notable market share.

Food production and processing in the United States is an area of outstanding quality, unmatched by the Orient. The most obvious example of food quality control is the safety of foods in the United States. There are 290 million people in this country who eat a total of about 870 million meals a day, or 318 billion meals annually. A benchmark study made by the Center for Disease Control analyzed the numbers and causes for food outbreaks across the country for an entire year. They found 460 reported outbreaks of food poisoning, in which an outbreak was defined as two or more people becoming ill from the same food eaten at the same time. The 460 figure represents the number of people who were reported by Public Health Departments, doctors, and hospitals to have become ill from foods during the year, but does not include those who became ill and who went to their own physician for treatment or waited without assistance until they recovered. Of course, such data is unavailable. Working only with the proven data, the 460 figure, expressed as percent product failure, indicates

$$(460 \times 100)/318,000,000,000 = 0.000000147\%$$

It is difficult to find another country which has achieved that kind of record for any product in any industry.

If the record for food quality is so superb, why this book? Because the need to improve quality control is unending. The safety appears adequate, but there are always improvements possible in net weight control, product color, flavor, keeping qualities, production cost, absence of defects, productivity, etc. There still is little factual quality information provided at the graduate level of business schools, and much of the information available to the management of food processing companies is supplied by the professionals in quality control who are likely to read this book. Perhaps management may find the time to read it as well.

## VARIABILITY

We live in a world of variability. The person who first used the expression "like two peas in a pod" was not looking very carefully. There are no two people exactly alike—even so-called twins. Astronomers tell us that in this vast universe, there are no two planets alike. Two man-made products which are "within specifications" may seem to be the same, but on closer inspection, we find that they are not identical.

It is generally known that perfection is not possible. You know it, your friends know it, children know it; but the Chief Operating Officer of many companies does not admit to it. He says that there is no reason why all products in a properly maintained production line—with adequately trained and motivated workers, the right raw materials, expert supervisors, and quality control employees who know what they are doing—cannot be perfectly uniform, with no defects, and with no variation. While we must accept the fact that variability does exist, there are methods to control it within bounds which will satisfy even the Chief Operating Officer. You will find that:

- Statistical tools are available
- Processes can be controlled
- Line people are not necessarily responsible for poor quality
- Management, and only management can improve quality

## QUALITY CONTROL PROGRAMS

The Shewhart control-chart technique was developed in 1924, and has been in use continuously since then. Perhaps the only fundamental change in the Shewhart chart was the simplification evolved by mathematical statisticians by which control charts could be simply determined using the range of observations, rather than the more time-consuming calculations for standard deviations for each subgroup. Evaluation of the statistical approach of Shewhart was published in 1939 by W.E. Deming, who later (1944) defined "constant-cause systems, stability, and distribution" in simple terms to show how a control chart determined when a process was in statistical control. After over 50 years, these principles are still valid, and are the basis for most of the successful quality control programs in use today.

One of the philosophies attributed to Dr. Deming is that judgment and the eyeball are most always wrong. $X$-bar, $R$ and $p$ charts are the only evidence that a process is in control. Failure to use statistical methods to discover which type of cause (common or system cause; and special or assignable cause) is responsible for a production problem generally leads to chaos; whereas statistical methods, properly used, direct the efforts of all concerned toward productivity and quality.

Dr. Deming has stated that 85% of the causes of quality problems are faults of the system (common causes) which will remain with the system until they are reduced by management. Only 15% of the causes are special or assignable causes specific to an individual machine or worker, and are readily detectable from statistical signals. Confusion between common and assignable causes leads to frustration at all levels, and results in greater variability and higher costs—the exact opposite of what is needed. Without the use of statistical techniques, the natural reaction to an accident, a high reject rate, or production stoppage is to blame the operators.

The worker is powerless to act on a common cause. The worker has no authority to sharpen the definition and tests that determine what constitutes acceptable quality. He cannot do much about test equipment or machines which are out of order. He cannot change the policy or specifications for procurement of incoming materials, nor is he responsible for design of the product.

Several quality control leaders have each developed a formalized program consisting of several steps. It is difficult to look at a summary of these steps to determine which system is best for a given company, since the programs must be tailored to each particular situation. Note how even these recognized authorities disagree on certain measures. A summary of the steps suggested by these quality control authorities follows. They are not complete descriptions, but serve to differentiate the emphasis of these programs.

## J.M. Juran

1. Establish quality policies, guides to managerial action.
2. Establish quality goals.
3. Design quality plans to reach these goals.
4. Assign responsibility for implementing the plans.
5. Provide the necessary resources.
6. Review progress against goal.
7. Evaluate manager performance against quality goal.

## W.E. Deming (*Quality* Magazine Anniversary Issue 1987)

1. Create constancy of purpose toward improvement of product and services.
2. Adopt the new philosophy: we are in a new economic age.
3. Cease dependence on mass inspection as a way to achieve quality.
4. End the practice of awarding business on the basis of price tag.
5. Constantly and forever improve the system of production and service; the system includes the people.
6. Institute training on the job.
7. Provide leadership to help people and machines do a better job.
8. Drive out fear.
9. Break down barriers between departments.
10. Eliminate slogans and targets for zero defects and new productivity levels.
11. Eliminate work standards and management by objectives.
12. Remove barriers that rob people of their right to pride of workmanship.
13. Institute a vigorous program of education and self-improvement.
14. Put everybody in the company to work to accomplish the transformation.

## Armand V. Fiegenbaum

1. Agree on business decision at the boardroom level to make quality leadership a strategic company goal and back it up with the necessary budgets, systems, and actions. Each key manager must personally assess performance, carry out corrective measures, and systematically maintain improvements.
2. Create a systemic structure of quality management and technology. This makes quality leadership policies effective and integrates agreed-upon quality disciplines throughout the organization.
3. Establish the continuing quality habit. Today's programs seek continually improving quality levels.

## Tom Peters

1. Abiding management commitment.
2. Wholesale empowerment of people.
3. Involvement of all functions—and allies of the firm.
4. Encompassing systems.
5. Attention to customer perceptions more than technical specifications.

## P.B. Crosby (*Quality is Free* by P.B. Crosby)

1. Management commitment.
2. Quality improvement team.
3. Quality measurement.
4. Cost of quality evaluation.
5. Quality awareness.
6. Corrective action.
7. Establish an Ad Hoc committee for the Zero Defects Program.
8. Supervisor training.
9. Zero defects day.
10. Goal setting.
11. Error cause removal.
12. Recognition.
13. Quality councils.
14. Do it all over again.

## M.R. Hubbard (N. Cal. Institute of Food Technologists October 1987)

1. Select an area within the company in need of assistance, using Pareto or political procedures.
2. Using statistical techniques, calculate process capability and determine control limits.
3. Establish sampling locations, frequency, size, and methods of testing and reporting. Study process outliers and determine assignable causes of variation at each process step.
4. Correct assignable causes by modifying the process, and calculate improved process capability. Report to management dollars saved, improved productivity, reduced scrap, rework, spoilage, product giveaway, overtime, etc.
5. Repeat steps 3 and 4 until no further improvements are apparent.
6. Design experiments to modify the process to further improve productivity, and follow by returning to step 2, using the most promising test results.
7. Move on to another area of the company (another line, another function, another department) until the entire company is actively pursuing quality programs.

8. Where possible, install quality attribute acceptance sampling plans as a safeguard for quality in the process and in the finished product. Expand this into a company-wide audit system.

## Total Quality Management Practices (General Accounting Office 1991)

1. Customer-defined quality.
2. Senior management quality leadership.
3. Employee involvement and empowerment.
4. Employee training in quality awareness and quality skills.
5. Open corporate culture.
6. Fact-based quality decision-making.
7. Partnership with suppliers.

## Hazard Analysis Critical Control Point (Department of Health and Human Services—FDA,1994–2002)

1. Identify food safety hazards.
2. Identify critical control points where hazards are likely to occur.
3. Identify critical limits for each hazard.
4. Establish monitoring procedure.
5. Establish corrective actions.
6. Establish effective record keeping procedures.
7. Establish verifying audit procedures.

## Computer Integrated Management (approximately 1987)

Computer integrated management (CIM) is a system designed to control all phases of a food process by the use of computers. The goal is to utilize computer power in product design, engineering, purchasing, raw material control, production scheduling, maintenance, manufacturing, quality control, inventory control, warehousing and distribution, cost accounting, and finance. By integrating the databases and commands of the individual computer systems throughout all of these stages, it should be possible to improve the efficiency of production planning and control, decrease costs of each operation, and improve both process control and product quality. The goal is to optimize the entire system through the use of computerized information sharing.

## ISO 9000 Standards (International Organization for Standardization revised 2000)

1. ISO 9000. Quality Management and Quality Assurance Standards— Guidelines for Selection and Use.

2. ISO 9001. Quality Systems—Model for Quality Assurance in Design/
   Development, Production, Installation, and Servicing.
3. ISO 9004. Quality Management and Quality System Elements—Guidelines.

## Malcolm Baldridge National Quality Award (U.S. Department of Commerce 1987, and revised annually)

The seven categories on which these quality awards are based have been revised over the years. Three years have been selected at random as examples.

| 1990 Examination categories | 1995 Examination categories |
| --- | --- |
| 1. Leadership | Leadership |
| 2. Information and analysis | Information and analysis |
| 3. Strategic quality planning | Strategic planning |
| 4. Human resources utilization | Human resource development and management |
| 5. Quality assurance of products and services | Process management |
| 6. Quality results | Business results |
| 7. Customer satisfaction | Customer focus and satisfaction |

The 2002 emphasis shifted further from product quality control toward business performance excellence. The Examination categories for 2002 were:

- Leadership
- Strategic planning
- Customer and market focus
- Information and analysis
- Human resource focus
- Process management
- Business results.

Although these business goals are admirable, they have failed to emphasize the goal of quality control.

## Six-Sigma (Motorola 1979) (see Chapter 16)

- Recognize
- Define
- Measure
- Analyze
- Improve
- Control
- Standardize
- Integrate.

The Six-Sigma process is often called the DMAIC system, referring to the steps 2–6 above. The system is explained in detail in Chapter 16.

## Other Quality Programs

Since the 1980s, several additional techniques have been offered with the goal of improving quality control programs. For the most part, they are business process tools rather than statistical quality control techniques, but have often had favorable effects on both productivity and product quality. Some of the many examples: Total Quality Management, Teams, Reengineering, Benchmarking, Empowerment, Continuous Improvement, Quality Function Deployment, Computer Applications, Six-Sigma, Computer Controlled Processes, Computer Analyses of Data, Computerized SPC, Real Time Manufacturing, Expert Systems, etc.

# PROBLEMS WITH TOOL SELECTION

The difficulty in selecting the correct statistical tools for problem solving might be explained using the analogy of selecting the correct tools for a painting project. Assume that we wish to paint a rod. We are faced with the following choices:

   5 finishes: flat, satin flat, satin, semigloss, gloss
   3 solubilities: oils, water, organic solvents
   5 types: lacquers, enamels, stains, primers, fillers
   4 spreaders: air and pressure spray, roll, brush, gel brush

Although all of these possibilities are not necessarily combinable, there is a possibility of at least

$$5 \times 3 \times 5 \times 4 = 300 \text{ combinations of kinds of paints and applicators}$$

This does not include all paint types, or the myriad of shades available. Nor does it consider the formulations for floor, ceiling, deck, wall, furniture, concrete, metal, glass, antifouling, etc. Nor does it include subclasses of metals: iron, galvanized, copper, aluminum, etc. Knowing which tools are available may not solve our painting problem. Any paint will cover wood, concrete, or some metals, but will it peel, blister, fade, discolor, weather, corrode, flake, or stain? Will it leave brush marks, or a slippery surface? Is it toxic? Does it have an unpleasant odor? Will it resist a second coat? In short, knowing which tools are available is necessary; and knowing the proper use of these tools is absolutely imperative.

# QUALITY CONTROL TOOLS

The following is a list of the more common statistical tools used in quality control applications. These will be covered in greater detail later. These tools have specific applications in industry, and care should be taken to select the proper ones. Computers provide a convenient and speedy method of converting large volumes of data into charts and summaries with uncanny accuracy, but the use of the incorrect program or the selection of an incorrect tool in a program can lead to confusion. Table 1-1 outlines some of the most common applications of statistical techniques for quality control in the food industry.

| Statistical tool | Use |
| --- | --- |
| Acceptance sampling plans | Evaluate product attribute quality |
| Analysis of variance | Establish significance of difference between two sets of data |
| Cause and effect chart | Display sources of problems |
| Confidence limits | Determine reliability of sample results |
| Check sheet | Locates major problems |
| Control charts | Early detection of process variability |
| Cusum chart | Cumulative subgroup difference plot |
| Design of experiments | Provide valid data with minimum test |
| Evolutionary operation | Short cut response surface testing |
| Flow chart | Defines process |
| Histogram | Process frequency distribution |
| Pareto chart | Display frequency of problem areas |
| Probability distributions | Summarize large groups of data |
| Process capability | Level of yield uniformity |
| Regression analysis | Determine mathematical relationships between two sets of variables |
| Sample size probabilities | Select sample size |
| Scatter diagram | Shows relationship of variables |
| Statistical inference | Significance of data difference |
| Taguchi method | Specification and tolerance technique |
| Trend chart | Systematic drift analysis |

Seven basic tools have been used successfully in food quality control programs for decades, and in all likelihood will remain as the foundation for future quality needs in the industry. Over many years, there has been general agreement (see *Quality Progress*, June–December 1990) that these seven tools should be in every quality control program. They are discussed in some detail later on. The following is a list, a brief explanation, and a simplified example of each.

- Flow chart
- Cause and effect diagram

**Table 1-1.   Use of SQC Techniques in the Food Industry**

**Statistical techniques**

A.   Experimental design:   Factorial, ANOVA, regression, EVOP, Taguchi
B.   Control charts:   X-bar, R, p, np, c
C.   Acceptance sampling:   Attributes MIL STD 105E, variables MIL STD 414
D.   Diagrams:   Pareto, cause and effect
E.   Special sampling:   Skiplot, cusum, scatter diagram, flow chart, histogram,
      check sheet
F.   Special charts:   Sequential, trend analysis. Consumer complaints

| Applications | Use technique |
| --- | --- |
| Product design | A |
|    Specifications: Product, ingredient, packaging | |
|    Sensory evaluation | |
|    Function, shelf life | |
| Process design | A,B |
|    Process specification, process capability | |
| Vendor | B,C |
|    Selection, qualification, control | |
| Incoming quality | B,C |
|    Materials, supplies | |
| Process specification conformance | B,C,E,F |
|    Sort, clarify, wash, heat, filter, cool, mill, other | |
|    Package integrity, code, fill, appearance | |
|    Microbiology | |
| Product specification conformance | B,C |
|    Sensory—color, flavor, odor | |
|    Structure, function | |
| Storage and distribution | B,C |
| Consumer complaints | B,D |
| Audit | A,B,C |
|    Laboratory control | |
|    Process, product, field performance | |
| Product, process improvement | A,D,F |

- Control chart (variable and attribute)
- Histogram
- Check sheet
- Pareto chart
- Scatter diagram.

## 1. Flow Chart

A picture of a process, using engineering symbols, pictures, or block diagrams, which indicates the main steps of a process (Figure 1-1).

## 2. Cause and Effect Diagram

A pictorial representation of the main inputs to a process, problem or goal, with detailed sub-features attached to each of the main inputs (Figure 1-2). (Also referred to as Ishikawa or fishbone diagrams.)

## 3. Control Chart (Variable and Attribute)

A graph of a process characteristic plotted in sequence, which includes the calculated process mean and statistical control limits (Figure 1-3).

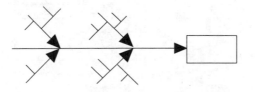

**Figure 1-1.**   Flow chart.

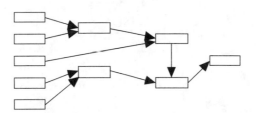

**Figure 1-2.**   Cause and effect diagram.

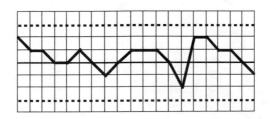

**Figure 1-3.**   Control chart.

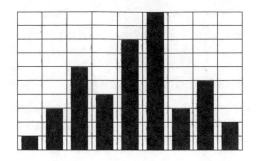

**Figure 1-4.** Histogram.

**Figure 1-5.** Check sheet.

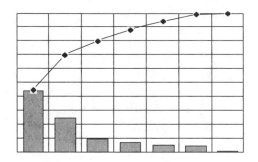
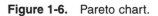

**Figure 1-6.** Pareto chart.

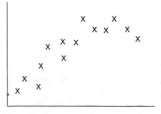

**Figure 1-7.** Scatter diagram.

## 4. Histogram

A diagram of the frequency distribution of a set of data observed in a process (Figure 1-4). The data are not plotted in sequence, but are placed in the appropriate cells (or intervals) to construct a bar chart.

## 5. Check Sheet

Generally in the form of a data sheet, used to display how often specific problems occur (Figure 1-5).

## 6. Pareto Charts

A bar chart illustrating causes of defects, arranged in decreasing order. Superimposed is a line chart indicating the cumulative percentages of these defects (Figure 1-6).

## 7. Scatter Diagrams

A collection of sets of data which attempt to relate a potential cause (X-axis) with an effect (Y-axis) (Figure 1-7). Data are collected in pairs at random.

# 2    Food Quality System

## THE FORMALIZED QUALITY SYSTEM

As a company grows, the need for formal departmental operating procedures and reports generally produces a large volume of standard manuals. The Quality Department in a food manufacturing company may be the last one to assemble a written system. There are perhaps more excuses than reasons for this:

- The products change from year to year, and someone would have to be retained on the Quality Department payroll just to keep up with the paperwork. (Unrealistic!)
- The food industry is regulated by federal agencies (Food and Drug, Commerce, Department of Agriculture, and others) and by state and local agencies (Weights and Measures, Public Health, and others). Therefore, there is no need to further formalize quality procedures. (Untrue!)
- A food processing company could not remain in business unless its quality systems were adequate. It might be risky to change the existing system. (Head in the sand!)
- It is necessary to remain flexible in the food business so that the company can take advantage of new developments quickly. A formalized system tends to slow things down. (Absence of a system may bring things to a standstill!)

These excuses might be applied equally to other departments within a food company (accounting, personnel or distribution), but companies generally have rigorous formalized procedures for these departments. The Quality Department has frequently been overlooked in this respect, partly because it is a relatively new discipline. The "Food Processing Industry Quality System Guidelines" was prepared for the American Society for Quality Control in 1986. Prior to that time, the Society, the professional organization dedicated to promotion of quality control in industry, had no recommendations specifically for the food industry.

Perhaps a more common reason for the lack of system emphasis of quality is that the techniques of statistical quality control are not well understood by upper management. Until the 1980s, few colleges or universities offered degrees in statistical quality control. In fact, few even offered classes in this subject. As a direct result, quality control was far too often mistakenly considered to be concerned with inspection, sorting, sanitation management, and monitoring the retorting process for low acid canned foods.

Quality control in the food industry, under these conditions, quite naturally was regarded as a cost center which contributed to overhead, rather than as a potential profit center which contributed to savings. From a series of successes at home and abroad in quality control—quality improvement, process improvement, productivity improvement, reduction in cost of scrap, rework, and product giveaway—the reputation of properly organized and operated quality control departments has gradually changed from a "cost generator" to a "profit center."

A quality system which starts at the product concept and development stage has the greatest opportunity to reduce new product costs. The cost to remedy design failures is minimized when these shortcomings are detected at the concept or prototype stages of development. The cost to remedy failures rises exponentially at the later stages (pilot plant, production run, market distribution). A documented system can assure that geographically separated divisions of a company know how to produce uniform product quality, and are capable of communicating process improvement information between plants. Such a system provides an effective tool for training new employees both within and outside of the quality department. Perhaps most important, a documented quality system can be created to emphasize continuously the twin goals of attainment of uniform quality and profit improvement.

## QUALITY SYSTEM GUIDELINES

Chapter 1 outlined nearly a dozen approaches to developing a quality system. A more detailed discussion of some of the more recent system guidelines follows. It should be emphasized that the seven basic tools of quality should be included in each of these systems.

Six-sigma (see Chapter 8) is based on counting defects, and using this data to rate quality control. Efforts to reduce the number of defects are centered around inputs from all levels of employment within the company. Management and employees all train in the statistics involved, the techniques of production inspection and product improvement. It is most effective in hardware industries where defects can be remachined or sorted and scrapped. For the most part, this industry's standards are self-imposed by the manufacturers or their customers. Because of legal standards (and complete unwillingness by consumers to accept any food defects) the six-sigma method has little application to the food industry.

HACCP (see Chapter 5) is centered on food production industries, and is based on defect prevention. Critical path diagrams are generally clear flow diagrams which can be understood by production personnel, thus contributing to their effectiveness. Unlike most hardware industries where some defective product is considered normal (though undesirable), in the food industry critical defects are not acceptable, and in many areas not permitted by law.

ISO 9000:2000 (see below) is an excellent tool to enforce control of quality. It is effective in the hardware industries. Food industries may be required to adhere to this standard in order to export their products to many countries which demand ISO certification.

TQM, PDAC, and many other pseudonyms are based on detailed standards (often legislated) which are achieved by the use of statistical quality controls. The principles may not be understood at all levels of employees, but these programs can be highly effective for preventing defects, improving quality and lowering process costs.

## Food Processing Industry Quality System Guidelines

The generic guidelines for quality systems developed by the American National Standards Institute (ANSI Z1.15) provides an excellent basis for establishing effective quality control systems, but is geared more toward hardware manufacture than food processing. A committee of food quality experts, chaired by Sydney Pearce, restructured this standard for use by the food processing industry, and published the guidelines in 1986. This document covers the following:

1. *Administration* includes quality policy, objectives, quality system, planning quality manual, responsibility, reporting, quality cost management, and quality system audits. Each of these subjects is covered in detail. For example, quality system provides for individual policies, procedures, standards, instructions, etc. covering: ingredients, packaging, processing, finished product, distribution, storage practices, vendor/contract processors relations, environmental standards, sanitation, housekeeping, pest management/control, shelf life, design assurance, recall, quality costs, user contacts, complaint handling and analysis, corrective action, motivational and training programs and others.

2. *Design assurance and design change control* contains 12 subsections, such as concept definition, design review, market readiness review.

3. *Control of purchased materials*—an excellent summary of supplier certification requirements, such as specifications, system requirements, facility inspection, assistance to suppliers.

4. *Production quality control* contains 24 detailed requirements under the following subheadings:
   - Planning and controlling the process
   - Finished product inspection
   - Handling, storage, shipping
   - Product and container marking
   - Quality information.

5. *User contact and field performance* includes product objective, advertising, marketing, acceptance surveys, complaints and analysis.
6. *Corrective action* covers detection, documentation, incorporating change, product recall, and non-conforming disposition.
7. *Employee relations*—selection, motivation, and training.

## Good Manufacturing Practice (GMP)

Although not one of the *statistical* tools of quality control, GMPs belong in every food quality control system. The Code of Federal Regulations (CFR 21 Part 110, GMP) provides excellent definitions and criteria which determine if the product has been manufactured under conditions which make it unfit for food; or if the product has been processed under insanitary conditions resulting in contamination with filth; or is otherwise rendered injurious to health. It contains detailed requirements for avoiding these possibilities in the following general areas:

*Personnel*—Disease control, cleanliness, education and training, and supervision.
*Plant and grounds*—Proper equipment storage, maintenance of surrounding property, effective systems for waste disposal, space for equipment placement and storage of materials, separation of operations likely to cause contamination, sanitation precautions for outside fermentation vessels, building construction to permit adequate cleaning, adequate lighting, ventilation and screening.
*Sanitary operations*—Building and Fixtures: maintenance, cleaning and sanitizing to prevent contamination. Special precautions for toxic sanitizing agents. Pest control. Food contact surfaces: sanitation procedures.
*Sanitary facilities and controls*—Water supply, plumbing, sewage disposal, toilet facilities, handwashing facilities, rubbish and offal disposal.

The Code then follows with specific GMP regulations for equipment and for process controls.

*Equipment and utensils*—Design, materials and workmanship shall be cleanable, protected against contamination, and shall be nontoxic, seamless, and properly maintained. (Some specific types of equipment are referred to: holding, conveying, freezing, instrumentation, controls, and compressed gases.)
*Processes and controls*—Adequate sanitation in receiving inspection, transporting, segregating, manufacturing, packaging, and storing. Appropriate quality control operations to insure that food is suitable for human consumption and that packaging materials are safe and suitable. Assigned responsibility for sanitation. Chemical, microbial and extraneous material testing. Rejection of adulterated or contaminated material.
  • Raw materials—Shall be inspected for suitability for processing into food. Stored to minimize deterioration. Wash and conveying water to be of adequate sanitary quality. Containers shall be inspected for possible contamination or deterioration of food. Microorganism presence shall not be at a level which might produce food poisoning, and shall be

pasteurized during manufacturing to maintain a safe level. Levels of toxins (such as aflotoxin), or presence of pest contamination or extraneous material to be in compliance with FDA regulations, guidelines or action levels. Storage of raw materials, ingredients or rework shall be protected against contamination, and held at temperature and humidity which will prevent adulteration. Frozen raw materials shall be thawed only as required prior to use and protected from adulteration.

- *Manufacturing operations*—Equipment, utensils and finished food containers to be sanitized as necessary. Manufacturing, packaging and storage to be controlled for minimum microorganism growth, or contamination. (A number of specific suggestions for physical factors to be controlled: time, temperature, humidity, water activity, pH, pressure, flow rate. Controls for manufacturing operations are also suggested: freezing, dehydration, heat processing, acidification, and refrigeration.) Growth of undesirable organisms shall be prevented by refrigeration, freezing, pH, sterilizing, irradiating, water activity control. Construction and use of equipment used to hold, convey or store raw materials, ingredients work in process, rework, food or refuse shall protect against contamination. Protection against inclusion of metals or other extraneous material shall be effective. Adulterated food, ingredients or raw materials shall be segregated and, if reconditioned, shall be proven to be effectively free from adulteration. Mechanical manufacturing steps such as washing, peeling, trimming, cutting, sorting, inspecting, cooling, shredding, extruding, drying, whipping, defatting, and forming shall be performed without contamination. Instructions are offered for blanching, with particular emphasis on thermophilic bacteria control.

  Preparation of batters, breading, sauces, gravies, dressings and similar preparations shall be prepared without contamination by effective means such as: contaminant-free ingredients, adequate heat processes, use of time and temperature controls, physical protection of components from contaminants which might be drawn into them during cooling, filling, assembling and packaging.

  Compliance may be accomplished by a quality control operation in which critical control points are identified and controlled during operation; all food contact surfaces are cleaned and sanitized; all materials used are safe and suitable; physical protection from contamination, particularly airborne, is provided; and sanitary handling procedures are used.

  Similar requirements are listed for dry mixes, nuts, intermediate moisture food, dehydrated foods, acidified foods, and ice-added foods. Lastly, the regulation forbids manufacturing human and non-human food grade animal feed (or inedible products) in the same areas, unless there is no reasonable possibility for contamination.

- *Warehousing and distribution*—Storage and transportation of finished foods shall be protected against physical, chemical and microbial contamination, as well as deterioration of the food and the container.

Food processing companies with sufficient staff might consider incorporating the GMP regulation in the quality control manual, and conducting routine audits to assure conformance. Consulting firms are available to perform periodic GMP inspections for smaller organizations. In either case, a file of satisfactory audits could prove invaluable in the event of suspected product failure resulting in litigation.

It should be noted that the FDA regulations for Good Manufacturing Practice is modified from time to time, and it is necessary to periodically review quality control procedures to insure compliance.

## ISO Standards (International Organization for Standardization, Revised 2000)

The International Organization for Standardization in Geneva, Switzerland, began to develop a series of standards to describe an ideal, generic quality system in the late 1970s. Based on the British quality standard, the initial intent was to clarify contracts between suppliers and their customers. It became apparent after a few years, that companies which were registered for compliance with these standards would not require most supplier audits, since customers could be assured of product which would meet their specifications. Some countries have expanded this concept to require that imported goods must be produced under ISO standards. Although the ISO requires that all standards be reviewed and updated every five years, it was expected that changes would be gradual. The 1994 changes, for example, included relatively minor format and wording, a greater emphasis on documented procedures (including the quality manual), a few small additions to management responsibility and staffing, addition of a quality planning document, and a few definitions such as verification and validation.

There are three Standards in the revised ISO 9000 series: 9000, 9001, and 9004. The 1994 standards 9002 and 9003 have been discontinued, and their coverage has been consolidated into the 9001 standard. An organization may be certified on the basis of compliance with 9001. The revised 2000 standards focus on customer satisfaction rather than products, continual product or service improvement, top management commitment (development and improvement of the quality management system), and emphasis on "continuous value added processes" rather than a list of "quality elements." Another new requirement is monitoring customer satisfaction information as a measure of system performance. A significant change is the recognition of statutory and regulatory requirements.

- ISO 9000:2000 Quality Management Systems—Fundamentals and Vocabulary. Covers explanations of definitions and fundamental terms.
- ISO 9001:2000 Quality Management Systems—Requirements. Procedures to meet customer satisfaction and regulatory requirements. Conformance to this standard alone can be certified by an external agency. Major requirements are outlined below.
- ISO 9004:2000 Quality Management Systems—Guidelines for performance improvements, based on maintaining customer satisfaction.

ISO 9001–2000 Introduction.    The following are outlines of the salient features of the second edition of ISO 9001 which may apply to the food industry. The standard consists of eight clauses. The first three clauses cover a number of definitions:

Quality management principles
The process approach
Relationship to ISO 9004
Compatibility with other management systems
Scope of the standard
Non-generic organization exclusions
Maintenance of currently valid registrations
Terms and definitions.

There are eight Quality Management Principles used in both ISO 9001:2000 and ISO 9004:2000:

1. Customer-focused organization—understand customers' present and future needs.
2. Leadership—create and retain management direction, and an environment which focuses on achieving objectives.
3. Involvement of people—continually use employees at all levels to provide input of their expertise.
4. Process approach—manage activities as processes is an effective use of resources.
5. System approach to management—recognize interrelated processes as a system to achieve objectives.
6. Continual improvement—create a permanent system for improving the organization's objectives.
7. Factual approach to decision making—resolutions of objectives are best reached by analysis of data.
8. Mutually beneficial supplier relationship—contributes to the ability of all parties to improve value.

The next five clauses (#4–8) are outlined below. They replace the 20-clause structure of the old standard 9001.

(Clause 4) Quality Management System—system, quality manual, and control of documentation.
(Clause 5) Management Responsibility—top management planning, communication, focus on customer.
(Clause 6) Resource Management—human resources, facilities, work environment, infrastructure, training.
(Clause 7) Product Realization—planning, monitoring, control of measuring equipment, design, customer-related processes from receipt of order through delivery.

(Clause 8) Measurement, Analysis, and Improvement—monitoring processes and customer satisfaction, control of defectives, and analysis of data.

Details of these five sections follow. (Some parts of Clause 7—Product Realization—may not be applicable to all companies, and exclusions are permitted for this section only.)

## Quality Management System (Requirement Clause #4)

4.1. General Requirements

A documented quality system which is continually reviewed and improved. Identification of the process sequence, and methods by which they can be monitored and verified to be working properly.

4.2. General Document Requirements

4.2.1. Control of Documents

A Quality manual which includes quality objectives, policy, procedures, and records.

4.2.2. Quality Manual

All procedures should be documented, along with flow charts. Separate volumes are permitted. Descriptions of interaction between processes of the quality management system.

4.2.3. Control of Documents

Written system for approving reviewing, updating, and identifying current issues of documents. Applicable documents shall be available at points of use.

4.2.4. Control of Records

Record retention program including specified holding period for each, protection, indexing record description, retention time, and location.

## Management Responsibility (Requirement Clause #5)

5.1. Management Commitment

Commitment of senior management to continuous system reviews and improvement. Statement of quality policy and quality goals (customer, statutory, and regulatory).

5.2. Customer Focus

Senior management plan to either meet with customers or designate a contact individual to determine customer satisfaction and the need for improvement.

5.3. Quality Policy

Establishment and review of quality objectives and a system to review those objectives. Contains a commitment to comply with requirements and continually improve the quality management system.

5.4. Planning

5.4.1. Quality Objectives

Definition of measurable quality consistent with quality policy objectives. Statement of resources available to meet these objectives.

5.4.2. Quality management system planning

Top management shall ensure that resources are planned and identified to meet requirements of quality system integrity.

5.5. Responsibility, Authority, and Communication

5.5.1. Responsibility and Authority

Identification of key duties to ensure that responsibilities and authorities of senior management and other department personnel are defined and communicated. Organization structure information shall make clear employee reporting system.

5.5.2. Management Representative

Employee appointed by top management to oversee that processes needed for quality management system are established, implemented, and maintained. Customer requirements promoted and system performance reported to top management.

5.5.3. Internal Communication

Communications are to report the effectiveness of the quality management system.

5.6. Management Review

5.6.1. General

Quality system review of current effectiveness and possible improvement. Shall include possible changes to quality policy and quality objectives.

5.6.2. Review Input

Specific process review

Audit results

Customer feedback

Process performance and product conformity

Preventive and correction action status

System changes

Recommended improvement.

5.6.3. Review Output

Specific output review

Effectiveness of management system

Improvement of product or services

Required resources for improvements.

## Resource Management (Requirement Clause #6)

6.1. Provision of Resources

Provide adequate resources to enhance customer satisfaction and implement required and improved processes.

6.2. Human Resources

6.2.1. General

Need for trained and experienced staff to perform indicated tasks which affect quality. Evaluation of competence shall be evaluated on the basis of education, training, skills, and experience.

6.2.2.  Competence, Awareness, and Training
The organization shall determine personnel competence for work affecting quality, and provide necessary training. Ensure that the personnel are aware of the importance of their activities.

6.3. Infrastructure
Provide workplace, equipment, and employee service support.

6.4. Work environment
Human and physical conditions managed for product and service requirements.

## Product Realization (Requirement Clause #7)

7.1. Planning of Product Realization
Establish product quality objectives and requirements
Plan all processes required for products and services
Plan inspection and test procedures
Identify reports and records related to product quality.

7.2. Customer-Related Processes

7.2.1.  Determination of Requirements Related to Product
Identify customer requirements for product or service, including legal requirements, delivery requirements, warranties (generally not applicable to most food industries).

7.2.2.  Review of Requirements Related to Product
Define and record customer requirements
Review resource capability before accepting order
Establish procedure for informing relevant sections of company regarding requirement changes.

7.2.3.  Customer Communication
Define customer procedure for communicating product information, such as complaints, inquiries, specification changes.

7.3. Design and Development

7.3.1.  Design and Development Planning
Identification of design, development and review stages.

7.3.2.  Design and Development Inputs
Identification of inputs relative to legal requirements, standards, safety, packaging, recycling, and labeling.

7.3.3.  Design and Development Outputs
Drawings, specifications: time, temperature, pH, concentration, bacterial levels, governmental regulations—all checked to insure the product meets requirements.

7.3.4.  Design and Development Review
Identify procedure and records for design review as development progresses. Identify problems and propose necessary actions. Record progress. Maintain records of review stages.

7.3.5. Design and Development Verification
Plans to include methods for verifying that design meets requirements. Records of the results shall be maintained.

7.3.6. Design and Development Validation
Produce product or service to establish that it meets requirements. Review customer feedback. Plans for further development. Records of the results of validation shall be maintained.

7.3.7. Control of Design and Development Changes
Review, verify, and record changes.

7.4. Purchasing

7.4.1. Purchasing Process
Specifications of supplied material are clearly stated to supplier. Suppliers evaluated and selected on their ability to meet requirements. Records of selection shall be kept.

7.4.2. Purchasing Information
Purchase order verification prior to sending to supplier. Include product description, requirement for approval, quality management system.

7.4.3. Verification of Purchased Product
Receiving inspection process. Procedure, documentation, record retention.

7.5. Production and Service Provision

7.5.1. Control of Production and Service Provision
Product characteristics specified
Test procedures
Testing equipment and maintenance
Availability of testing equipment
Procedures for release and delivery.

7.5.2. Validation of Processes
(Hardware industries oriented. Probably no application to food industry.) Covers processes of monitoring and measurement after product or service has been delivered.

7.5.3. Identification and Traceability
Manufacturing date and line code identification.
Food industry—data for potential recall or shelf-life.

7.5.4. Customer Property
Hardware industry practice where customer supplies part for further processing and return as finished product. Procedures for identifying, verifying and protecting product during operations.

7.5.5. Preservation of Product
Storage, identification, packaging, handling and stock rotation.

7.6. Control of Monitoring and Measuring Devices
Identification of which tests need to be performed, and which test equipment is to be used.

Maintenance and accuracy of devices—calibration and adjustment.
Equipment safeguards against damage or deterioration.

## Measurement, Analysis, and Improvement (Requirement Clause #8)

8.1. General

Specify measuring and monitoring techniques to be utilized.

Itemized list of statistical techniques.

Process for continual product and process analysis and improvement.

8.2. Monitoring and Measurement

    8.2.1. Customer Satisfaction

        Monitoring Customer Perception of Requirements.

    8.2.2. Internal Audit

        Required to demonstrate system's ability to meet requirements of ISO 9000.

        Auditor is independent of process audited, competent to understand the process, and impartial.

        External or internal auditors are acceptable.

        Define and document criteria, scope, and conduct of audits.

    8.2.3. Monitoring and Measurement of Processes

        Establish suitability of process and product control through examination of product failures, customer complaints.

    8.2.4. Monitoring and Measurement of Product

        In-process and final inspection procedures.

        Records of tests completed before delivery, with evidence of conformity with acceptance criteria.

        Product or service release for delivery only if tests prove satisfactory.

8.3. Control of Nonconforming Product

Documented system for assurance that non-acceptable product is not used in the process or shipped to customer. System to include procedures for removing or retrieving faulty product.

8.4. Analysis of Data

Collection and analysis of data to evaluate the quality system.

Includes data from suppliers as well as customer satisfaction.

Based on analyses, plans are made to improve.

8.5 Improvement

    8.5.1. Continual Improvement

        Based upon quality reports, failure analysis. Analysis of customer satisfaction and non-conformance.

        Improve quality management system: quality policy, quality objectives, audit results, analysis of data corrective and preventive action.

    8.5.2. Corrective Action

        Identify, record, and repair problem areas.

        Evaluation of effectiveness of action taken.

Provide documented procedures to review nonconformities and customer complaints, causes of nonconformities, implementing action needed, and recording results.

8.5.3. Preventative Action
Evaluate problems—actual and possible.
Determine need for corrective action to eliminate causes of nonconformities.
Review and document effectiveness of corrective action.

## MALCOLM BALDRIDGE NATIONAL QUALITY AWARD

The Malcolm Baldridge National Quality Award is presented by the President of the United States to a limited number of winners in manufacturing, service, and small companies each year. The primary purpose of the award is to recognize U.S. companies for business excellence and quality achievement. Additionally, the award has encouraged thousands of organizations to undertake quality improvement programs. As mentioned in the previous chapter, the Malcolm Baldridge National Quality Award categories have undergone significant changes since inception. The early guidelines focused on the applicant's *total quality system*. After a few years, this was modified to recognize *business excellence and quality achievement*. Either approach provides a useful format for evaluating and improving a company quality system. However, it is suggested that companies use the 1990 Baldridge Award Application Guidelines as a model if they are not yet satisfied that they have achieved a detailed and effective quality system. Category 5 of this early guideline, titled Quality Assurance of Products and Services, covers seven of the basic requirements:

1. Design and introduction of quality products and services, emphasizing how processes are designed to meet or exceed customer requirements.
2. Process and quality controls which assure that the products and services meet design specifications.
3. Continuous improvement techniques (controlled experiments, evaluation of new technology, benchmarking) and methods of integrating them.
4. Quality assessments of process, products, services and practices.
5. Documentation.
6. Quality assurance, quality assessment and quality improvement of support services and business processes. (Accounting, sales, purchasing, personnel, etc.)
7. Quality assurance, quality assessment and quality improvement of suppliers. (Audits, inspections, certification, testing, partnerships, training, incentives and recognition.)

Once a company is satisfied that an effective quality system has been established, it might be desirable to examine the Baldridge Award criteria of subsequent years to see if the goals of business excellence and quality achievement are also being met.

## TOTAL QUALITY MANAGEMENT (TQM)

TQM is more of a quality philosophy than a quality system. TQM is a goal in which every individual in every department of an organization is dedicated to quality control and/or quality improvement. In addition, this web of management may encompass suppliers and distributors as well. Some plans include the customers as partners in the quality efforts. Each organization has the freedom to define "total," "quality," and "management" in any way they believe will enhance the quality efforts within their scope.

One company instituted a program to train all of their employees with team techniques (see below) for solving quality problems and improving product quality. They referred to this activity as their TQM process.

Some companies have tried to avoid the stigma of reports of failed TQM programs by using other terminology: "The Universal Way," "Quality Improvement Process," "Total Quality Service," "Quality Strategies," "Leadership Through Quality," "Total Systems Approach," "Quality Proud Program," "Customer Driven Excellence," "Total Quality Commitment," "Total Quality Systems," "Statistically Aided Management."

*Food Technology* magazine, published by the Institute of Food Technologists, had several articles with widely varied definitions of TQM. One all-encompassing description of TQM states that it is a way of managing business, and is based on the American Supplier Institute approach:

- Incorporate leadership, cooperation, partnerships, trust.
- Aim for long-term goals of the business.
- Concentrate on improving processes to improve results.
- Deploy means and goals to all levels of the organization.
- Use statistical quality control tools, quality function deployment, Taguchi methods, and other quality tools.

*Quality Progress* magazine, a publication of the American Society for Quality Control, devoted an entire issue (July 1995) to 13 articles which described and evaluated TQM. Each paper was required to submit a brief definition of TQM. Table 2-1 indicates the differences of opinion between the authors.

The observation that there is no consensus for a TQM definition should not lessen its virtues. Applying as many quality control principles in as many areas as possible is a desirable goal for any organization. Some companies have built successful TQM programs (using their own definition) by proceeding stepwise; others have attempted to formulate the entire concept at once, and implement it in stages. There probably is no one best way.

**Table 2-1.  TQM System Components**

| TQM system components | Author's definitions |  |  |  |  |  |  |  |  |  |  |  |  |
|---|---|---|---|---|---|---|---|---|---|---|---|---|---|
|  | 1 | 2 | 3 | 4 | 5 | 6 | 7 | 8 | 9 | 10 | 11 | 12 | 13 |
| Business or management philosophy | X |  |  |  | X |  |  |  |  |  |  |  |  |
| Continuous (mindset for) improvement |  |  |  |  | X | X | X | X |  | X |  | X | X |
| Continuous redesigned work process |  |  |  |  |  |  |  | X |  |  |  |  |  |
| Customer focus/driven | X |  |  |  | X | X |  | X | X |  |  | X | X |
| Customer/supplier partnership |  |  |  | X |  |  |  |  |  |  |  |  |  |
| Delight customers |  |  |  | X |  |  |  |  |  |  |  |  |  |
| Develop improvement measure |  |  |  |  |  |  |  |  |  |  | X |  |  |
| Develop with time |  |  |  | X |  |  |  |  |  |  |  |  |  |
| Employee cooperation |  |  |  |  |  |  |  |  | X | X |  |  |  |
| Evaluate continuous improvement |  | X | X |  |  |  |  |  |  |  |  |  |  |
| Failure mode effects analysis |  |  |  |  |  | X |  |  |  |  |  |  |  |
| Goal alignment |  |  |  |  |  | X |  |  |  |  |  |  |  |
| Guarantee survival | X |  |  |  |  |  |  |  |  |  |  |  |  |
| Improve customer satisfaction |  |  | X |  |  | X |  |  |  |  |  |  |  |
| Improve financial performance | X |  |  |  |  |  |  |  |  |  |  |  |  |
| Improve management system | X |  |  |  |  |  |  |  |  |  |  |  |  |
| Improve materials and services supplied |  |  |  |  |  |  |  |  |  |  |  | X |  |
| Learning architecture |  |  |  |  |  |  | X |  |  |  |  |  |  |
| Meet customers' needs | X |  |  |  |  |  |  |  |  |  |  |  |  |
| Meet employees needs | X |  |  |  |  |  |  |  |  |  |  |  |  |
| Meet owners' needs | X |  |  |  |  |  |  |  |  |  |  |  |  |
| Mission driven |  |  |  |  |  |  |  | X |  |  |  |  |  |
| Plan/do/check/act |  |  |  |  |  |  |  |  |  |  | X |  |  |
| Reduce costs |  |  |  | X |  |  |  |  |  |  |  |  |  |
| Reengineering |  |  |  | X |  |  |  |  |  |  |  |  |  |
| Responsible individuals |  | X |  |  |  |  |  |  |  |  |  |  |  |
| Responsive customer service |  |  |  |  |  |  |  | X |  |  |  |  |  |
| Self-directed teams |  |  |  | X |  |  |  |  |  |  |  |  |  |
| Set targets for improvement |  |  |  |  |  |  |  |  |  |  |  | X |  |
| Statistical process control |  |  |  |  |  | X |  |  |  | X |  |  |  |
| Strategic architecture |  | X |  |  |  |  |  |  |  |  |  |  |  |
| Teamwork and trust |  | X |  |  |  |  |  | X |  |  |  |  |  |
| Use of quality tools |  |  |  |  |  |  |  |  |  | X |  |  |  |

## TEAM QUALITY SYSTEMS

Teams generally represent a subsystem of quality control and improvement systems. They rarely work well without a centralized quality control structure since they are usually task oriented. Team concepts and operation will be covered in some detail in Chapter 14.

## COMPUTER NETWORK QUALITY SYSTEMS

Where operational cost/benefit analyses can justify the installation of computer systems on processing lines, opportunities for improved quality control accompany production gains. In addition to creating a system which can graphically show current progress of every step of a process, track the quantities of ingredients, supplies, work in process, and output, these same data signals can be harnessed to operate and control valves, burners, feeders, pumps, mixers and other equipment. Additionally, quality control can have the ability to use real-time data for instantaneous control, to collect larger quantities of more meaningful and accurate data, to provide high speed calculation and display of control charts, quality costs and other quality reports, and to instantly create and access data storage. Computer systems are discussed in greater detail in Chapter 15.

## SUMMARY

The many quality system approaches outlined in Chapter 1 are still used in the food industry. There will never be complete agreement on *the one best way*, but recent trends have emphasized the use of portions of the older philosophies, along with several new ones. Those considered most effective at present include the systems explained above.

- Seven basic tools
- Food processing industry quality system guidelines
- Good manufacturing practice
- ISO 9000 Standards
- Malcolm Baldridge Quality Award
- Total quality management
- Team quality systems
- Computer network quality systems.

## Policy

A quality system is all but useless without an understanding of the company quality policy. If the company policy on product packaging is to provide inexpensive and functional sealed plastic film envelopes, then the quality control manager might be doing the company a costly disservice to insist on precise printing and color control of these envelopes. A verbal understanding of policy is not enough. The policy should be in writing, signed by a company official, and reproduced as the front page of the quality manual.

If a policy statement is unavailable, it may be difficult to obtain one because of organizational peculiarities or sensitive personal relationships between department managers. One of the more effective methods for generating a policy document is the preparation by the quality manager of a clearly labeled draft policy which is forwarded to the company president for revision and approval through whatever channels are considered necessary. This provides the president with the necessary guidelines, and still leaves him with complete authority to select the details.

A simple policy statement might read:

The Midwest Food Company is dedicated to supplying our customers with a competitive quality line of standard products, and a superior line of choice quality products under the Gourmet label. All of our products shall be shelf stable for one year, safe, nutritious, and flavorful, and shall compare favorably with the quality level of similar products found in our marketing area. They shall conform in all respects to the standards and regulations of federal, state, and local governments. The responsibility for quality control shall rest jointly with the managers of purchasing, production, distribution, and quality control.

Major specifics might be included:

The Quality Control Department Manager shall establish and maintain product and process specifications. He shall be responsible for preparing and distributing routine quality reports, exception reports, quality improvement reports, cost reduction proposals, and shall perform monthly audits to evaluate all phases of the quality control system, including sanitation, product performance, house-keeping, and quality training. Quality control meetings shall be chaired weekly by the Chief Executive Officer, and attended by managers and lead foremen of all departments.

For some companies, more specific quality statements may be considered desirable, but generally, the longer and more complicated a policy statement is made, the less likely it will be read and followed. The details are best left for inclusion in the quality manual. Where there is likely to be a potential for misunderstanding, it might be necessary for the policy statement to include the functional quality responsibilities for each department. The drawbacks to details of this nature are related to possible personality conflicts, turf battles, and restriction of the contributions of capable employees.

## Quality Control Manual

If a quality manual were prepared to conform to the best recommendations of all of the respected quality professionals, it would be so voluminous as to be useless.

And yet, it is essential that all of the critical elements of a manual be prepared, updated, and—most important—used. One suggestion to minimize the size without compromising the contents is to divide the manual into a set of numbered "submanuals":

| MANUAL NO. | SUBJECT |
|---|---|
| Q1 | Policy, Organization, Interrelationships |
| Q2 | Research and Pilot Plant Quality Control |
| Q3 | Material Quality Control |
| Q4 | Production Quality Control |
| Q5 | Product Performance |
| Q6 | Corrective Action |
| Q7 | Microbiological Procedures |
| Q8 | Packaging Material Test Methods |
| Q9 | Product Specifications |
| Q10 | Quality Audit Procedures |
| Q11 | Reports and Forms |
| Q12 | Government Regulations |
| Q13 | Product Recall System |
| Q14 | Special Quality Requirements |

An excellent guide to content and wording for an overall quality manual is Military Standard 9858A, obtainable from the Government Printing Office. Although it is hardware-industry oriented, it includes most of the basics and many details which might be overlooked.

The Secretary of Defense cancelled MIL Q 9858A as of October 1996, and adopted the ISO 9000 system in its place. The basic principles offered in MIL Q 9858A are similar in most respects to the ISO 9000 standards, differing chiefly in the following areas:

a. MIL Q 9858A has generally been the responsibility of the Quality Control Department; whereas ISO 9000 cuts across organizational lines, requiring input and responsibility from all departments.

b. ISO 9000 quality systems require constant review and modification. In fact the standard itself has a scheduled review and modification requirement.

c. ISO 9000 extends beyond the production quality concept, and includes product design. It also covers, as previously noted, a number of activities which are more common to hardware manufacturers than food processors.

Test methods, other than those specifically listed above, may be included under other headings, or may be compiled into a separate manual. Primarily, these tests are concerned with raw materials, process, and product. Additionally, sample size, selection and frequency may be chosen from the above classifications, particularly if the techniques are complex. By preparing separate manuals to cover such subjects as sampling and testing, training of new employees is often simplified.

## Product Research and Development

From a legal viewpoint, one of the most important requirements in development of a new or improved product is documentation. This protects the company in the event of patent or copyright litigation. For the scientist or engineer, accurately prepared records provide a route map for others to follow, and frequently suggest other developments for products, processing, or packaging to be explored at a later date.

Many creative food technologists prefer to work in an unstructured and informal atmosphere, and for small companies or short-term projects, formalized procedures are not required. Guidelines are intended to assist workers, not to bind them into a mold which restricts the inquisitive mind. Perhaps the systematizing of Research and Development (R&D) should be formalized more as an outline or checklist, rather than as a rigorous procedure. This permits flexibility but provides the safeguards which reduce the possibility of overlooking important development steps.

1. Quality goal of R&D—The quality system should assure that specifications are evolved which conform to marketing needs, manufacturing practices, and regulatory requirements, and cover all aspects of materials, process, packaging, storage and use. These specifications might include the following:
   - Physical and sensory properties
   - Product function and nutritive values
   - Process equipment and process rates
   - Packaging and packaging equipment
   - Composition
   - Microbiological limitations
   - Shelf life
   - Labeling and coding
   - Regulatory requirements.
2. *Methods and standards*—A checklist for methods and standards might best be considered as a flexible guide, since most developmental projects have special requirements that will not fit a rigorous mold. To avoid the possibility of overlooking important requirements, such a list should be reviewed at the start of any new development project:
   - Sampling and statistical analysis
   - Analytical testing methods
   - Calibration and maintenance of analytical equipment
   - Shipping tests and evaluation
   - Tolerance evaluation and review
   - Hazard and failure mode effect analysis

- Specifications for material, process, product, package
- Consumer test design and evaluation
- Patent potential evaluation
- Outline of quality system for initial production
- FDA, USDA, EPA, and local regulations.

3. *Documentation*—Progress reports usually are considered the basic documents for most development studies, and their presentation should be a planned procedure. Haphazard reporting may impede the efforts by requiring disrupting staff meetings to explain the reasons for failures or the direction of current research efforts. At the very least, milestone reports should be made (where applicable) at six stages:

- Product concept
- Prototype development
- Pilot plant trial
- Qualification testing
- Field evaluation
- Market readiness.

4. *Reviews*—Reports at the above six stages are intended to generate critical reviews by involved personnel. A system should be evolved by which criteria for acceptance of progress can be established, permitting work to proceed to the next step. This should avoid unnecessary backtracking and interruption of the research process. A universal checklist is impractical since each research project tends to be unique, but it is unwise to use this fact as an excuse to avoid detailed documentation. Perhaps a master list for qualification testing review might illustrate the scope of such a report.

- Critical material
- Product function and composition
- Product stability—nutrient, sensory, function
- Packing performance
- Process variability
- Process response—drift, interactions, errors
- Shipping
- Cost analysis.

## Material Control

Over the years, "Just-in-Time" manufacturing systems have set goals to lower costs by reducing inventories and eliminating rejects, rework and downtime by accepting only defect-free raw materials in quantities no larger than those which were sufficient to keep the plant operating. Obviously, it is costly to produce satisfactory product from defective raw materials by sorting at the receiving dock or correcting on the line, and much effort has been made to assist vendors in production and delivery of defect-free field crops, food additives, and packaging materials. The key word here is "assist." The farmer may be unaware of the critical nature of the sugar-acid ratio of his fruit in a jelly-making process. The packaging film supplier may be unaware of the relationship of film thickness uniformity and shelf life of a product. A fraction of a percent of gravel in unprocessed grain may abrade your raw material cleaning equipment. By working with suppliers, a manufacturer can often reduce the quality costs of both parties. Improved materials reduce downtime, permit reduction of "Just-in-Case" inventories, and should improve profits.

Control of materials starts, at least theoretically, at the product development stage. It becomes a real consideration during supplier (also referred to as "vendor") selection, and eventually evolves into a daily consideration. Selection may be based on marginal criteria such as reputation, past performance, or reputation. Where material specifications are clearly written, they may eliminate suppliers because of their inability either to reach or maintain the requirements. This should emphasize the need to be realistic when preparing specifications. They need not be more strict than is actually needed in the process, since this may rule out alternate sources, and may often result in unnecessary added material cost to achieve an unrealistic level of quality.

By working together, those responsible for quality control, for purchasing and for production can work with their counterparts in the supplier's organization to prepare realistic purchase specifications and contractual requirements. With a clear understanding of the rules, suppliers may be willing to provide certification documents with each shipment, thus avoiding the necessity for major incoming inspection facilities at the processing plant. Under these conditions, suppliers may also be willing to accept financial responsibility for production losses due to shipment of nonconforming raw material resulting in line stoppages, rejected or spoiled product, lost packaging materials, repacking, and overtime costs. This has long been a practice for large-volume food processors, and should be equally applicable to smaller companies as well.

Assessing the quality of materials from suppliers should include:

- Inspection of supplier quality control program.
- Evaluation of supplier specifications.
- Examination of supplier facilities and production procedures.
- Evaluation of performance through periodic audits.

Although one of the goals of supplier–purchaser relationships is the elimination of incoming material inspection, it is unlikely that 100% achievement can be accomplished without serious risk. For both attribute and variable inspection plans commonly used, switching plans are available which permit reduced inspection when a supplier has provided satisfactory material over a specified number of lots. This can substantially reduce the costs of material inspection while providing assurance that only useable material will be accepted. Ultimately, this might lead to a "Just-in-Time" supplier program, should this be considered desirable.

In the event of discovery of nonconforming incoming material, quality control purchasing and manufacturing departments should all be informed immediately to avoid the possibility of losses from production downtime or spoilage. Unless properly controlled, rejected materials can find their way onto the production floor. The simplest of controls include four distinct steps:

1. Document the identification of the shipment, the inspection results, disposition recommended, and (when completed) final disposition.
2. Label nonconforming material clearly with distinctive markings and easily read warnings not to use.
3. Segregate the shipment to prevent the possibility of mixing with satisfactory product. If contaminated, isolate the shipment from edible storage areas.
4. Dispose of promptly. If destruction of the material is deemed necessary, it should be observed by quality control personnel.

Under even the most satisfactory conditions, minimal receiving inspection is always required for compliance with current FDA Good Manufacturing Practices Regulations, other governmental regulations, and company policies. Supermarket chain stores and large food product distributors consider food processors as their suppliers, and are likely to require specific receiving inspection procedures at the processing plant as part of their overall quality systems. Regulatory or contractual considerations notwithstanding, it is always imperative to inspect at the receiving dock for evidence of potential problems with sanitation, housekeeping, pest control, adulteration or fitness for use. This represents a major departure from the practices of the hardware manufacturing industries: the end-product of the food processor is eaten!

## Production Quality Control

The major effort in quality control is devoted to the production line, and it must be designed around process capability studies. These will be discussed in detail in a later chapter. A process capability study determines whether the process can meet the desired product specifications, where sampling points should be established, and the frequency of the sampling cycles. In addition, the direction for process change to improve quality and productivity is frequently highlighted during process capability studies. The necessity for periodically conducting confirming studies is often overlooked. When a process is running comfortably in control for a period of time, a repeat process capability study may indicate the need for new control limits, and perhaps new product specifications and new process targets. Normally, this is the basis for on-line product quality improvement.

It is next to impossible to continuously control a process without written procedures. Word-of-mouth instructions from a supervisor to an operator are invariably modified, misunderstood, or forgotten. Control cannot be accomplished when a process operator explains his unusual procedures with excuses such as: "I've always found it works better this way"; or "I thought that is what I was instructed to do"; or "It didn't seem to be that important"; or, worst of all, "Nobody ever explained that to me." Documented procedures are difficult to establish, and they require continuous upgrading and correction; but they are essential to providing a system which can be controlled to the same standards, day after day. A second major benefit from written procedures is their availability for providing uniform training to all new employees.

No one person should be required to take on the total responsibility for writing production operation procedures. They are less likely to be adhered to if they are forced onto the production department by an "outsider," even if they have been prepared under the direction of top company management. To ease the introduction and acceptance of procedures, all affected supervisory or management employees should have the opportunity to review drafts as they are developed, and should be given the responsibility for offering modifications to a draft before the final document is issued for approval and use.

Occasionally one might expect management to balk at the concept of providing written procedures of proprietary operations to employees who might allow them to fall into the hands of competitors. It may be possible to alleviate these fears by the use of codes for critical temperatures, pressures, flows, or special equipment. Or where possible, such critical data might be left blank on the procedures, but posted on the equipment for use by the employee. It is not unusual for employees with proprietary process knowledge to leave one company, with or without documents, for employment with a competitor. Nevertheless, the advantages of written process procedures far outweigh the disadvantages.

Control of operations may include several functions, such as:

1. In-process inspection of materials, product, and package.
2. Process verification.
3. Status identification of raw materials, partially processed and finished product (ordinarily performed by use of a tag system; see Figure 2-1.)
4. Test equipment calibration. This would include meters, gages, media, reagents, metal detectors, check-weighers, production instrumentation, etc.

**Figure 2-1.** Suggested forms for controlling reagent standardization and for controlling nonconforming materials.

5. Nonconforming material identification, isolation, and release. An accurate history of nonconforming material must be maintained for the life of the product. Corrective action recommendations should be made a part of this system.
6. Document control: procedures, specifications, test methods, formulations, quality reports.
7. Material identification, rotation, and protection from infestation or physical damage.

Perhaps some emphasis on test-equipment calibration is advisable. There is a natural tendency to program a computer, install a micrometer, or calibrate a scale, and then walk away from the project believing that the job is finished. It is not. Others may change the calibration; vibration or wear might induce drift, or corrosion may affect readings. The reliability of the test equipment is as important as the quality control system itself. The simplest calibration system should encompass the following:

1. Verification of accuracy before permitting use. All new measuring and testing equipment, chemicals, reagents, scale weights, microbiological media and secondary standards should be verified or calibrated against certified standards, preferably having a known relationship to national standards.
2. Periodic verification, to be performed at defined intervals.
3. Control over the verifications and standardizations. This shall include tagging of equipment, where practical, with the date tested and the next inspection date shown on the tag. Additionally, a complete up-to-date record of the status of each measuring and testing item shall be maintained. The analytical methods used for standardization should also be subjected periodically either to collaborative tests or internal monitoring by control charts.

The final step in production quality control is finished-product inspection. Prior to the common use of statistical quality control, this single step was often considered to be a quality control system. Experience (and common sense) have shown that finished-product inspection is generally an audit of all of the quality subsystems that were used to assist in creating the product. The quality has already been built into the product at this stage, and nothing can be done other than accept, rework, or scrap. On the other hand, this is an important step, and should not be overlooked when constructing a quality system.

However, it should be emphasized again that finished-product inspection is not quality control. It is an inefficient way to attempt to assure consistent quality, and for many reasons it is ineffective. The individual responsible for finished-product testing can easily succumb to boredom, and fail to see defects. Or previously undiscovered flaws may be present but not observed, since the inspector has not been conditioned to look for them. But final inspection cannot be ignored—it is the last chance to discover quality problems before the consumer discovers them.

Finished-product testing can include examination for whatever attributes are included in the specification, but the crucial test should be performance of the product. The product should look, smell, dissolve, and taste right. It should be tested in the same manner as the customer or consumer would normally use it. If it is a teabag, the paper should not disintegrate in the cup—the customer is not interested in the number of grams of force required to separate the seam, but only that the tea leaves remain inside the bag.

Another class of finished-product inspection is acceptance testing to assure that production lots have met quality requirements. This may be accomplished either by 100% testing or by lot sampling, to be discussed in Chapter 5.

Storage, handling, and shipping may adversely affect the quality of food products, with the possible exception of some wine and cheeses. Lengthy storage permits settling of some emulsified liquids, and may be responsible for damage from corrosion, infestation, or degradation. Damage from improper handling and shipping would include leaking or deformed containers, scuffed labels, or crumbled product. Abnormally high-temperature storage of "commercially sterile" canned goods may result in thermophilic bacterial spoilage. Temperature control during storage of frozen or refrigerated products may be critical. Control of the quality during this phase of the product life starts with audit of shipping procedures and of stored items. Systems development for storage, handling and shipment generally rest with the distribution manager, but quality problems associated with these areas can be highlighted by quality control personnel, and perhaps some suggestions for quality improvement may be uncovered and suggested.

Product coding and case marking may seem like a simple housekeeping chore, but it can assume monumental proportions in the event of a product recall. Whether the problem in a recall is relatively small, such as forgetting to include a minor ingredient in a short run, or accidentally labeling the orange-flavored product "mandarin orange flavored," or whether it is a major product quality failure, proper coding can mean the difference between calling back a truckload or two from a small area, or bringing back product from the entire distribution area for many days of production because the code could not be identified.

Governmental requirements for coding are the minimum to be considered. If possible, all information should be included in codes: date, shift, line and plant of origin, plus batch, and product description. By the use of the Julian Date Code (the number of the day in the year) and a combination of letters and colors, a great deal of information can be inserted in small space. The month of the year can be designated as 1 through 9 for January through September, and the last three months might use either O, N, D or inverted 1,2,3; and the year could be expressed as a single digit; thus the date portion of the code need not exceed five characters. Universal product codes, label edge scoring codes, and other ingenious schemes might also be used where they do not conflict with regulations.

Finally, production quality records shall be considered. Since these tend to become voluminous in a short time, some procedures need to be established for retaining only those records which are actually useful. The general rule is to retain records for no longer than the normal life of the product. It is strongly suggested that this rule be modified somewhat. In order to retain skeletal data for use in comparing product quality progress over several years, and to retain records of prior formulations and methods in the event they may some day be used again, one scheme retains all records selected at random from one day per month for a period of one year after the normal product life, and then reduces older records by selecting at random one month per year for retention for two years. If these quality records are retained in a computer, there is no reason to destroy any of them, since they may be stored on inexpensive disks. But if the records are on paper, a system such as the one described is recommended. Following is a schematic representation of the records retention system for a product with maximum life of two years.

The goals of a quality records retention system need to be clearly established if they are to be effectively reached (Table 2-2). And there are many such goals:

1. Provide evidence that prescribed tests and inspections were made.
2. Identify personnel responsible for performing quality tasks in the event that they need to be questioned.
3. Accumulate rejections, scrap and rework data for calculations of quality costs (for cost reduction efforts).
4. Identify product, and trace materials, equipment, operations, and date of production.
5. Establish database for providing reports to management of quality performance, action taken, and recommendations for product, materials, process or cost improvement.

As with any system, success is based upon effective management. Unless the quality manager constantly reviews the goals of his programs, and takes action, when necessary, they tend to be overlooked, dooming the quality function to that of a paperwork generator. This no more apparent than in the collection of quality records. They contain much valuable information but, to be of use to the

Table 2-2.  A Records Retention system

| Year | Complete file | Retain one day's files per month for year no | Permanent file one day per year |
|---|---|---|---|
| 1 | 1 | — | — |
| 2 | 1,2 | | |
| 3 | 2,3 | 1 | — |
| 4 | 3,4 | 1,2 | — |
| 5 | 4,5 | 2,3 | 1 |
| 5 | 5,6 | 3,4 | 1,2 |
| 6 | 6,7 | 4,5 | 1,2,3 |
| 7 | 7,8 | 5,6 | 1,2,3,4 |

company, they must be studied, analyzed, and reported in digest form with recommendations for constant productivity improvement.

## Departmental Relationships

The quality control system cannot function in a vacuum. It must be integrated into the entire fabric of the company. Unfortunately, there is no magic formula to accomplish this. Cooperative relationships between departments cannot be expected to operate because of a presidential edict, and there is little to be gained by writing such relationships into the quality manual. In fact, telling other departments how to function is bound to feed some amount of antagonism into a company. The amount may be significant, or it may be negligible, depending upon personalities of those involved. And here is the heart of the matter. A food processing company is often thought of as a steel and concrete building, filled with noisy equipment and fragrant vapors, fed by truckloads of raw material and generating truckloads of finished product. It certainly is more than that: it functions only because there are people within it—all with diverse personalities, needs, intellectual abilities and goals. Men and women are all prone to jealousies and anger, pride and despair, energy and lassitude, ambition and despondency, cooperativeness and selfish independence.

Is it possible to create a quality policy which will lead all of these individuals into one common direction? Perhaps so. If it is true that only management can create a system under which quality control works, then it would seem logical to expect that a system could be evolved in which all departments participate to contribute to quality control. The following oversimplified example might suggest a way around the problem.

Let us consider a company in which the purchasing department is expected to supply the production line with least-cost packaging material. In the same company, the quality control department is expected to control the quality of the packaging at some predetermined level. Finally, the production department is expected to keep the packaging lines operating without interruption in order to minimize costs, eliminate a second shift or overtime. If, under these suppositions, purchasing supplies inexpensive but unreliable packaging, production will suffer numerous downtime episodes because of malfunction of the packaging material, and quality control will reject vast quantities of substandard production, thus requiring production to schedule additional shifts to make up for lost product. Each department, exclusive of the other two, may have done its job in exemplary fashion, but the total effort was a costly quality failure.

No amount of arguments between the departments will alleviate the problem. No department can be expected to tell another department how to run its affairs without adding to the conflict. Now let us bring enlightened management into the picture. If the company president decides the time has come to correct these costly problems, he might arrange a solution which overcomes the special needs of the individual departments. Purchasing, quality, production, and finance shall meet once each week to select jointly a packaging supplier whose material cost and quality level would result in the highest output, at the specification quality level, and at the least cost. We now have a system of purchasing packaging materials which overcomes personalities and departmental goals. This is a company goal-oriented system, and has a far better chance of success.

The problem and its solution have perhaps been oversimplified. It might, for example, be worthwhile to include other departments such as research or engineering to consider alternative package design or production methods or equipment which might use cheaper packaging material successfully. The principle of management systems is a valid one, however. It is equally applicable to the production, shipping, commodity procurement, and possibly personnel departments. It cannot completely overcome personality conflicts, but it can control them to a large extent.

This chapter is directed at the quality control department, not the president of the company. So far it has not supplied the quality manger with a solution to the conflicting departmental goals. Teaching quality control methods to the company president can only be accomplished in small steps, since his priorities are usually directed elsewhere; but small steps are certainly not undesirable.

It would be a near miracle to provide a company with a new and effective complete quality control program in a single step. However, one can be developed in phases. If, in the example above, the purchasing manager could be approached and shown how he could become a company hero by reducing the costs due to packaging problems on the line, then the first step toward interdepartmental quality operations would be launched.

In a similar fashion, solutions to quality problems on the line which are discovered through analysis of quality reports can be discussed with management in other departments for their corrective action. This will assist in cementing relations with the quality department and in gradually developing the company-wide quality team approach, frequently referred to as "total quality control."

## Product Performance

The objective for a company is to make a profit. The objective for the manufacture of a product by that company is to satisfy the consumer's needs and expectations. The consumer provides the final—and most important—audit of the product quality, and contact with the consumer by the quality control department is the only way to really understand if the product quality is satisfactory. The contacts may be direct through form letters, coupons, mailings, or meetings. These techniques may be administered by outside consultants, internal sales and marketing departments, or by quality control.

Whenever practical, potential users of a proposed new product should be contacted to determine its intended use and practical requirements. Once a product is in the market, care must be taken to review dissemination of product information to the public. It should reflect the needs of the consumer, as well as truthful representations of the performance, quality and safety. The advertising should be reviewed for accuracy and conformance to regulations. Statements such as "twice the strength," "instantly soluble," "will keep forever in an unopened container," "sugar-free," "preferred by 80% of the product users," "organically produced," "reduced calories," and "guaranteed fresh" all require supportable evidence from the technical staff of either the research department or the quality control department. The quality system should provide the mechanics for reviewing marketing strategies where such advertising is proposed.

If nothing else, contact with the consumer in the form of a consumer complaint must be handled quickly and with care. It has been claimed that a single consumer complaint may represent somewhere between 20 and 200 other consumers who had the same negative experience, but merely changed brands rather than writing to the manufacturer. Each company may have its own preferred procedures for responding to the complaint, but in all cases, a copy of the complaint and the response should be forwarded to quality control so that efforts to prevent recurrences can be started immediately. Complaints may fall into many different categories, depending upon the product. They may be concerned with flavor, odor, foreign material, appearance, net volumes or net weights, presumed health effects, deterioration, etc.

Complaints which fall within the product specification may be classified as nuisance complaints, and handled by replacement of product to the consumer. Where a product has apparently been produced outside of specifications, attempts should be made to obtain the package code date so that an immediate investigation can be performed. Remembering the 20 to 200 unwritten complaints (see above), this one might represent a serious quality problem, perhaps even a recall. Finally, where a product complaint indicates health hazards, all information should immediately be turned over to the company's insurance handler or legal department. No direct contact with the customer by other company personnel is advisable because of possible legal liability considerations. However, close contact between technical and legal is essential in order to establish the facts quickly.

Individual responses to consumer complaints are not the end of the story. Complaints should be compiled into categories by month, and cumulative for the past 12 months (do not use cumulative data for the current calendar year—this can be misleading). The data should be listed by number of complaints and by percent. If the company is unfortunate enough to have a large volume of complaints each month, control charts can be constructed to highlight both trends and emergencies. In any event, use of percentage figures will highlight unusual situations. All data should be considered confidential, and not distributed throughout the company except as required. A sample of such a compilation (from a fictitious company) is shown in Table 2-3.

An improvement over simple tabulation of complaints would be to include the sales figures as well as the complaints. Graphs showing types of complaints by product by month along with the number of units sold during that period may show that seasonal increases in complaints are accompanied (adjusting for lead

**Table 2-3.  Consumer Complaint Record**

| Complaint description | 12-month cumulative | | Jan | | Feb | | Mar | | Apr | |
|---|---|---|---|---|---|---|---|---|---|---|
| | No. | % | No. | % | No. | % | No. | % | No. | % |
| Weak | 82 | 25 | 5 | 17 | 6 | 14 | 3 | 8 | | |
| Too yellow | 132 | 41 | 16 | 55 | 12 | 29 | 22 | 60 | | |
| Bitter | 28 | 9 | 2 | 7 | 1 | 2 | 3 | 8 | | |
| Sticky | 18 | 6 | 0 | 0 | 1 | 2 | 0 | 0 | | etc. |
| Glass | 1 | — | 0 | 0 | 0 | 0 | 0 | 0 | | |
| Dirt, sand | 54 | 17 | 5 | 17 | 4 | 10 | 9 | 24 | | |
| Unsealed | 7 | 2 | 1 | 4 | 18 | 43 | 0 | 0 | | |
| Total | 322 | 100 | 29 | 100 | 42 | 100 | 37 | 100 | | |

*Note:* The abnormal figure for unsealed in February.

time) by increases in sales. To further improve the interpretation of this chart, the data might be plotted as number of complaints per thousand (or hundred thousand, million ...) per month, which would de-emphasize the seasonal nature of complaints. Simple computer programs are available which eliminate mathematical computations. It might be noted that seasonal complaints are not completely tied into seasonal sales data. It has been observed that in the colder and wetter parts of the sales areas, particularly during the winter months, the volume of complaints tend to rise over those from milder climates. This is probably due in part to some bored and perhaps irritable people, indoors and isolated by the weather, searching for amusement by writing letters of complaint for real and imagined product failures. Still, all complaints should be considered as legitimate until proven otherwise.

Another class of psychological complaint frequently occurs when a new product is introduced. Consumers with a negative attitude find change disagreeable and, though it is difficult to explain why they would try a new product in the first place, they may tend to complain about it because "it is different, therefore no good." Other consumers may be confused by a new or novel package, flavor, or product application. More often, however, the unfortunate truth is that there were still a few "bugs" in the new product which had not been discovered before the initial sales period. One can count on 100% inspection performed by the consumer to discover these failings. The more optimistic outlook regarding new product complaints is that inevitably, the complaints per number of units sold decreases over time.

For those classes of complaints which generate fairly large numbers, it is possible to calculate the standard deviation, and construct three-sigma control charts to signal an alarm whenever an abnormally high number occurs. On the other hand, it would be far more logical to spend the time and effort to eliminate the causes for those classes of complaints, rather than to construct control charts.

## Corrective Action

We shall define corrective action as creating a change in development, production, or distribution of the product to eliminate the risk of a quality failure. Note that repair or rework of a defect does not fully comply with this definition. Corrective action requires periodic assessment of quality failures by product, by failure mode, and by material suppliers. The analysis of specific failures may require an evaluation of the quality system to prevent recurring failures.

Reviews of quality reports to management are the usual starting place for corrective action, but such requests might be originated by any department or individual within the company. Under these circumstances, the request should be written, since this encourages the originator to think the problem through, and provide sufficient information to launch an investigation. Even at this early stage, the projected cost of corrective action should be considered to determine the economic justification of further analysis and change.

Agreeing with Dr. Deming's principles of quality control, the Food Processing Industry Quality System Guidelines recommends designating the responsibility for the quality problem as management-controllable or operator-controllable. If the operator is in control, it must be established that the operator knows what is expected of him, that he is able to determine how well he is conforming to those expectations, that he is able to adjust the process if he is not conforming, and that he is motivated to achieve what is expected. Management factors would include systems considerations related to design, materials, equipment, manufacturing, calibration and inspection methods, training, and personnel.

Statistical analysis of causative factors may lead to possible solutions. Once the important variables have been identified, appropriate controls should be established to provide manufacturing with the assurance that they can maintain those variables within established limits. The new controls must then be added to the list of audited tests and test data.

In the event of a serious or catastrophic quality failure, a withdrawal or recall of finished product must be considered, and it is mandatory that the quality system includes a series of procedures for quickly responding to such failures. Corrective action decisions regarding failures include evaluation of all manufactured product which might be affected by the same quality problem. Production and distribution records should be available for determining where the product is stored: on the production floor, the factory warehouse, in transit, in outside warehouses, retail distribution system, or in the hands of the consumer. Procedures for locating and retrieving suspect product should be ready for immediate implementation after whatever approvals required by company policy have been obtained.

A recall at any level is a serious event. Every food processing company is expected to have a recall system primed and ready to go at any time. Undetected "six-sigma events" which are beyond the control of management do occur: shattered mixer paddle, exploding light bulb with defective shield, incorrectly certified materials, corroded valve seat, disintegrated contact sensors, atmospheric contamination, and even sabotage. This is the one program, more than any other, which must be formalized, documented, and maintained current. Considerable assistance is available to set up such a program. The Code of Federal Regulations Number 21, Part 7, covers FDA regulations. The Consumer Product Safety Commission, CPSC Act 1972, Section 15 also refers to recalls. The Grocery Manufacturers of America, Washington, DC has published an excellent pamphlet which describes a recall system for food products. Once a system has been established, it should be periodically tested by performing a mock recall. This will quickly verify that either the system is workable, or that there are deficiencies in either the system or in the record keeping procedures in various departments, or perhaps in outside warehouses. These deficiencies must be immediately remedied, and tested again at a future mock recall.

Company policies differ on the desirability of written reports of quality failures, such as recalls. Some companies may prefer that no written material be prepared; others use recall records as training materials for certain employees. It is suggested that those responsible for company legal matters be consulted on this matter. A concise report covering steps taken in the event of a recall may be required by the legal representative as defense material to be used, should litigation arise as a result of product quality failure.

## Quality Personnel

In classical industrial engineering system studies, the four "ms" considered are men, materials, methods, and money. So far in this discussion of quality systems, we have only considered the last three items. And yet, the best quality system will not function without people who are carefully selected, trained, and motivated. Regardless of which department is responsible for recruitment of quality control personnel, the responsibility for identifying the characteristics of a satisfactory quality employee must be assumed by quality control management. Quality control need not be staffed by the same type of outgoing and personable individual required by, for example, the sales department. Nor should the qualifications rigorously insist on the often mistaken need for a stereotype of an unbending police-type. Such character traits as dedication and honesty are required, along with alertness and a proficiency in at least basic mathematics. Mechanical aptitudes are often useful as well. The process for recruitment and selection of employees differs from company to company, but the final approval should always remain with the quality manager. Vacancies may be filled by promoting from within, reviewing resume files, newspaper and trade journal advertising, employment agencies, and other channels. Consequently, the manager is expected to have a list of candidate employee qualifications available for guidance. The final selection of employees should be based on their capability and experience, or their potential to qualify for the job.

Large corporations may have training departments which can quickly indoctrinate a new employee with company policies and procedures. Large or small, however, a company quality manager is still responsible for training new employees in their new assignment. Company policies and procedures manuals are useful for training newcomers to any department. The special training required for quality department newcomers is highly specialized. It must be general in nature, but at the same time, specifically oriented to the company's operations and quality policies. Study of the quality manuals is perhaps the quickest (and most uniform) method for training quality control personnel. Quality control personnel from other industries may not be aware of the critical nature of quality in the food industry. If a lawn mower is delivered with its blade's cutting edges installed backwards, it can be repaired or returned for replacement. Not so with foods: if a food product contains a major defect which might affect the health of the purchaser, it cannot be repaired , and might cause serious consequences. Many hardware industries deal with slow production lines of expensive products which are controlled by indirect nondestructive test methods; most food processing plants operate high-speed processes, requiring different quality testing techniques. All quality employees should be required to take refresher courses geared to introducing new concepts and to the eradication of complacency, and elimination of "short cuts."

The need for training of line and supervisory employees in the principles of quality performance is best handled by personnel other than those from quality control. The production employee would willingly agree that the principles of quality workmanship would make the quality control department "look good." But at the same time, the possibility of reducing production rates by using careful and precise quality methods might make the production line supervisor "look bad." Of course this is not necessarily true, but the problem exists, nevertheless. By utilizing trainers from other departments, or from outside the company, this apparent conflict is reduced. Obviously, the best control of quality performance is achieved when both the production personnel and the quality control personnel are all effectively trained in the principles of quality effort.

Motivation is another key to the success of quality control, and there is no one formula for providing this tool. It is important for quality employees to understand the consequences of a slip-shod performance. It is equally important for the line and supervisory personnel to be aware of this. A thorough knowledge of the principles of quality effort, as mentioned above, is required for quality control; but without the motivation to actually perform under these principles, the results will be less than acceptable. Motivational tools are numerous, and many are effective. Unfortunately, most are short-lived, and when applied a second time, there is frequently a disinterest, or even an antagonism exhibited on the part of the employees. There is one motivational tool which seems effective in practically all situations: recognition of achievement. How this is best applied must be carefully thought out. Setting up a reward system consisting of a dinner with the plant manager may work once or twice, but it may not be considered a reward by many. Worse, it is difficult to bring the program to an end or to replace it with another reward. (And this hardly considers the plant manager's long-term feelings about such an arrangement.) Recognition can be as simple as a kind word from the company manager to an employee at his workstation. Financial awards, vacations, promotions, pronouncements at the Christmas Party or in the company newspaper, plaques, gifts, posters—these are some of the many ways to recognize efforts and to provide motivation, some of which are more effective than others. Whatever method is selected, it should be made clear to the recipient and all of his fellow workers that the award is made in recognition of outstanding performance—it definitely is not a bribe to do the job correctly.

# 3    Control Charts

## THE IMPORTANCE OF CHARTING

The ancient saying, "one picture is worth a thousand words," is somewhat time-worn, but far from worn out. In a world where statistics and columns of numbers are not well understood, a picture of a process may be far easier to grasp than a quality summary laced with standard deviations, averages, equations, and numerical computations. The control charts used in statistical quality control represent a picture of a process. When used and continuously updated on the production floor, charts represent a moving picture of the process.

The frequent misuse, or occasional fraudulent misrepresentation of charts and graphs, has placed them under suspicion. For one example, let us suppose we are considering the increase in the cost of changing raw product for a process from standard grade to choice. In preparing a report on this project, a simple bar chart (Figure 3-1) may be used to compare the cost in 1998 and the cost in 2003 when the new material is intended to be used. The dollar column on the left of the chart below represents the 1998 cost; the column on the right shows the year 2003 cost at about double the amount.

To emphasize the difference, we may choose to represent the increase in costs by a three-dimensional picture. Here, as before, the vertical stack height of the year 2003 is double the height of the 1998 dollars, but by doubling the depth and

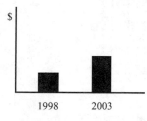

**Figure 3-1.**  Bar chart comparing costs for two years.

width as well, the appearance of the new costs is dramatically (and untruthfully) magnified. The year 2003 dollars are not the same dollars as those of 1998 (Figure 3-2).

Another commonly used trick of reaching unwarranted conclusions by manipulating graphs is the magnifier principle, frequently practiced in newspapers to prove a point. Although it might be useful occasionally, extreme caution should be taken when employing this technique. For example, in Figure 3-3 the left graph shows the level of petty cash in an office safe over the years. The enlarged graph on the right has selected the most recent week with grossly magnified axis values, suggesting that petty cash is skyrocketing. In truth, it is most likely fluctuating in the general range where it has wandered over the past few years.

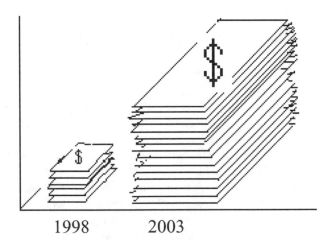

**Figure 3-2.**   Three-dimensional diagram exaggerating costs for year 2000.

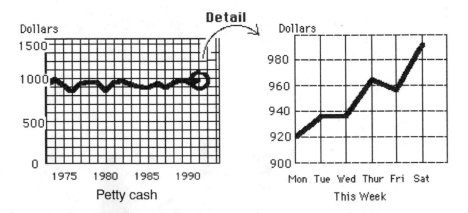

**Figure 3-3.**   Misleading effect of an enlarged portion of a chart.

## THE PIE CHART GAME

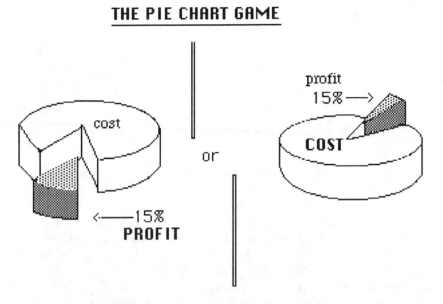

**Figure 3-4.**  Misleading effect of a pie chart.

Pie charts (Figure 3-4) may be used to represent percentages or portions of a population, and are sometimes useful. They are difficult to interpret if many sections are shown, and they can be misleading if they are portrayed in three-dimensional perspective. Note in Figure 3-4 how a 15% portion in the front of the pie can emphasize the profit of a proposal in the left pie slice; by contrast, note how the same 15% placed in the rear of the pie (and with a change in type size and font) suggests that the costs of this proposal are prohibitive compared to the meager profits.

On the other hand, charts and graphs are valuable tools for presenting statistical data. Of particular interest are the control charts used universally to present quality data. They are sufficiently simple to interpret so that misunderstandings are avoided. Regardless of the type of control chart, they all contain a few fundamental characteristics (Figure 3-5). They contain upper and lower control limits within which all observations will lie if the process is in control. They contain a center line which is usually considered to be the target value for the process. And they generally show numbers along the vertical axis to define the values of the control limits and of the observations. Beyond these basics, the charts may be tailored to suit the requirements of the operation.

If one were to ask for the basis of statistical quality control, there might be considerable discussion; but the $X$-bar and $R$ chart would rate very high on the list. These two charts are easy to prepare, simple to understand, and extremely useful in locating problems and even in suggesting possible solutions. They are ideal tools for discovering ways to improve product quality and process control, and they can drastically reduce scrap and rework while assuring the production of

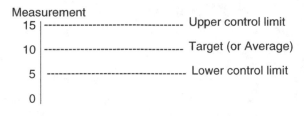

**Figure 3-5.** Characteristics of a control chart.

only satisfactory product. They are used for controlling every step of a production process, for the acceptance or rejection of lots, for product improvement studies and for early detection of equipment or process failures. An example of an X-bar control chart taken from an orange-flavored dry drink mix packaging line at the filler discharge shows the citrate concentration as determined periodically by the quality control laboratory (Figure 3-6). (The average line has been omitted for clarity.) At nine o'clock, a point appears just above the control limit. Since, by chance alone, this may happen as infrequently as three times per thousand observations and still represent a process in control, no action is indicated. As the chart subsequently shows, the next few points are indeed within limits, and the process is allowed to proceed. At 12 o'clock, another point is out-of-limits, and the recheck point immediately following shows, without doubt, that the process is far out of control. The line was stopped, and it was determined that an operator had inadvertently dumped a drum of lemon drink mix into the hopper during the noon shift relief change. The citrate control limits for the lemon mix product are 6 to 8 units, which would explain the increase. Additional subsequent evidence obtained by the laboratory confirmed the mistake by product color analysis, and flavor evaluation.

In addition to X-bar and R charts, a group of charts loosely defined as *attribute* charts are used for control of defect analysis. They are particularly useful for controlling raw material and finished product quality, and analyzing quality comments in consumer letters. An attribute is a characteristic of a product, a process, or of any population, which can be counted or tallied, but which otherwise cannot be described in incremental numbers. It would be more meaningful to define an attribute as a characteristic which is either satisfactory or unsatisfactory; go-or-no-go; defective or nondefective; good or bad; heavy or light; etc. The only numbers which can be applied are the number or percent of the satisfactory or unsatisfactory units. These charts (which will be considered later on) are the p charts (fraction or percent defective), np charts (number of defectives), and c or u charts (number of defects). First, we shall explore the X-bar and R charts.

To start with, the X-bar and R charts are used for control of variables. Variables are measurements which are expressed in discrete numbers: inches, pounds, pH units, angstroms, percent solids, degrees centigrade, etc. In the case of a leaking container, a variable measurement would be the rate at which gas flows through the leak in cubic centimeters per hour. An attribute measurement, on the other

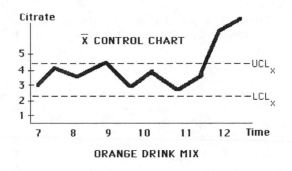

**Figure 3-6.** *X*-bar control chart for orange drink mix.

hand, would be the number of leakers or nonleakers in a batch or in some time period.

*X*-bar is usually written as $\overline{X}$, and is the average value of several measurements, each of which is called *X*, and generally identified by the means of subscripts: $X_1$, $X_2$, $X_3$, etc. As we shall learn later (in Chapter 4, Fundamentals), some distributions of data may be lopsided (skewed) because of peculiarities of the process from which the data has been obtained. It will also be explained that taking averages of data from a process, regardless of the distribution of the data, tends to reduce these irregularities, resulting in a smooth distribution. Hence, a graph of a series of *X*-bars shows a smoothed representation of the process being measured. This makes it easier to determine when the process is out of control, since smoothed data will lie in a relatively narrow range.

A possible weakness of an $\overline{X}$ chart is that individual points on it are represented by averages of data which might contain wide ranges of values, but which are masked by the very smoothing effect that makes these charts so desirable. To overcome this difficulty, the range of data from which each average was obtained is also required, and in turn, its control values (upper and lower) must also be calculated.

## PROCEDURE FOR CONSTRUCTING *X*-BAR AND *R* CHARTS

The preceding verbose description is a good argument for the statement that a picture is worth a thousand words. With that in mind, we shall proceed to construct an actual control chart (Figure 3-7). A first example will be a control chart for headspace in one-pound jelly bean jars packed on one of the six lines in the plant. Our concern is that insufficient headspace is likely to crush the product when the lid is applied, and that excess headspace will allow the product to abrade during transport. Experience has shown that a headspace of about 3/32 or 4/32nds of an inch is satisfactory, but plant operating people are not certain of what any of the lines are capable of producing. We will work with Line #3 to start. A preliminary chart can be prepared from about 100 headspace measurements. For a more

**Table 3-1.  *X*-bar and *R* Chart Calculations.
Headspace—One Pound Jelly Bean. Line No. 3**

| Sample number | | | | | Average ($\overline{X}$) | Range ($R$) |
|---|---|---|---|---|---|---|
| 1 | 2 | 3 | 4 | 5 | | |
| 7 | 10 | 8 | 2 | 8 | 7 | 8 |
| 6 | 9 | 3 | 4 | 3 | 5 | 6 |
| 6 | 7 | 2 | 6 | 4 | 5 | 5 |
| 4 | 8 | 5 | 7 | 7 | 6.2 | 4 |
| 10 | 3 | 5 | 6 | 4 | 5.6 | 7 |
| 8 | 11 | 9 | 6 | 6 | 8 | 5 |
| 5 | 7 | 5 | 6 | 7 | 6 | 2 |
| 9 | 5 | 8 | 7 | 6 | 7 | 4 |
| 6 | 5 | 4 | 5 | 5 | 5 | 2 |
| 1 | 3 | 1 | 0 | 3 | 1.6 | 3 |
| 3 | 4 | 6 | 5 | 4 | 4.8 | 3 |
| 5 | 3 | 6 | 3 | 3 | 4 | 3 |
| 5 | 6 | 8 | 9 | 7 | 7 | 4 |
| 7 | 7 | 8 | 7 | 6 | 7 | 2 |
| 7 | 7 | 6 | 7 | 7 | 6.8 | 1 |
| 7 | 7 | 7 | 9 | 9 | 7.8 | 2 |
| (add columns) Totals | | | | | 93.8 | 61 |
| (divide by 16) Averages | | | | | 5.86 | 3.81 |

$A_2$ for sample of $5 = 0.577$
$D_4$ for sample of $5 = 2.115$ (from tables)
$D_3$ for sample of $5 = 0$

realistic working chart, it is customary to obtain 30 subgroups of five samples
for a total of 150. (Refer to the data and calculations in Table 3-1 as we go through
the following steps.)

1. When Line #3 is operating satisfactorily, select about 100 jars off the line
   before the lids are applied. All jars must be selected in sequence, and kept
   in order until measured.
2. Measure the headspace in each jar, in order, and record in groups of five,
   as shown in Table 3-1. (Sample 1 measures 7/32, and is recorded in the
   first column as "7." Sample 2 measures 10/32, and is recorded in the second column as 10. Continue until reaching sample 6, which measures
   6/32 and is recorded in the first column under the first sample taken.
   Eventually 16 rows of five columns will be completed. This totals 90 samples—close enough to the 100 generally required for a first attempt. The
   reason for the five columns will be explained shortly.

3. Calculate the averages for each row. For example, $7 + 10 + 8 + 2 + 8 = 35$. Divide 35 by $5 = 7$. Record under "Average" heading.
4. Record the ranges for each row under the "Range" heading. For example, in row 1 subtract 2 from 10 and list 8 under Range.
5. Add all of the averages (93.8 total) and divide by the number of rows (16) to determine the grand average (5.86).
6. Add all of the ranges (61 total) and divide by 16 to determine the average range (3.81).
7. Refer to Appendix Table A-8, "Factors for Computing Control Chart Limits," and find 0.577 under column A2 for sample size 5. Note that the values for samples 2 through 4 are quite large and decrease in substantial quantities; but starting with sample of size 5, these numbers decrease at a much smaller rate. This is the reason for combining samples in groups of five. If more precision is required at some later date, the test may be repeated with larger groupings and more samples to obtain 30 subgroups.
8. Using the same table, find 2.115 under column D4, and 0 for D3. We now have prepared all of the data necessary to calculate the control limits.
9. Calculate the upper and lower control limits from the formulas as shown.

$$UCL_X = \overline{\overline{X}} + A_2\overline{R} = 5.86 + 0.577 \times 3.81 = 8.06$$
$$LCL_X = \overline{\overline{X}} - A_2\overline{R} = 5.86 - 0.577 \times 3.81 = 3.66$$
$$UCL_R = D_4\overline{R} = 2.113 \times 3.81 = 8.05$$
$$LCL_R = D_3\overline{R} = 0 \times 3.81 = 0$$

10. Plot headspace against sample number, in which the sample number refers to the average for each set of 16 sample groups.
11. Draw the upper and lower control limits (8.03 and 3.64), and the grand average (5.84). This completes the *X*-Bar Control Chart
12. Plot headspace against sample number in which the sample number refers to the range for each subgroup of 5 samples.
13. Draw the upper and lower control limits (8.06 and 0) and the average range (3.81). This completes the Range Chart (Figure 3-8).

There are several points of interest in these simple charts. Note that in the control chart for averages, even in this short burst of production, one of the points (number 10) is out of limits for headspace on the low side. Whether this is a desirable or undesirable measurement might be worth knowing. Some effort should be made to find out what might have been responsible for this apparent out-of-control jar. Perhaps the beans are oversized, there is less breakage, or the jar is overfilled. There is always the possibility that one of the filling machine heads has some peculiarity which might explain the abnormality. In any event, if the answer is found, action should be taken to assure that this will not occur again (if undesirable), or to assure that it will always occur (if a lower headspace is desirable). The fact that the remainder of the measurements are in control should not be surprising. After all, the limits for control were defined by the data which is plotted.

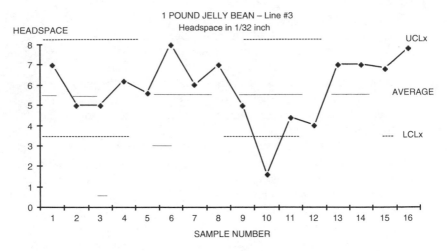

Figure 3-7.   *X*-bar control chart for jelly bean jars.

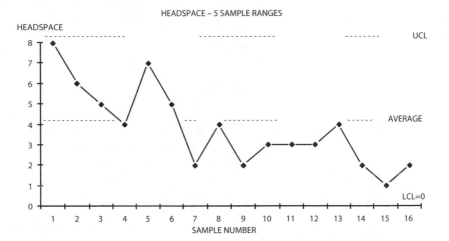

**Figure 3-8.**   Range chart.

One conclusion which may be drawn from this initial effort is that the "satisfactory" limits of 3/32 to 4/32, used informally in the past as specifications, are not attainable under this process as it now exists. The current process can run successfully between 3.6/32 and 8/32, and if the tighter specifications are truly important to the quality of the product, considerable work lies ahead to improve this process.

Having developed this initial control chart, the limits should be used every day for a while, plotting headspace averages of five samples taken off the production line every 30 min. It should not take long to find out if the limits are realistic. They should be refined as more data is gathered, until it can be decided whether process improvement is needed. Should that point be reached, simple improvements, which management is aware of but has been putting off, should be evaluated and installed. If improvement beyond this is required, a series of experiments should be designed to investigate other means of reaching the specification goals. To measure the effects of improvements, continue to plot the five sample averages, and if a shift is suspected, prepare a new control chart showing the tighter control limits.

It might be noticed that the range control chart seems to have a downward trend. Since all of the values are within the control limits, it is probable that this unusual pattern has occurred by chance alone. On the other hand, it might be worth an initial investigation to find out if there is some special cause responsible for what appears to be more uniform weight ranges.

The chart for averages (Figure 3-7) shows that the headspace is higher than it should be for optimum performance; the range chart (Figure 3-8) shows more discouraging information. It indicates that even though the average headspace could be reduced to a more acceptable level, the range between readings is still going to be very high. At the moment, the range is plus or minus 4/32, and the present hoped-for specification limit is very much less at plus or minus 0.5/32. It now becomes obvious that there is work to be done to improve the performance of this line, and that it should probably start by attempting to remove the wide range of headspaces currently produced. After that, the target level should also be reduced.

In subsequent chapters on process capacity and process control, there are detailed discussions on the meaning and use of quality control charts such as the above.

## PROCEDURES FOR CONSTRUCTING ATTRIBUTE CHARTS

There are four commonly used attribute charts, and each has a specific use. They are generally easier to construct and to use on a routine basis, though they occasionally lack the power of variable charts to spot problem areas quickly. One of their major advantages is the simple nature of the concept. They are easily explained to line workers and management alike. Most everybody understands the meaning of "percent defective" or "number of defects in the lot." The mathematics do not require tables. They can be applied to systems where measurements consist of pass-fail and variable measurements are not possible or are difficult to obtain. These are the charts (Table 3-2):

1. *p chart with constant lot size*: Used to determine control of percent defective units, and to establish whether the process is in control for the day (week, month). "Constant" means "within 20%."
2. *p chart with variable lot size*: Usually intended to control percent defective units where the number of units varies from sample to sample. Determines if a process is in control for each lot's control limits.
3. *np chart (also known as m chart)*: Used to control the number of defective products in each lot, and to assure that the process is in control. Requires constant lot size.
4. *c chart*: Used to determine if the number of defects in a single product is within control limits. Final inspection.

**Table 3-2. Control Charts**

| Symbol | Name | General use | Application | Based on | Lot | Sample |
|---|---|---|---|---|---|---|
| X/R | Shewhart | Locate assignable cause of process shift | Process control by variables (inches, ml, pH, ounces, count, etc.) | Normal distribution | Constant | Constant and arbitrary |
| p | Fraction or percent defective constant lot | Detect process change | Process control by attributes: leaks, flavor, blemishes, miscounts, seeds, dents, etc. | Binomial distribution | Constant (within 20%) | Constant, 100% or sample |
| p | Fraction or percent defective variable lot | Detect process change | Process control by attributes | Binomial distribution | Varies | Varies 100%, average or stabilized |
| np | Number of defectives | Detect process change | Process control by attributes | Binomial distribution | Constant | Constant |
| c | Number of defects constant sample | Accept or reject item | Process control by attributes | Poisson distribution | Constant or continuous | Constant |
| u | Number of defects, variable sample | Accept or reject item | Process control by attributes | Poisson distribution | Varies, or one unit | Varies |

**Table 3-3.  Comparison of Attribute Charts**

| | $P$ charts | $np$ chart | $c$ or $u$ chart |
|---|---|---|---|
| Chart name | Fraction or % defective | Number of defectives | Number of defects |
| Examine for | Defective unit | Defective unit | Defects in unit |
| No. of items in sample | $n$ | $n$ | $n$ (usually 1) |
| Defectives in lot | $np$ (or $m$) | $np$ (or $m$) | — |
| Defects in lot | — | — | $c$ |
| Average fraction defective | $\overline{P} = \dfrac{\Sigma m}{\Sigma n}$ | $\overline{P} = \dfrac{\Sigma m}{\Sigma n}$ | — |
| Average defectives | — | $\overline{m} = \dfrac{\Sigma m}{N}$ | — |
| Average defects | — | — | $\overline{c} = \dfrac{c}{N}$ |
| Control limits | | | |

$$CLY_p = \overline{p} \pm 3\sqrt{\frac{\overline{p}(1 - \overline{p})}{n}}$$

$$CLY_m = \overline{m} \pm 3\sqrt{\overline{m}(1 - \overline{p})}$$

$$CLY_c = \overline{c} \pm 3\sqrt{\overline{c}}$$

In Table 3-2 there are three classes of attributes considered: fraction defective ($P$), number of defectives ($np$), and number of defects ($c$, $u$). The mathematics involved in finding the averages is simply a matter of dividing the total number of non-acceptables by the number of samples. The calculation for control limits is based on this figure, and again is an easy calculation. The formulae are shown in Table 3-3.

## Examples
Examples of the three basic types of charts have been selected from different industries to illustrate the broad application of these techniques. The first example is the construction of a $p$ chart for fraction defective.

## Example 1    Percent defectives chart: $p$ chart with constant lot size
In a plant producing drums of dill pickle chips, a day's production varies from 476 to 580 drums (Table 3-4). Over the years, the major types of defective drums have been found to be leaks, contamination, discoloration, mislabeled drums, off flavor, overweight and underweight. In all, there are 18 different types of defects which have been observed. The

**Table 3-4.  Defective Pickle Drums—100% Inspection**

| Sample no. | Date | No. of drums $n$ | Defectives $m$ | Fraction defective $p$ |
|---|---|---|---|---|
| 1 | 4 Oct | 502 | 18 | 0.036 |
| 2 | 5 | 530 | 13 | 0.025 |
| 3 | 8 | 480 | 13 | 0.027 |
| 4 | 9 | 510 | 15 | 0.029 |
| 5 | 10 | 540 | 21 | 0.039 |
| 6 | 11 | 520 | 17 | 0.040• |
| 7 | 12 | 580 | 28 | 0.048• |
| 8 | 15 | 475 | 10 | 0.021 |
| 9 | 16 | 570 | 23 | 0.040• |
| 10 | 17 | 520 | 16 | 0.031 |
| 11 | 18 | 510 | 15 | 0.029 |
| 12 | 19 | 536 | 22 | 0.041• |
| 13 | 22 | 515 | 18 | 0.035 |
| 14 | 23 | 480 | 12 | 0.025 |
| 15 | 24 | 548 | 24 | 0.044• |
| 16 | 25 | 500 | 11 | 0.022 |
| 17 | 26 | 515 | 19 | 0.037 |
| 18 | 29 | 520 | 16 | 0.031 |
| 19 | 30 | 485 | 13 | 0.027 |
| 20 | 31 | 520 | 14 | 0.027 |
| 21 | 1 Nov | 515 | 12 | 0.023 |
| 22 | 2 | 545 | 25 | 0.046• |
| 23 | 5 | 515 | 16 | 0.031 |
| 24 | 6 | 505 | 13 | 0.026 |
| 25 | 7 | 518 | 15 | 0.029 |
| 26 | 8 | 484 | 12 | 0.025 |
| 27 | 9 | 520 | 22 | 0.042• |
| 28 | 12 | 535 | 22 | 0.041• |
| 29 | 13 | 518 | 14 | 0.027 |
| 30 | 14 | 554 | 26 | 0.045• |
|  | Total | 15,565 | 515 |  |

$$\bar{p} = \frac{\sum m}{\sum n} = \frac{515}{15,565} = 0.033 \,(\text{or } 3.3\%)$$

$$\bar{n} = \frac{15,565}{30} = 519$$

$$\frac{(580 - 475)}{519} \times 100 = 20\% \text{ variation}$$

plant manager believes that on the average, his operation produces no more than 2.5% defectives (or .025 fraction defective), based on the number of customer complaints he receives over the year. Using this meager evidence, he has set product specifications at 0 to 0.040 fraction defective (or 4% defective) drums. Because of a serious quality complaint from a major customer, a quality consultant is called in to find out if the specifications are being met. The consultant is charged with the responsibility of a) determining if the specifications are realistic, b) if the process is in control. If needed, the consultant is requested to make suggestions on how to improve quality and to maintain it at a higher level. The relatively few drums produced each day permit a test program in which the output is 100% inspected. If any defects are found in a drum, it is counted as a defective; if more than one defect is found in a single drum, it is still counted as a single defective. Data for 30 successive production days are presented in Table 3-4.

As shown by the data marked with •, there are many drums (nearly a third of the production) which equal or exceed the plant manager's arbitrary 4% defective specification limit. The consultant now proceeds to calculate the average fraction defective and average number of drums checked. In addition, the variation of sample sizes is also checked to see if a "constant size" chart will suffice.

The average number of defective drums is not the 2.5% which the plant manager had believed, but is actually 3.3% during this 30-day run. These numbers (0.033 and 519) are all that are required to calculate the control limits for the process.

Upper control limit

$$\bar{p} + Y_p = \bar{p} + 3\sqrt{\frac{\bar{p}(1 - \bar{p})}{n}}$$

$$= 0.033 + 3\sqrt{\frac{0.033(1 - 0.033)}{519}}$$

$$= 0.033 + 0.023 = 0.056$$

Lower control limit

$$\bar{p} - Y_p = 0.033 - 0.023 = 0.010$$

Expressing these findings in terms of percent defective (moving the decimal point two places to the right), the control limits are 1% to 5.6% defectives, and the average is 3.3% defective. The consultant is now ready to plot the 30 day's data on a control chart (Figure 3-9) to show the data in pictorial form. The chart shows that the process is within the statistical control limits, with average defective of each day's drums ranging from about 2% to nearly 5%. The arrow points to the 4% maximum specification limit which the plant manager had optimistically hoped not to exceed. A more realistic specification would be 6% maximum defective drums. The chart more clearly pictures the large number of drums exceeding the present specification than does the list of numbers in the inspection sheet on page 59.

Having achieved this overall view of the plant's capabilities, and finding them less than satisfactory, the next step is to attempt to improve the process. We will

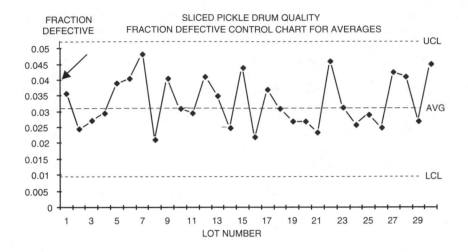

**Figure 3-9.** Fraction defective control chart, constant lot size.

not follow this example to its conclusion, but will indicate the general direction the consultant will probably take. He will now pinpoint the causes for the defective drums, using a control chart for *defects*, rather than the broader concept of *defectives* shown in the above chart.

A chart for defects will show some periods where defects are very numerous at or near the upper control limit. The causes should be immediately identified, and new procedures installed to be sure that the problems are not repeated. Conversely, the defect chart also will likely show that on some days the plant is capable of producing fewer defects than others. By identifying the reasons for these improvements, additional procedure changes can be installed which will ensure that they will be a permanent part of the operation.

Although the processing line is the first place to look for these improvements, there are other areas which may be equally as responsible for defects, and the following possibilities should be explored.

- Purchase a better quality of raw materials.
- Work with the drum supplier to eliminate substandard drums.
- Engineer more gentle handling systems to reduce line damage.
- Identify hazardous and critical control points so that national inspection locations can be installed.
- Improve the training of processing personnel.
- Revise maintenance schedules so that equipment (scales, slicers, meters, mixers, thermometers, etc.) will function more reliably.
- Study the quality drift cycles to precisely specify sampling schedules.

When the improvements are all in place, and the improvement rate in the defect chart is leveling off, then it is time to return to the fraction defective control chart for averages, and to calculate the new process averages. Realistic specifications may now be written, and the quality control procedures revised to reflect the changes of the program. To continue the process at the new quality level, the fraction defective control chart should become a part of the daily process routine, with immediate response to any adverse signal.

### Example 2 Percent defectives chart: *p* chart with variable lot size

A small shrimp breading plant produced frozen shrimp in various types of packages. One of the first steps in the line was deveining the raw shrimp and packing them in intermediate cases for refrigeration and further processing. In an effort to control the deveining process, the plant manager attempted to use a *p* chart for attributes, but found that the widely fluctuating lot sizes produced meaningless charts. A lot of 10 cases with 3 defectives calculated to 30% defectives; a lot of 40 cases with 3 defectives calculated to 7.5% defectives. These two figures made no sense when plotted on a graph with the same control limits. It became apparent that a variable lot size requires variable control limits. There seemed to be no way to control the case output size, since they devein all of the raw shrimp received each day, and that quantity depends entirely on the luck of the local shrimp boats. To simplify the classification, an arbitrary system of defect quantities was devised. Any case containing more than 6 shells over 1/16 inch, 4 tails, 10 veins over 1/4 inch, and several other criteria, was considered a defective case. Over a period of 10 days, the data for the number of cases produced, and the number of defectives were tabulated as shown in Table 3-5.

The first calculation is determining the average fraction defective. (Note that this cannot be done until after the tenth day—one of the weaknesses of the attribute charting technique is that it is slow to set up initially, and equally slow to evaluate the results of attempted process improvements.)

$$\bar{P} = \frac{\sum m}{\sum n} = 38/420 = 0.0905$$

### Table 3-5. Defective Shrimp Cases—100% Inspection

| Day | No. of cases *n* | Defectives *m* | Fraction defective *p* |
|---|---|---|---|
| 1 | 40 | 4 | 0.100 |
| 2 | 30 | 2 | 0.067 |
| 3 | 60 | 4 | 0.067 |
| 4 | 60 | 7 | 0.117 |
| 5 | 50 | 5 | 0.100 |
| 6 | 40 | 2 | 0.050 |
| 7 | 10 | 3 | 0.300 |
| 8 | 50 | 6 | 0.120 |
| 9 | 40 | 3 | 0.075 |
| 10 | 40 | 2 | 0.050 |
| | $\Sigma$ 420 | 38 | |

The upper control limit for the fraction defective is defined as:

$$\text{CL}\,Y_p = \bar{p} \pm 3\sqrt{\frac{\bar{p}(1 - \bar{p})}{n}}$$

$$= 0.0905 + 3\sqrt{\{0.0905(1 - 0.0905)\}/n}$$

$$= 0.0905 + 3\sqrt{0.0824n}$$

Since $n$ varies for each day, the calculation for the upper control limit is performed each day, using this formula. The results are shown in Table 3-6 under the column, Control Limit UCL.

Plotting the fraction defectives for each day is performed in the same way as for a constant lot size (see Figure 3-9). The difference arises when the control limits are plotted. The lower limit is so close to zero that a lower control limit would not be particularly useful.

The upper control limits are plotted for each sample size. Thus, the sample size for day 1 is $n = 40$. The corresponding upper control limit is 0.227. Continuing, the control limits for each day are plotted along with that day's fraction defective until all 10 points are completed.

Interpreting the chart in Figure 3-10 discloses that for the relatively small sample sizes for the shrimp example, the control limits are very liberal, and suggest that the quality level would have to be atrocious before a day's production is out of limits. As a matter of fact, the fraction defective average of 0.0905 (or 9%) is already an index of very poor quality performance, and yet even at this level, the operation appears to be in statistical control. As was the case with other control charts, when the sample size increases, the upper control limit changes less and less. This is clearly shown by comparing the distance on the vertical axis between $n = 10$ and $n = 20$, with the distance between $n = 50$ and $n = 60$. Most of the days sampled exhibited defective levels clustered near the 0.0905 average except for the spike data of Day #7. Again, even though this day showed a distressing level of 30% defectives, the numbers involved (3 out of 10 cases) are so small as to allow the control limits to indicate

**Table 3-6. Defective Shrimp Cases—100% Inspection: Calculation of Control Limits**

| Day | No. of cases $n$ | Defectives $m$ | Fraction defective $p$ | Control limit UCL |
|-----|------------------|----------------|------------------------|-------------------|
| 1   | 40               | 4              | 0.100                  | 0.227             |
| 2   | 30               | 2              | 0.067                  | 0.248             |
| 3   | 60               | 4              | 0.067                  | 0.202             |
| 4   | 60               | 7              | 0.117                  | 0.202             |
| 5   | 50               | 5              | 0.100                  | 0.212             |
| 6   | 40               | 2              | 0.050                  | 0.227             |
| 7   | 10               | 3              | 0.300                  | 0.363             |
| 8   | 50               | 6              | 0.120                  | 0.212             |
| 9   | 40               | 3              | 0.075                  | 0.227             |
| 10  | 40               | 2              | 0.050                  | 0.227             |

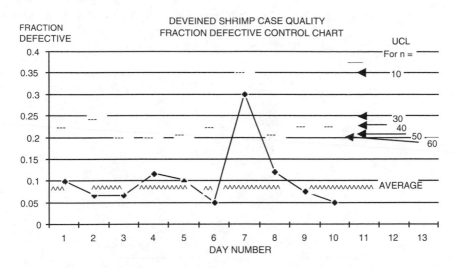

**Figure 3-10.**   Fraction defective control chart, variable lot size.

the process is in control. As the sample sizes increase, this type of chart analysis becomes more powerful. For example, at $n = 500$, the UCL drops to about 0.13. This is quite close to the average value, and provides a fairly tight control level.

The control chart for fraction defective with a variable lot size (Figure 3-10) is cumbersome to use, difficult to interpret, slow to respond to changes, and clumsy to explain. With all of these difficulties, it would be wise to avoid use of this chart whenever an alternative is available. There are two possibilities. The easiest way might be to change the sampling plan to a constant size subgroup of 30 cases selected throughout the day's production, with 100% inspection on those days when too few total cases are produced. Another way would be to select a single major defect (such as presence of veins), and in a subgroup of 5 cases per day, selected at uniform intervals throughout the day, measure the length of veins in millimeters. From this, a variables control chart can be prepared to assist in locating those conditions which produce an excessive number of these defects. As improved procedures are developed, they may continue to be monitored by control chart techniques, and the quality improvement program may be shifted to another type of defect.

**Example 3   Number of defectives chart or *np* chart (also known as *m* chart)**
An olive processing plant has had a long-established quality control system using percent defective control charts (*p* charts) to evaluate the quality level of production. However, they wish to start a quality improvement program, and find that grouping all of the defects into one category, specific quality problems are not highlighted. In the past, any jar containing blemished pieces, pit fragments, missing pimientos, leaking seal, low brine, and/or other defect was classified as a single defective. For the purposes of this new program, the company has selected presence of pit fragments as a major defect which shall be reduced or eliminated. A constant-sized sample of 500 jars uniformly spaced within each production run shall be examined for pit fragments, and an np chart for number of defective jars prepared. The first ten days' data are shown in Table 3-7.

**Table 3-7.   Data for Defective Olive Jars**

| Day | Sample size (n) | Number of defectives (m) or np |
|-----|-----------------|-------------------------------|
| 1 | 500 | 14 |
| 2 | 500 | 14 |
| 3 | 500 | 17 |
| 4 | 500 | 10 |
| 5 | 500 | 13 |
| 6 | 500 | 8 |
| 7 | 500 | 24 |
| 8 | 500 | 11 |
| 9 | 500 | 8 |
| 10 | 500 | 10 |
| $N = 10$ | $\Sigma n = 5000$ | $\Sigma m = 129$ |

The calculations based on the data establish the average number of defectives, the average fraction defective, and the control limits.

Average number of defectives

$$\overline{m} = \frac{\Sigma m}{N} = \frac{129}{10} = 12.9$$

Average fraction defective

$$\overline{m} = n\overline{p} \qquad \overline{p} = \frac{\overline{m}}{n}$$

$$= 12.9/500 = 0.026$$

Control limits

$$CL = \overline{m} \pm 3\sqrt{\overline{m}(1 - \overline{p})}$$

$$= 12.9 \pm 3\sqrt{12.9(1 - 0.026)}$$

$$= 12.9 \pm 10.6$$

A control chart may now be constructed using these figures:

Average number of defectives = 12.9
Upper control limit = 12.9 + 10.6 = 23.5
Lower control limit = 12.9 − 10.6 = 2.3

Figure 3-11 provides a good starting point to improve the quality of the product insofar as pits are concerned. It shows that the pit level centers around 12.9%, and that the quality level stays fairly close to that level. On day 7, some abnormality in the process caused the defects to rise out of limits on the high side, and operating conditions for that day should be examined closely for clues. Those familiar with the industry would know the most likely places to look—a change in raw material, maintenance programs on the pitter equipment, line speeds, and new line workers. If the "special cause" for this unusual day's production can be found, steps should be taken to insure that it cannot happen again. After a few more days of production with this process modification, a new chart should be prepared with its associated closer control limits. Any time a point falls outside of limits, it should be analyzed to improve the process. At the same time, any observation which drops below the lower control limit should be studied with equal care, since it too may point to directions for quality improvement.

To see the effect of eliminating the problem responsible for the out-of-control level of day #7, replace the 24 with 12.9, and calculate a new set of control points. The average would be reduced slightly to 11.8, and the upper control limit would now be 22.0. The significance of this little exercise is that it demonstrates how removal of just this one special cause of product quality variation can reduce the average defects by about 1%, and reduce the upper control limit by 1.5%. It should be obvious that continuing with this quality control chart, and eliminating other special causes as they are discovered will appreciably reduce the defects caused by the presence of pits. Once the procedure reaches an equilibrium, and further improvement appears unlikely, the next most important defect (perhaps blemishes) may be explored in a similar manner. An impatient manager might find this process painfully slow, but it can be a very effective quality improvement procedure to follow.

**Example 4    Number of defects chart: constant size sample *c* chart**
In contrast to all of the above charts where the quality issue is the number of *defectives*, the *c* chart is concerned only with the number of *defects*. The number of *defects* is

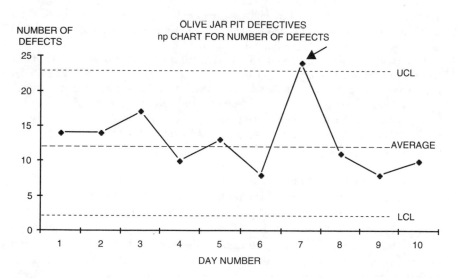

**Figure 3-11.** Number of defectives (*np*) control chart.

**Table 3-8.  Data for Syrup Batch Defects**

| Sample number | Date | Number of defects ($c$) |
|:---:|:---:|:---:|
| 1 | Feb. 3 | 15 |
| 2 | 4 | 21 |
| 3 | 5 | 27 |
| 4 | 6 | 18 |
| 5 | 7 | 39 |
| 6 | 10 | 10 |
| 7 | 11 | 18 |
| 8 | 12 | 22 |
| 9 | 13 | 13 |
| 10 | 14 | *18* |
|  |  | $\Sigma c = 201$ |

$$\bar{c} = \frac{\Sigma c}{N} = \frac{201}{10} = 20.1$$

$$\mathrm{CLY}_c = \bar{c} \pm 3\sqrt{\bar{c}}$$

$$= 20.1 \pm 3\sqrt{20.1}$$

$$= 20.1 + 13.4 = 33.5 \text{ UCL}$$

$$20.1 - 13.4 = 6.7 \text{ LCL}$$

measured for a subgroup of fixed size, and very often is concerned with a sample of size 1. The process to be measured may be a continuous one in which samples of a standardized number of units are selected at a prescribed constant interval. Or a process producing a constant lot size may be checked with a standardized sample size. The sample of size 1 refers to a process in which the end product is a single sample: a batch of corn syrup, a side of beef, a vat of chocolate, a bin of blended tea.

For our example, we shall select a batch of flavored syrup produced by a bottler of carbonated fruit drink. Ordinarily, a single batch of fruit-flavored syrup is produced each day to be diluted and packed on the following shift. The defects commonly found during quality control evaluations are: foreign matter, temperature above 49°F, poor color, pH over 5.3, weak flavor, concentration incorrect (for each of 22 ingredients), incorrect viscosity, excessive bacteria (6 tests), yeast, or molds. The following data (Table 3-8) was obtained over ten production days.

Again, we find one of the batches out of control in the initial chart. With the large number of attributes entered into the data, it is apparent that at least one of them is seriously out of control. If possible, the factor responsible should be isolated and studied under its own control chart until a reason for the variation has been isolated. At that point, new procedures should be initiated to prevent further occurrence of this defect in a day's batch. If the identity of the cause defect cannot be easily determined, control charts for small groups of defects should be prepared until the problem is uncovered.

## Example 5    Number of defects chart: variable size sample *u* chart

The *u* chart is a special case of the *c* chart, and has limited use in the food processing industry. It is applied to products where the number of units in the sample subgroup vary—and they may vary from a subgroup of 1 to perhaps a dozen units. The defects are fed into an equation to reduce the data to the number of nonconformities in a subgroup. The control limits are calculated in the same manner as for the *c* chart, except that the control limits are divided by the square root of the number of units in the subgroup. At best, this type of control chart serves as a score card to let management know how the quality was for some past period of time. It is slow to react to quality control problems, and it is recommended that other types of control charts be used before considering the *u* chart. Even in the automotive manufacturing industry where the *u* chart is used in an attempt to control the number of defects in fenders or bumpers, it has limited effectiveness. It might be used as a blunt weapon to brandish at parts suppliers in an effort to assist them in product improvement, but it carries few if any suggested cures. With this bleak description and the limitations of the *u* chart, no example will be presented here.

It is difficult to generalize on the use and effectiveness of attribute charts of all types since there are so many possible applications in so many different food processing industries. Generally speaking, they are slow to react, and tend to be constructed in such a way as to include a number of quality problems under one format. They do provide management with an easily understood picture of how the quality has been controlled over some period of time. There are, of course, some applications where precise variable measurements of a process are neither possible nor desirable, and attribute charts are satisfactory. Under some circumstances they can suggest areas where the process should be modified to better control the quality level, or in some instances, to actually improve the quality level. The one generalization which might be made is: wherever possible, select a variable control chart instead of an attribute chart to control process quality.

This chapter has not covered all of the charts which are available for use in quality control applications. Many of them have specialized uses for special industrial applications. Many are complicated, and are perhaps best understood by the statisticians who have created them than by the plant operating and quality control personnel who must use them. Some mature industries tend to take short cuts with the classical chart techniques discussed above. This is somewhat risky. An example is the abandonment of range charts when experience has shown that the averages chart is sufficiently indicative of process quality. Sooner or later, one would expect to miss an out-of-control period of production when, for example, a worn machine abruptly spews out widely varying product which somehow continues to average out within the control limits. This is not to suggest that some experimentation with the tried and true control chart technique should not be attempted. But it is strongly recommended that experimentation be delayed until after all plant processes have been exposed to the simpler procedures long enough to ensure that all the processes are under excellent control, and that further quality improvements using these principles are unlikely to be uncovered.

# 4     Fundamentals

There are four distributions of data which form the basis for statistical quality control: Binomial, Poisson, Normal and Student's *t*. Common to these distributions are the concepts of measures of central tendency and measures of dispersion.

## ANALYSIS OF DATA

### Measures of Central Tendency

In order to illustrate a large quantity of data (population) with a single number, perhaps the best single measure is the arithmetic mean. This is simply the total of all values, divided by the number of items.

$$\text{Mean}(\mu) = \frac{X_1 + X_2 + X_3 + \cdots + X_N}{N} = \sum \frac{X}{N}$$

Another illustration of the data is called the Median. When the numbers are arranged in order of magnitude, the Median of the set of data is the middle number if $N$ is odd, and the average of the two middle numbers if $N$ is even. For example:

$$2, 3, 7, 9, 21, 22, 45, 67, 68, 69, 72$$

$$\text{Median} = 22$$

$$4, 6, 9, 12, 24, 45, 56, 58, 59, 63, 68, 69$$

$$\text{Median} = (45 + 56)/2 = 50.5$$

As can be seen from these illustrations, neither the mean nor the median has to be one of the numbers observed in the set of data.

**Table 4-1. Database, Set A: Net Weights, 184-g Jars of Pistachios**

| Sample | $X_1$ | $X_2$ | $X_3$ | $X_4$ | $X_5$ | Total | Mean |
|--------|-------|-------|-------|-------|-------|-------|------|
| 1 | 187 | 190 | 183 | 185 | 188 | 933 | 186.6 |
| 2 | 186 | 182 | 183 | 185 | 189 | 925 | 185 |
| 3 | 187 | 189 | 192 | 192 | 188 | 948 | 189.6 |
| 4 | 190 | 189 | 189 | 187 | 187 | 942 | 188.4 |
| 5 | 188 | 188 | 183 | 187 | 183 | 929 | 185.8 |
| 6 | 187 | 185 | 187 | 190 | 185 | 934 | 186.8 |
| 7 | 194 | 189 | 194 | 186 | 187 | 950 | 190 |
| 8 | 183 | 188 | 187 | 187 | 183 | 928 | 185.6 |
| 9 | 189 | 192 | 194 | 183 | 188 | 946 | 189.2 |
| 10 | 188 | 186 | 193 | 191 | 185 | 943 | 188.6 |
| 11 | 187 | 187 | 188 | 182 | 183 | 927 | 185.4 |
| 12 | 168 | 188 | 190 | 190 | 183 | 939 | 187.8 |
| 13 | 186 | 189 | 194 | 186 | 192 | 947 | 189.4 |
| 14 | 189 | 184 | 194 | 188 | 190 | 945 | 189 |
| 15 | 190 | 186 | 189 | 191 | 187 | 943 | 188.6 |
| 16 | 188 | 189 | 185 | 185 | 183 | 930 | 186 |
| 17 | 185 | 188 | 193 | 188 | 188 | 942 | 186.4 |
| 18 | 187 | 187 | 190 | 189 | 183 | 936 | 187.2 |
| 19 | 194 | 183 | 191 | 191 | 192 | 951 | 190.2 |
| 20 | 186 | 188 | 188 | 180 | 186 | 928 | 185.6 |

$$\Sigma X = 18{,}766 \qquad \bar{X} = 18{,}765$$

Tables 4-1 and 4-2 demonstrate calculation of the population mean by different methods. Both examples are based on the same data obtained by weighing the net contents of 184-g jars of pistachios taken in sequence from the production line. The data has been arbitrarily arranged in groups of five weighings. The first method for calculating the mean is based on the actual data; the second method uses deviations from the target weight of 184 g. Both produce the same mean. The mean of the population is known as $\mu$; the mean of the samples representing the population is $\bar{X}$.

## Measures of Dispersion

The mean and the median do not provide much information about the distribution of the data. Two different sets of data may have the same means, but the dispersion of the data about the two means may be totally different. The dispersion is better described by three other expressions: *Range, Variance, and Standard Deviation*.

The Range $(R)$ is found simply by subtracting the smallest value from the largest value found in the data. By itself, the range has limited use. This is because

### Table 4-2. Deviations from 184-g ($X - 184$). Jars of Pistachios

| Sample | $X_1$ | $X_2$ | $X_3$ | $X_4$ | $X_5$ | Total | Mean | Range |
|--------|-------|-------|-------|-------|-------|-------|------|-------|
| 1 | 3 | 6 | −1 | 1 | 4 | 13 | 2.6 | 7 |
| 2 | 2 | −2 | −1 | 1 | 5 | 5 | 1 | 7 |
| 3 | 3 | 5 | 8 | 8 | 4 | 28 | 5.6 | 5 |
| 4 | 6 | 5 | 5 | 3 | 3 | 22 | 4.4 | 3 |
| 5 | 4 | 4 | −1 | 3 | −1 | 9 | 1.8 | 5 |
| 6 | 3 | 1 | 3 | 6 | 1 | 14 | 2.8 | 5 |
| 7 | 10 | 5 | 10 | 2 | 3 | 30 | 6 | 8 |
| 8 | −1 | 4 | 3 | 3 | −1 | 8 | 1.6 | 5 |
| 9 | 5 | 8 | 10 | −1 | 4 | 26 | 5.2 | 11 |
| 10 | 4 | 2 | 9 | 7 | 1 | 23 | 4.6 | 8 |
| 11 | 3 | 3 | 4 | −2 | −1 | 7 | 1.4 | 6 |
| 12 | 4 | 4 | 6 | 6 | −1 | 19 | 3.8 | 7 |
| 13 | 2 | 5 | 10 | 2 | 8 | 27 | 5.4 | 8 |
| 14 | 5 | 0 | 10 | 4 | 6 | 25 | 5 | 10 |
| 15 | 6 | 2 | 5 | 7 | 3 | 23 | 4.6 | 5 |
| 16 | 4 | 5 | 1 | 1 | −1 | 10 | 2 | 6 |
| 17 | 1 | 4 | 9 | 4 | 4 | 22 | 4.4 | 8 |
| 18 | 3 | 3 | 6 | 5 | −1 | 16 | 3.2 | 7 |
| 19 | 10 | −1 | 7 | 7 | 8 | 31 | 6.2 | 11 |
| 20 | 2 | 4 | 4 | −4 | 2 | 8 | 1.6 | 8 |
| | | | | | | 366 | 73.2 | 140 |

Note: $\bar{X} = 184 + (366/100) = 187.66$.

the smallest or the largest value may be extremely far from the mean value, thus giving an exaggerated impression of the dispersion of data.

For this reason, the Variance $\sigma^2$ is more commonly used to measure dispersion of data. The variance is calculated by subtracting each item ($X$) in the distribution from the mean $\mu$, squaring each of the resulting figures, adding the results, and dividing by the number of items ($N$). Note that if the subtraction operation were not squared, the final figure would be zero, since half of the values are positive and half of the values are negative with respect to the mean. Another way to express this: the average of all of the deviations from the mean will always turn out to be zero. By squaring, all of the figures become positive.

Unfortunately, the Variance is expressed as the square of some number, and has limited direct use in quality control. For example, if all of the observations were weights of a distribution of packages, the variance calculated as above would be expressed in (grams)$^2$. To eliminate this complication, the square root of the variance is commonly determined, and the resulting figure is referred to as the Standard Deviation or $\sigma$. When using this formula to determine Variance and Standard Deviation, there are many opportunities for calculation errors because of the many subtractions required. To avoid this possibility, a simpler formula may be used, in which only a single subtraction is needed:

$$\sigma = \sqrt{\frac{\sum(X - \overline{X})^2}{N}}$$

When $N$ is large, it is not always possible or desirable to calculate $\mu$ and $\sigma$ from the entire population, and estimates $\overline{X}$ (the sample mean) and $s$ (the sample standard derivation estimate) obtained from a sample of $n$ items ($n < N$) from the population are used to approximate $\mu$ and $\sigma$. These estimates are calculated as follows:

$$\overline{X} = \frac{X_1 + X_2 + X_3 + \cdots + X_n}{n}$$

and

$$s = \sqrt{\frac{\sum(X - \overline{X})^2}{n - 1}} = \sqrt{\frac{\sum X^2 - \frac{(\sum X)^2}{n}}{n - 1}}$$

Note that $\overline{X}$, the sample mean, is generally close to the population mean $\mu$. Similarly, $s$, the sample standard deviation estimate, is close to the value of the population standard deviation $\mu$.

When working with frequency distributions, the formulas and the calculations are quite similar. Using the same data as in the previous pistachio examples, note the similarity of computations.

$$\overline{X} = \frac{fX_1 + fX_2 + fX_3 + \cdots + f_nX_n}{n} = \frac{\sum fX}{\sum f}$$

and

$$s = \sqrt{\frac{\sum f(X - \overline{X})^2}{n - 1}} = \sqrt{\frac{\sum fX^2 - \frac{(\sum fX)^2}{n}}{n - 1}}$$

where $n = \sum f$

$$\overline{X} = \frac{\sum fX}{\sum f} = \frac{18,766}{100} = 187.66$$

$$s = \sqrt{\frac{\sum f(X - \overline{X})^2}{n - 1}} = \sqrt{\frac{984.44}{99}} = 3.15$$

**Table 4-3.   Frequency and Standard Deviation Calculations for Pistachios**

| X | Frequency | f | fX | $X - \overline{X}$ | $f(X - \overline{X})^2$ |
|---|-----------|---|----|--------------------|------------------------|
| 180 | / | 1 | 180 | −7.66 | 58.6756 |
| 181 |  | 0 | 0 | −6.66 | 0 |
| 182 | // | 2 | 364 | −5.66 | 64.0712 |
| 183 | //////////// | 12 | 2196 | −4.66 | 260.5872 |
| 184 | / | 1 | 184 | −3.66 | 13.3956, |
| 185 | //////// | 8 | 1480 | −2.66 | 56.6048 |
| 186 | //////// | 8 | 1488 | −1.66 | 22.0448 |
| 187 | /////////////// | 15 | 2805 | −0.66 | 6.5340 |
| 188 | ///////////////// | 17 | 3196 | 0.34 | 1.9652 |
| 189 | /////////// | 11 | 2079 | 1.34 | 19.7516 |
| 190 | //////// | 8 | 1520 | 2.34 | 43.8048 |
| 191 | //// | 4 | 764 | 3.34 | 44.6224 |
| 192 | ///// | 5 | 960 | 4.34 | 94.1780 |
| 193 | // | 2 | 386 | 5.34 | 57.0312 |
| 194 | ////// | 6 | 1164 | 6.34 | 241.1736 |
|  |  | 100 | 18766 |  | 984.44 |

Range of weights = 194 − 180.

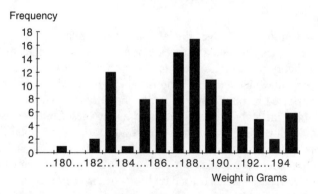

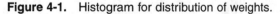

**Figure 4-1.**   Histogram for distribution of weights.

The "Frequency" columns in Table 4-3 are more usually assembled vertically in a format referred to as a histogram. The same data have been rearranged as shown in Figure 4-1 and Table 4-4.

$$\overline{X} = 184 + \frac{366}{100} = 187.66$$

$$s = \sqrt{\frac{\sum fX'^2 - \frac{(\sum fX')^2}{n}}{n - 1}} = \sqrt{\frac{2324 - \frac{(366)^2}{100}}{99}} = 3.15$$

**Table 4-4.   Range of Differences $= (X - 184)$**

| $X - 184$ | $f$ | $fX$ | $fX^2$ |
|---|---|---|---|
| $-4$ | 1 | $-4$ | 16 |
| $-6$ | 0 | 0 | 0 |
| $-2$ | 2 | $-4$ | 8 |
| $-1$ | 12 | $-12$ | 12 |
| 0 | 1 | 0 | 0 |
| 1 | 8 | 8 | 8 |
| 2 | 8 | 16 | 32 |
| 3 | 15 | 45 | 135 |
| 4 | 17 | 68 | 272 |
| 5 | 11 | 55 | 275 |
| 6 | 8 | 48 | 286 |
| 7 | 4 | 28 | 196 |
| 8 | 5 | 40 | 320 |
| 9 | 2 | 18 | 162 |
| 10 | 6 | 60 | 600 |
| | 100 | 366 | 2324 |

# PROBABILITY

Statistics and probability are so fundamentally interrelated that it is impossible to discuss statistics without an understanding of the meaning of probability. Probability theory makes it possible to interpret statistical results. Many of the statistical procedures involve conclusions which are based on samples, and these are always affected by random variation. Probability theory allows us to express numerically the inevitable uncertainties in drawing these conclusions.

When assessing the quality of a lot of goods, rarely is the total lot examined piece by piece. The costs associated with such a practice generally make it impractical, and it has been shown that 100% inspection is rarely 100% effective because of inaccuracies. These may be caused by limits of the test procedures, or boredom of the inspectors, often leading to near hypnosis which allows defects to pass unnoticed. Instead, a sample of the lot is usually tested, and the number of defective articles recorded. In order for this result to be meaningful, the sample should be representative of the lot, and selected in such a way that each article within the lot has an equal chance of being selected. This is referred to as a *random sample*.

Within a lot, if there are $r$ defective articles in a random sample of $n$ articles, then, in the selection of each article from the lot, the probability of obtaining a defective article is $p = r/n$, and the probability of obtaining a nondefective article is $q = (n - r)/n$. In the event that $p = 0$, there are no defective articles; should $p = 1$, all articles are defective.

For example, if 1/10 (or 10%) of the articles in a lot are defective, then 9/10 (or 90%) are nondefective. This may also be expressed as $p = 1/10$, and $q = 9/10$.

It should also be noted that a frequent practice is to report defects as percentages: a probability of $5/100 = 0.05$ may be referred to as a 5% probability.

Another simple concept of probability frequently used in procedures for chart control of quality is the expression $np$. If $p$ is the probability of a defective article in a single trial, then the expected number of defective articles in $n$ trials is $np$. For example: If there are 15 defective articles in a lot of 100 articles, then in a random selection of a single article from the lot, the probability of drawing a defective article is $p = 15/100 = 0.15$; and in drawing 20 articles from the lot, one might expect to obtain $np = 20(0.15) = 3$ defective articles. Note that these calculations refer to "the long run." It is entirely possible that a sample of 20 might contain no defectives at all, but "in the long run" the probability is that three defective articles would be found in each sample of 20.

In the course of controlling quality, observations of one type or another are required. These observations may be physical measurements of length, weight, volume, wavelengths, volts, etc.; or the observations might be merely counts, such as the number of products which were incorrect, too big, too small, scratched, dented, etc. The observations may be made over a period of time (a day, an hour ...); within a selected quantity (a production lot, a 30 gallon tank, a pallet load ...); until some required event occurs (failure of an electrical device, end of shelf life, melting point temperature, a satisfactory item is produced ...); or any other arbitrarily established limits.

The observations within the limits selected are referred to as "data." The data may be obtained from selected items ("samples") within those limits, or they may be obtained from all of the items (the "population"). The position, arrangement, or frequency of occurrence of the data within the population is defined as the "distribution" of data.

Depending upon the product, process, or system being examined, the data may be found within many different types of distributions, and analysis of the data cannot logically proceed without first understanding which distribution is present. Following are brief descriptions of distributions commonly encountered in quality control calculations.

1. Binomial—The probability that an event will happen exactly $x$ times in $n$ trials is calculated from a formula based on the binomial theorem $(a + b)^n$, also referred to as the Bernoulli distribution. The data conforming to these conditions define the resulting binomial distribution curve. This is useful in determining sampling plans and establishing control charts for defectives.
2. Poisson—An approximation to the binomial which is valid when the samples are large and probabilities are small. Used for developing defect control charts and for calculating sampling plan probabilities.
3. Normal—A common distribution found when the variable examined is the result of many causes which have a 50/50 chance of occurring. Average and Range control charts are based on this curve. Although the population may not be normally distributed, the averages of groups of samples selected from it generally follow a normal distribution.

4. $t$-Distribution—A somewhat nonnormal curve produced when comparing samples and population means when the population standard deviations are unknown and must be estimated from the samples.
5. $F$-Distribution—The distribution of the ratio of two estimates of variance.
6. Exponential—Logarithmic curves which describe events such as flavor loss, shelf life, or container failure.
7. Weibull—These may take many shapes, and are based on a single formula with three variables: shape, scale, and location. Used to study shelf life and product failure rates.
8. Chi-Square—As contrasted with $t$-distribution, the chi-square distribution is used when the standard deviations are known for the sample and for the population. These are then compared.
9. Others—Multinomial (a generalization of the binomial), hypergeometric (samples removed from population for testing are not replaced), uniform, Cauchy, gamma, beta, bivariate, normal, geometric, Pascal's, Maxwell.

## BINOMIAL DISTRIBUTION

Many quality control problems deal with a population in which a proportion $p$ of the individuals have a certain characteristic, and a proportion $q = 1 - p$ of the individuals do not. This is called a binomial population since each individual in it falls into one of only two classes. Sampling for defective articles in a lot where each article selected at random is either good or defective leads to a binomial distribution.

In a random selection of a sample size $n$ from a large lot, if $p$ is the probability of obtaining a defective item in each draw, and $q = 1 - p$ is the probability of obtaining a nondefective item, then the probability $P(r)$ that there will be exactly $r$ defective items in the sample size $n$ is the term in the binomial expansion of $(q + p)^n$ for which the exponent of $p$ is $r$; that is:

$$P(r) = \binom{n}{r} q^{n-r} p^r$$

where

$$\binom{n}{r} = \frac{n!}{(n-r)!\,r!}$$

(Value of $\binom{n}{r}$ may be found in Table A4 in the Appendix.)

A distribution whose terms are proportional to successive terms of the binomial expansion is called a binomial distribution. Its mean is $\mu = np$, and its standard

$$\text{deviation: } \sigma = \sqrt{npq}$$

A binomial probability distribution may be represented by a histogram. This is a figure obtained by plotting all possible values of $r$ along a base line, and erecting on each an equal-base rectangle whose height (and area) is proportional to the corresponding probability.

## Example

In a manufacturing process in which it is known that 10% of the manufactured articles are defective, a sample of five articles is taken. What is the probability that in the sample there are (a) exactly 0,1,2,3,4,5 defective articles, and (b) 2 or less defectives?

(a)

| No. of defectives $r$ | Probabilities $P(r)$ | |
|---|---|---|
| 0 | $P(r = 0) = \binom{5}{0}\left(\dfrac{9}{10}\right)^5\left(\dfrac{1}{10}\right)^0 = \left(\dfrac{9}{10}\right)^5$ | $= 0.59049$ |
| 1 | $P(r = 1) = \binom{5}{1}\left(\dfrac{9}{10}\right)^4\left(\dfrac{1}{10}\right)^1 =$ | $0.32805$ |
| 2 | $P(r = 2) = \binom{5}{2}\left(\dfrac{9}{10}\right)^3\left(\dfrac{1}{10}\right)^2 =$ | $0.07290$ |
| 3 | $P(r = 3) = \binom{5}{3}\left(\dfrac{9}{10}\right)^2\left(\dfrac{1}{10}\right)^3 =$ | $0.00810$ |
| 4 | $P(r = 4) = \binom{5}{4}\left(\dfrac{9}{10}\right)^1\left(\dfrac{1}{10}\right)^4 =$ | $0.00045$ |
| 5 | $P(r = 5) = \binom{5}{5}\left(\dfrac{9}{10}\right)^0\left(\dfrac{1}{10}\right)^5 = \left(\dfrac{1}{10}\right)^5$ | $0.00001$ <br> $1.00000$ |

(b)

$$P(r < 3) = P(r = 0) = P(r = 1) = P(r = 2) = 0.99144$$

Under the conditions of the above example, therefore, in repetitions of the sampling one would expect to obtain a sample without defects more than one half of the time, with one defect about one third of the time and 2 defects about 7 times in 100 samplings. (See Figure 4-2.)

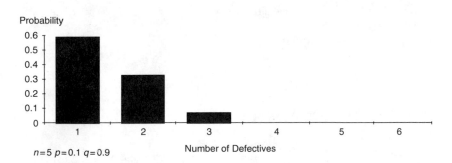

$n=5\ p=0.1\ q=0.9$

**Figure 4-2.**  Histogram of binomial distribution example.

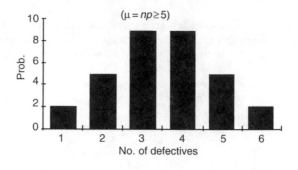

**Figure 4-3.** Histogram for $p = q - 0.5$.

In the event that the values of $p$ and $q$ approach equality (for example: $p = q = 0.5$), the histogram for any value of $n$ approaches symmetry around the vertical line erected at the mean. The total area included in the blocks of the histogram always equals 1. (See Figure 4-3.)

## Poisson Distribution

An approximation to the binomial distribution, the Poisson distribution, may be used when:

1. the probability of defectives is small (less than 0.1%);
2. the number of observations is large (more than 16);
3. the sample is small compared to the population ($<10\%$).

The Poisson distribution may be used to prepare control charts for defects, but perhaps has its greatest use in calculating probabilities for developing sampling plans. Much of the mathematics has been prepared in the form of a table which simplifies its use. (See Table A-1 in the Appendix.) This table gives the sums of the probabilities of $c$ or fewer defects for various values of $np' = c'$.

If $n$ is large and $p$ and $q$ are small so that $np$ is fairly small, that is:

$$(\mu = np \leq 5)$$

then the binomial probability:

$$P(r) = \binom{n}{r} q^{n-r} p^r$$

is given approximately by the equation:

$$P(r = c) = e^{-np} \frac{(np)^r}{r!} \quad (r = c = 0,1,2,...,n) \text{ and } e = 2.178...$$

A distribution where the frequencies are given by this formula is called a Poisson distribution, and is extremely useful in the inspection and control of manufactured goods where the proportion of defective articles $p'$ in a large lot is expected to be small. Following are several examples:

## Example 1

In a large shipment of articles 5% are known defective. What is the probability that in a sample of 100 articles from the shipment, there are 2 or fewer defectives?

$$\mu = np'' = 100(.05) = 5; \quad c = 2$$

| $c$ <br> $np'$ | No. of defectives | | | | | | | | |
|---|---|---|---|---|---|---|---|---|---|
| | 0 | 1 | 2 | 3 | 4 | 5 | 6 | 7 | 8 |
| (Mean 5.0) | 0.007 | 0.040 | 0.125 | 0.265 | 0.440 | 0.616 | 0.762 | 0.867 | 0.932 |

(From Table A-1.)

Therefore: $p(c = 2) = 0.125$ is the probability of 2 or fewer defectives. That is, the chances are between 12 and 13% that a sample of 100 will contain 2 or fewer defectives.

## Example 2

If the true process average of defectives (proportion defective) is known to be 0.01, find the probability that in a sample of 100 articles there are (a) no defectives, (b) exactly two defectives, (c) fewer than 5 defectives, (d) at least 5 defectives (i.e., 5 or more). Here, $\mu = np' = 100(0.01) = 1$; $c =$ number of defectives less than and including $c$. (It normally is the acceptance number; i.e., the maximum allowable number of defects in a sample of size $n$ to be acceptable.) We shall use the table to find probabilities for various values of $c$.

(a) $P(c = 0) = 0.368$
(b) $P[(c = 2) - (c = 1)] = 0.920 - 0.736 = 0.184$
(c) $P(c = 4) = 0.996$
(d) $P[1 - (c = 4)] = 1 - 0.996 = 0.004$ (total of all terms = 1)

## Example 3

A pump in a cannery has a failure, on the average, once in every 5000 hours of operation. What is the probability that (a) more than one failure will occur in a 1000-hour period, (b) no failure will occur in 10,000 hours of operation?

(a) $np' = 1000(1/5000) = 1/5 = 0.20$. Since the table gives the probability of one or less failures at $c = 1$, then the probability of more than one failure is represented by $P[1 - (c = 1)]$ since the total of all terms = 1. From the table,

$$P[1 - (c = 1)] = 1 - 0.982 = 0.018$$

(b) $np' = 10,000(1/5000) = 2$; $c = 0$.

From the table, $P(c = 0) = 0.135$.

**Example 4**

A purchaser will accept a large shipment of articles if, in a sample of 1,000 articles taken at random from the shipment, there are at most 10 defective articles. If the entire shipment is 0.5% defective, what is the probability that the shipment will be accepted?

$$np' = 1000(0.005) = 5; \quad c = 10$$

From the table, $P(c = 10) = 0.986$. Therefore, the shipment has a 98.6% probability of being accepted.

# THE NORMAL DISTRIBUTION

In quality control, the normal distribution has the widest application of any of the distributions. In the binomial and Poisson distributions, the random variable takes only a countable (or discrete) number of possible values. By contrast, there are many observations which involve a process of measurement, and in all such processes we are dealing with quantities which have an infinite number of gradations. This gives rise to the normal distribution; as the number of gradations ($n$) approaches infinity, the histogram resulting approaches a curve (see Figures 4-4 and 4–5). Early studies of the weights of groups of people, or lengths of corn stalks, or other naturally occurring phenomena produced distributions which approximated a bell-shaped curve. Because this curve is found so often in nature, it was assigned the name "Normal Curve."

A clingstone peach cannery may have a target drained weight of 19 ounces in a $2\frac{1}{2}$ size can. If the filling operation is in control, most of the production will center near the target weight, but the frequency will taper off from the average as the

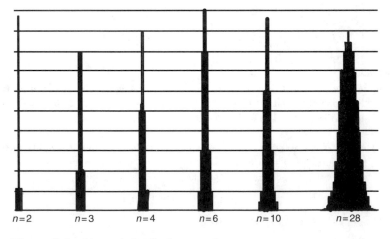

**Figure 4-4.** Normal distribution as $n$ increases.

deviations become wider. The need to establish attainable control weight limits on either side of the target weight is obvious, since the consumer feels cheated if the product weighs less than stated (and this could be in violation of the law), and overly excessive weights can be costly to the producer. The normal distribution is used to calculate the values of "less than stated" and "overly excessive weights."

The equation which defines the normal probability is:

$$P = \frac{1}{\sigma\sqrt{2\pi}}e^{(X-\mu)^2/2\sigma^2}$$

where $P$ is the relative frequency (or probability) with which a value of $X$ occurs and where $\mu$ and $\sigma$ are the mean and standard deviation, respectively ($\pi = 3.1459$ and $e = 2.178$). The value of $\sigma$ determines the concentration of data about the mean, as shown in Figure 4-5.

In order to understand the power of the normal curve and its application to quality control, let us consider some of its universal qualities. The areas under the normal curve represent probabilities of the normal distribution. As was the case of the histograms depicting the binomial distribution, the total area under the normal curve equals 1. Therefore, areas under the curve may be considered as proportions, probabilities, and (when multiplied by 100) as percents.

In the Figure 4-6, the area $A$ under the normal curve between two vertical lines at $X_1$ and $X_2$ represents the probability that a randomly drawn measurement $X$ will fall between $X_1$ and $X_2$. In algebraic terms: $P(X_1 \geq X \geq X_2) = A$.

To simplify calculations, it is customary to use tables to find the areas under the normal curve. Since there is an unlimited number of normal curves, the tables were constructed for universal use as follows: the deviation of any measurement $X$ from the mean is expressed in terms of standard deviations, that is,

$$Z = \frac{X - \mu}{\sigma}$$

Referring to Figure 4-6,

$$A_1 - A_2 = P(X_1 \geq X \geq X_2) = A$$

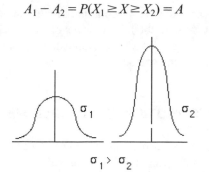

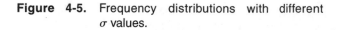

**Figure 4-5.** Frequency distributions with different $\sigma$ values.

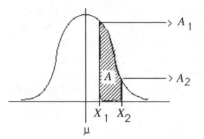

**Figure 4-6.** Area of probability between $X_1$ and $X_2$.

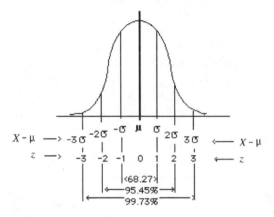

**Figure 4-7.** Areas under the normal curve.

and

$$Z_1 = \frac{X_1 - \mu}{\sigma} \quad Z_2 = \frac{X_2 - \mu}{\sigma}$$

Table A-2 in the Appendix is a compilation of areas under the normal curve to the right of valves of Z (or to the left of valves of $-Z$). Certain special areas are shown in Figure 4–7. These areas are designated by horizontal axis divisions expressed in the terms of the two scales: $X - \mu$ and Z.

## DISTRIBUTION OF SAMPLE MEANS

Many references are made to "parameters" and "statistics" when discussing normal distributions. The following definitions should make these differences clear:

*Parameters*—these are the values of $\mu$ (the mean) and $\sigma$ (the standard deviation) obtained from the *population*.

*Statistics*—these are the values of $\overline{X}$ and $s$, the estimates of $\mu$ (the mean) and $\sigma$ (the standard deviation) obtained from *samples*.

The relation between the population and samples taken from that population is of vital importance to establishment of quality control chart methods. That relationship is:

1. If measurements in a population are distributed normally, then averages of groups of measurements from that population will be found to be closer to the mean than the original measurements.
2. (and this is even more vital) If measurements of a population are NOT distributed normally, averages of groups of measurements from that population WILL approach a normal distribution.

This may be expressed in mathematical terms. When a lot or population is sampled, the sample mean is usually denoted as $\overline{X}$. Sampling generates a population of $\overline{X}$'s with a mean $\mu_{\overline{X}}$ and standard deviation $\sigma_{\overline{X}}$ of its own. If the population consists of $k$ sample means, $\overline{X}$, it can be shown that:

$$\mu_{\overline{X}} = \frac{\sum \overline{X}}{k} = \mu \qquad \sigma_{\overline{X}} = \sqrt{\frac{\sum (\overline{X} - \mu)^2}{k}} = \frac{\sigma}{\sqrt{n}}$$

To state this in words: if a set of measurements $X$ has a distribution with mean $\mu$ and a standard deviation $\sigma$, and if all possible samples of $n$ measurements are drawn, then the sample mean $\overline{X}$ will have a distribution which, for larger and larger $n$, approaches the normal distribution with mean $\mu$ and standard deviation $\sigma/\sqrt{n}$ This is known as the Central Limit Theorem, and states that for significantly large $n$ (at least 30), $\overline{X}$ will be approximately normally distributed, even if the original population of $X$s does not satisfy a normal distribution.

Figure 4-8 demonstrates the way in which the population distribution is narrowed as the sample size is increased from 4 to 16 measurements. Three curves are shown: the distribution of the population from which the samples are drawn; the distribution of sample means when samples of size 4 are selected; and the distribution of sample means when samples of size 16 are selected. With increasing $n$, the curve becomes taller and narrower. The areas under the curves will always be equal to one. Probabilities for $\overline{X}$ values can be found from Table A-2 in the Appendix by calculating

$$Z = \frac{\overline{X} - \mu}{\sigma_X} = \frac{\overline{X} - \mu}{\sigma/\sqrt{n}}$$

where $\sigma/\sqrt{n}$ is the standard deviation of the means and $\sigma_{\overline{X}}$ is the standard deviation of a sample of $n$ items.

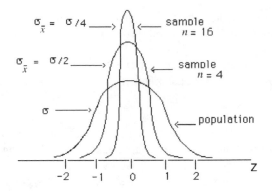

**Figure 4-8.** Comparison of population curve with curves for two sample sizes.

## Example 1

If the drained weights of canned peaches follow a normal distribution with $\mu = 19.0$ ounces and a standard deviation of $\sigma = 0.2$ ounce, what is the probability of a can selected at random having a drained weight (a) between 19.1 and 19.2 ounces; (b) between 18.7 and 19.1 ounces; and (c) less than 18.8 ounces?

(a)    $X_1 = 19.1; Z_1 = \dfrac{19.1 - 19.0}{0.2} = \dfrac{0.1}{0.2} = 0.5; A_1$ (from Table A-2) $= 0.3085$

   $X_2 = 19.2; Z_2 = \dfrac{19.2 - 19.0}{0.2} = \dfrac{0.2}{0.2} = 1.0; A_2$ (from Table A-2) $= 0.1587$

$$P(19.2 < X < 19.2) = A_1 - A_2 = 0.3085 - 0.1587 = 0.1498$$

Therefore, about 15% of the drained weights are between 19.1 and 19.2 ounces. (See Figure 4-9a.)

(b)    $X_1 = 18.7; Z_1 = \dfrac{18.7 - 19.0}{0.2} = \dfrac{-0.3}{0.2} = -1.5; A_1$ (from Table A-2) $= 0.0668$

   $X_2 = 19.1; Z_2 = \dfrac{19.1 - 19.0}{0.2} = \dfrac{0.1}{0.2} = 0.5$ (from Table A-2) $= 0.3085$

$$P(18.7 < X < 19.1) = 1 - (A_1 + A_2) = 1 - 0.3753 = 0.6247$$

Therefore, about 62.5% of the drained weights are between 18.7 and 19.1 ounces. See Figure 4-9b.

(c)                    $X = 18.8; Z = \dfrac{18.8 - 19.0}{0.2} = -1.0; A = 0.1587$

$$P(X < 18.8) = A = 0.1587$$

Therefore, between 15% and 16% of the cans have contents the drained weights of which are less than 18.8 ounces. See Figure 4.9c.

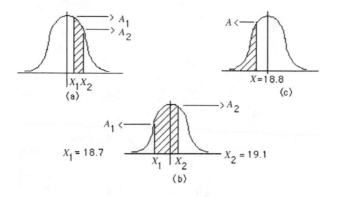

**Figure 4-9.**   Areas under the curve for Example 1(a)–(c).

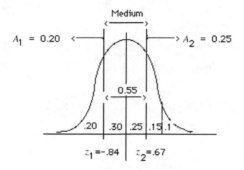

**Figure 4-10.**   Distribution chart, showing medium grade area
for Example 2.

## Example 2

In a packing plant grading Satsuma plums whose weights are normally distributed, 20% are called small, 55% medium, 15% large, and 10% very large. If the mean weight of all Satsuma plums is 4.83 ounces with a standard deviation of 1.20 ounces, what are the lower and upper bounds for the weight of Satsuma plums graded as medium?

Since $A_1 = 0.20$, from Table A-2 the corresponding value of $Z_1$ is $-0.84$, and for $A_2 = 0.25$ the nearest value of $Z_2$ from the table is 0.67. Then

$$Z_1 = -0.84 = \frac{X_1 - 4.83}{1.20} \quad Z_2 = 0.67 = \frac{X_2 - 4.83}{1.20}$$

Then, solving for $X_1$ and $X_2$ we get

$$X_1 = 4.83 - (0.84)(1.20) \qquad X_2 = 4.83 + (0.67)(1.20)$$
$$= 4.83 - 1.01 \qquad\qquad = 4.83 + 0.80$$
$$= 3.82 \text{ ounces} \qquad\qquad = 5.63 \text{ ounces}$$

Therefore, all fruits weighing between 3.82 and 5.63 ounces should be graded as medium (see Figure 4-10).

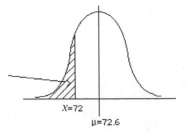

**Figure 4-11.** Example 3(a) calculation and curve.

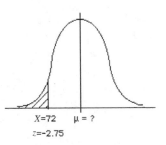

**Figure 4-12.** Example 3(b) calculation and curve.

## Example 3

Number 10 cans of peaches are supposed to hold 72 ounces. The cannery adjusts the filling machine to fill the cans with, on the average, 72.6 ounces. The distribution of fill weights is approximately normal with a standard deviation of 0.4 ounce.

(a) How many of 100,000 cans will contain less than 72 ounces?

$$\mu = 72.6, \quad \sigma = 0.4, \quad X = 72$$

$$z = \frac{X - \mu}{\sigma} = \frac{72 - 72.6}{0.4}$$

$$= 1.5, \quad P = 0.0668 \text{ (See Figure 4-11.)}$$

$$\text{Number of cans} = 100,000(0.0668)$$

$$= 6680$$

(b) The cannery considers this too many. Assuming that the standard deviation remains unchanged, what mean value should the cannery use for the machine setting if it wants no more than 300 out of the 100,000 cans to contain less than 72 ounces?

$$P = \frac{300}{100,000} = 0.003, \quad z = -2.75 \text{ (See Figure 4-12.)}$$

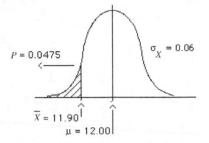

**Figure 4-13.** Example 4(a) curve.

$$-2.75 = \frac{72 - \mu}{0.4}$$

$$\mu = 72 + 2.75(0.4) = 73.1 \text{ ounces}$$

In other words, if the filling machine is set to deliver 73.1 ounces, the chances are that no more than 300 cans per 100,000 produced will contain less than 72 ounces.

## Example 4
A lot of 1000 fried chicken dinners has a mean weight of 12 ounces and a standard deviation of 0.6 ounce. (a) What is the probability that in a random sample of 100 of those dinners the average weight will be less than 11.90 ounces? (b) How large a sample must be taken to be 95% sure that the sample mean does not fall below 11.95 ounces.

(a)

$$\mu = 12; \quad \overline{X} = 11.90; \quad \sigma = 0.6; \quad \sigma_X = 0.6/\sqrt{100} = 0.06$$

$$Z = \frac{\overline{X} - \mu}{\sigma/\sqrt{n}} = \frac{11.90 - 12.00}{0.06} = -1.67 \quad \text{(See Figure 4-13.)}$$

From Table A-2, P = 0.0475 (see Figure 4-13).

(b) To be 95% sure that $\overline{X}$ is not less than 11.95 means that the area to the left of $\overline{X}$ =11.95 is $A = 0.05$. From Table A-2, when $A = 0.05$, $Z = 1.645$. Since $11.95 < 12.00$, $Z = -1.645$.

$$-1.645 = \frac{(11.95 - 12.0)}{0.6/\sqrt{n}} = \frac{-0.05\sqrt{n}}{0.6}$$

$$\frac{-0.5}{\sqrt{n}} = -1.645(0.6) = -0.9870$$

$$-\sqrt{n} = \frac{0.9870}{0.5} = 19.74; \quad n = (19.74)^2 = 389.7$$

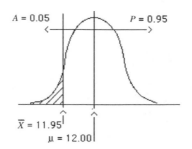

**Figure 4-14.** Example 4(b) curve.

Therefore, it will require a sample size of 390 to be 95% sure that the sample mean will not be less than 11.95 ounces. (see Figure 4-14.)

## NORMAL APPROXIMATION TO THE BINOMIAL DISTRIBUTION

If $n$ is large and $p$ is small, so that $\mu = np \le 5$, the Poisson distribution gives a good approximation of the probabilities of the binomial distribution. If $n$ is large and $\mu = np > 5$, the Poisson approximation is less satisfactory, and the normal curve may be used to find such probabilities. The closer together $p$ and $q$, and the larger the value of $n$, the better the approximation.

The binomial distribution is discrete, and its graphical representation consists of adjacent blocks, the areas of which represent the corresponding probabilities. Since the normal curve is continuous, in order to use it to approximate binomial probabilities, the area under the curve must include the block of the histogram centered at any value of r (the number of occurrences under consideration). To include the block centered at $r$, the value of $X$ to be used in the normal curve equation for the normal deviate must be adjusted by adding $\frac{1}{2}$ to or subtracting $\frac{1}{2}$ from the value of $r$. The procedure is illustrated in the following example.

**Example**
If 8% of a packaged product is known to be underweight, what is the probability that a random sample of 100 packages will contain (a) 14 or more underweight packages, (b) 4 or fewer underweight packages, (c) 5 or more underweight packages?

$$\mu = np = 100(0.08) = 8.0 \quad \text{and} \quad \sigma = \sqrt{npq} = \sqrt{100(0.08)(0.92)} = 2.71$$

(a)

$$r = 14; \quad X = r - \tfrac{1}{2} = 13.5$$

$$Z = \frac{13.5 - 8.0}{2.71} = \frac{5.5}{2.71} = 2.03$$

From the normal curve, $P = 0.0212$
Therefore, there is a little better than a 2% chance of this happening. (See Figure 4.15.)

(b) $r = 4; \qquad X = r + \frac{1}{2} = 4.5$

$$Z = \frac{4.5 - 8.0}{2.71} = \frac{-3.5}{2.71} = -1.29$$

$P = 0.0985$. There is about a 10% chance of this happening. (See Figure 4-16.)

(c) $\qquad\qquad r = 5; \qquad X = r - \frac{1}{2} = 4.5$
$\qquad\qquad Z = -1.29$ from part (b)
$\qquad\qquad P = 1 - 0.0985 = 0.9015$

There is about a 90% chance of this happening. (See Figure 4-17.)

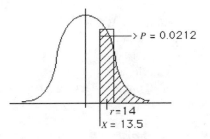

**Figure 4-15.** Example (a) curve.

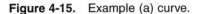

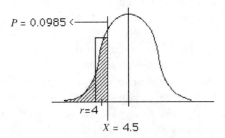

**Figure 4-16.** Example (b) curve.

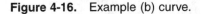

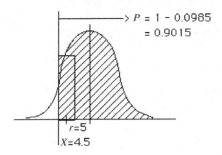

**Figure 4-17.** Example (c) curve.

## *t*-DISTRIBUTION

In all of the above examples using the normal distribution, the standard deviation has been known. More usually, the variability of a population is not known, and in order to determine the standard deviation, it is necessary to take a sample from the population and estimate the standard deviation $\sigma$. Under these conditions, the normal distribution is no longer applicable, and the *t*-distribution is more appropriate.

The sample mean $\bar{X}$ provides a satisfactory estimate of the population mean $\mu$, but the standard deviation is estimated using the expression:

$$\text{Estimate of standard deviation} s = \sqrt{\frac{\sum (X - \bar{X})^2}{n-1}} = \sqrt{\frac{\sum X^2 - \frac{(\sum X)^2}{n}}{n-1}}$$

where $\bar{X}$ is the mean of the sample, and $n$ is the number of items in the sample.

The quantity $(n-1)$ is called the number of degrees of freedom. It represents the number of independent relationships existing among the $(X-\bar{X})$ values. Since $(X-\bar{X}) = 0$, there are only $n-1$ independent differences.

The curves representing the *t*-distributions are symmetrical and bell-shaped, but not normal and somewhat flatter, with greater dispersion than those of the normal distribution. Their shapes depend on the number of degrees of freedom, and approach the normal as $n$ becomes increasingly larger.

Values of $t$ depend on the number of degrees of freedom, and are shown in Table A-3 for various probabilities. Note that the *t*-values in the table for infinite degrees of freedom ($df = \alpha$) are identical with the $Z$-values of the normal curve, and the entries in any column are obviously approaching the corresponding $Z$-values. From a practical standpoint, the dividing line between a "small sample" and a "large sample" is sample size of 30. With sample size 30 or over, the normal curve table or $t$ table may be used; below 30, the $t$ table should be used.

Formulas similar to those for the normal distribution are obtained for the *t*-distribution by replacing $\sigma$ by $s$ and $z$ by $t$; that is:

$$t = \frac{\bar{X} - \mu}{s_{\bar{X}}} = \frac{\bar{X} - \mu}{s/\sqrt{n}} \quad (\text{with } n-1 \; df)$$

satisfies a *t*-distribution with $n-1$ degrees of freedom where $\bar{X}$ is the sample mean, and as in the normal curve procedure,

$$s_{\bar{X}} = \frac{s}{\sqrt{n}}$$

## CONFIDENCE LIMITS FOR THE POPULATION MEAN

The $t$-distribution is convenient for use in estimating limits for the population mean. If we solve the following equation for $\mu$,

$$\pm t = \frac{X - \mu}{\dfrac{s}{\sqrt{n}}}$$

we find:

$$\mu = \overline{X} \pm \frac{ts}{\sqrt{n}}$$

which provides lower and upper limits for the population mean $\mu$ with a degree of confidence depending upon the value selected for $t$. These are referred to as the confidence limits for $\mu$.

To find the 95% confidence limits for $\mu$, select the value of $t$ from Table A-3 in the Appendix in the column headed 0.05 with $n - 1$ degrees of freedom (often denoted by $t_{n-1,0.05}$). Similarly, 99% confidence limits would be found by using the $t$-value found in the column headed 0.01 (i.e., $t_n - 1,0.01$).

### Example 1
A cannery is supposed to be filling cans with 19.0 ounces of fruit. To check conformance of a new machine with this standard, a random sample of 5 cans is selected from the new machine's production, and its mean is calculated. If the drained weights of the sample are 18.6, 18.4, 19.2, 18.3, and 19.0, can one be 95% sure that the production standard is being maintained on this machine? (This is equivalent to asking whether 19.0 is between the 95% confidence limits found for $\mu$.)

| $X$ | $(X - \overline{X})$ | $(X - \overline{X})^2$ |
|------|------|------|
| 18.6 | −0.1 | 0.01 |
| 18.4 | −0.3 | 0.09 |
| 19.2 | 0.5 | 0.25 |
| 18.3 | −0.4 | 0.16 |
| 19.0 | 0.3 | 0.09 |
| 93.5 | 0 | 0.60 |

$$\overline{X} = 93.5/5 = 18.7$$

$$s = \sqrt{\frac{0.60}{4}} = \sqrt{0.15}$$

$$s_{\overline{X}} = \frac{s}{\sqrt{n}} = \frac{\sqrt{0.15}}{\sqrt{5}} = \sqrt{0.03} = 0.1732$$

$$\pm\, t_{4,0.05} = \frac{18.7 - \mu}{0.1732}$$

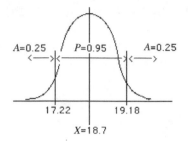

**Figure 4-18.** Example 1 curve.

From Table A-3,

$$t_{4,0.05} = 2.776$$

Therefore,

$$\mu = 18.7 \pm (2.776)(0.1732) = 18.7 \pm 0.48 = 18.22$$

and 19.18 and since 19.0 is between these two 95% confidence limits, the machine is conforming (see Figure 4-18).

## Example 2

The drained weights (in ounces) of a random sample of 9 cans of cherries are as shown. Find the 95% confidence limits for the mean weight of the entire lot.

| $X$ | $(X - \bar{X})$ | $(X - \bar{X})^2$ |
|---|---|---|
| 12.1 | 0 | 0 |
| 11.9 | −0.2 | 0.04 |
| 12.4 | 0.3 | 0.09 |
| 12.3 | 0.2 | 0.04 |
| 11.9 | −0.2 | 0.04 |
| 12.1 | 0 | 0 |
| 12.4 | 0.3 | 0.09 |
| 12.1 | 0 | 0 |
| 11.7 | −0.4 | 0.16 |
| 108.9 | 0 | 0.46 |

$$\bar{X} = \frac{108.9}{9} = 12.1$$

$$s = \sqrt{\frac{0.46}{8}} = 0.24; \quad s_{\bar{X}} = \frac{s}{\sqrt{n}} = \frac{0.24}{\sqrt{9}} = 0.08$$

95% *confidence limits for* $\mu$: $\bar{X} \pm t_{8,0.05} s_{\bar{X}}$

$$12.1 \pm (2.306)(0.08) = 12.1 \pm 0.18 = 11.92 \quad \text{and} \quad 12.28$$

Therefore, one can be 95% certain that 11.92 ounces to 12.28 ounces will contain the mean of the lot from which the sample came.

## STATISTICAL HYPOTHESES—TESTING HYPOTHESES

Frequently we are required to make a decision about populations based on data obtained from samples. For example, we may be interested in knowing if one lot is better than another, or if one production line manufactures better material than another. In order to reach such a decision, it may be useful to start with guesses or assumptions about the two populations. Such an assumption is known as a *statistical hypothesis*. A statistical procedure or decision rule which leads to establishing the truth or falsity of a hypothesis is called a *statistical test*. Decision rules enable the investigator to attach to his decisions probability statements about possible outcomes if the experiment were to be repeated many times under the same conditions. Decision rules are also referred to as *tests of significance*, or *tests of hypotheses*.

The hypothesis to be tested, often denoted by $H_0$, is called the *null hypothesis* since it implies that there is no real difference between the true value of the population parameter and its hypothesized value from the sample. For instance, if we wish to determine whether one process is better than another, we would formulate the hypothesis that there is no difference between the two processes, and that whatever differences were observed were merely due to fluctuations in sampling from the same population. Any hypothesis which differs from $H_0$ is called an *alternate hypothesis*, and is denoted by $H_1$.

In Figure 4-19, the yield for process A is designated by the shaded portion labeled A. After a change is made in the process, the results are plotted in the portion of the chart labeled B. The null hypothesis Ho states that if the B process falls within the shaded area, then it is no different than process A. The alternate hypothesis $H_1$ states that if the process should fall outside of the shaded area, then the two processes are different. This very simplistic example has limited usefulness since it does not consider probabilities, but it should illustrate the meanings of the two types of hypotheses.

To test the null hypothesis, one assumes that it is true and examines the consequences of this assumption in terms of a sampling distribution which depends on it. If we should find that the results obtained from a random sample are vastly different than those expected under the hypothesis, we would conclude that the observed differences were significant, and we would reject the hypothesis. As an example: if 100 tosses of a coin produced 83 tails, we would be likely to reject the hypothesis that the coin is fair.

The *level of significance* of a statistical test defines the probability level $\alpha$ which is the critical value in decision-making. It is the dividing line between accepting and rejecting. In the event that the calculated probability is less than $\alpha$, the hypothesis is considered false, and the result is termed *significant*. It is customary to consider a result significant if the probability is less than $\alpha = 0.05$, and to term it *highly significant* if the calculated probability is less than 0.01.

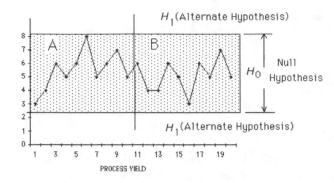

**Figure 4-19.** Null hypothesis for process A and process B.

An *alternate hypothesis* is any hypothesis other than $H_0$, and is denoted by $H_1$. If $H_1$ is adopted as the alternative hypothesis, then rejection of $H_0$ may be regarded as a decision to accept $H_1$.

If the area of only one tail of a curve is used in testing a statistical hypothesis, it is called a *one-tailed test*. Similarly, if the areas of both tails are used, the test is referred to as *two-tailed*. Instead of calculating probabilities, one can compare the calculated values of $Z$ with those shown in Figures 4-20 and 4-21. A calculated value which is greater than those shown indicates rejection of the null hypothesis.

The values for $Z$ may be selected from the table of normal curve areas. The critical values for $Z$ most frequently used are for probabilities of 10%, 5%, and 1%. These critical values for $Z$ for both one- and two-tail tests are tabulated below:

| $\alpha$ (Level of Significance) | 0.10 | 0.05 | 0.01 |
|---|---|---|---|
| $Z$ for one-tailed test | −1.28 | −1.645 | −2.33 |
| | or +1.28 | or +1.645 | or +2.33 |
| $Z$ for two-tailed test | −1.645 | −1.96 | −2.58 |
| | and +1.645 | and +1.96 | and +2.58 |

In testing statistical hypotheses there is no absolute certainty that the conclusion reached will be correct. At the 5% level, we are willing to be wrong once in 20 times, and at the 1% level, once in 100 times. If we are not willing to chance a wrong conclusion this often, we must select an even smaller value for $\alpha$. Two types of incorrect conclusions are possible. If the hypothesis is true but the sample selected concludes that it is false, we say that a *type 1 error* has been committed. The probability of committing a type 1 error is the relative frequency with which we reject a correct hypothesis, and this is precisely equal to the significance level $\alpha$.

If it happens that the hypothesis being tested is actually false, and if from the sample we reach the conclusion that it is true, we say that a *type 2 error* has been committed. The probability of committing a type 2 error is usually denoted by $\beta$. For a given number of observations it can be shown that if $\alpha$ is given, $\beta$ can be determined, and that if $\alpha$ is decreased, $\beta$ is increased, and vice versa. If we wish to decrease the chances of both types of error at the same time, it can be

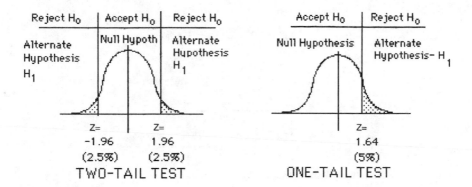

Figure 4-20.  Normal distribution, 5% level of significance.

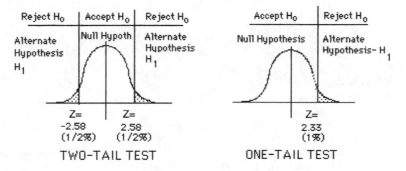

Figure 4-21.  Normal distribution, 1% level of significance.

accomplished only by increasing the sample size. $\alpha$ is also called the *producer's or sellers risk*; and $\beta$ is termed the *consumer's or buyer's risk*. (See Figure 4-22.)

### Hypotheses and Decisions Reached from Sample

| Decision from sample | $H_0$ true | $H_0$ false |
|---|---|---|
| Reject $H_0$ | Type 1 error | correct |
| Accept $H_0$ | correct | Type 2 error |

### Example 1

A company produces frozen shrimp in packages labeled "Contents 12 ounces." A sample of 4 packages selected at random yields the following weights: 12.2, 11.6, 11.8, and 11.6 ounces. At the 5% level, is the mean of the sample significantly different from the label claim of 12 ounces?

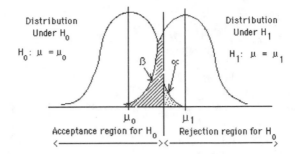

**Figure 4-22.** Relation of $\alpha$ and $\beta$.

Null Hypothesis                    Alternate hypothesis

$H_0: \mu = 12\,\text{ounces}$          $H_1: \mu \neq 12\,\text{ounces}$

| $X$ | $(X-\bar{X})$ | $(X-\bar{X})^2$ |
|------|------|------|
| 12.2 | 0.4 | 0.16 |
| 11.6 | −0.2 | 0.04 |
| 11.8 | 0 | 0 |
| 11.6 | −0.2 | 0.04 |
| 47.2 | 0 | 0.24 |

$$s = \sqrt{\frac{\sum(X-X)^2}{n-1}} = \sqrt{\frac{0.24}{3}} = \sqrt{0.08}$$

$$s_{\bar{X}} = \frac{s}{\sqrt{n}} = \frac{\sqrt{0.08}}{\sqrt{4}} = \sqrt{0.02} = 0.14$$

$$t_{0.05}\ (3\ \text{df}) = t_{3,0.5} = 3.182$$

$X = 47.2/4\ = 11.8$ ounces
$X \pm t_{3,.05}\,s_{\bar{X}}\ = 11.8 \pm 3.182(0.14)$
$\qquad\qquad = 11.8 \pm 0.45$

12.0

11.35        11.8        12.25

or

$\mu\ \pm\ t_{3,.05}s_{\bar{X}}\ = 12.0 \pm 0.45$

11.8

11.55        12.0        12.45

Therefore, there is no significant difference between $\bar{X}$ and $\mu$.

## Example 2
A new machine in a cannery is supposed to be filling cans of pears with a standard weight of 19.0 ounces. To check conformance with the standard, a sample of five cans is selected at random from the production line and its mean is calculated. The drained weights of the samples are shown below. Is the sample average significantly different from the standard at the 5% level?

| X | $(X-\bar{X})$ | $(X-\bar{X})^2$ |
| --- | --- | --- |
| 18.6 | −0.1 | 0.01 |
| 18.4 | −0.3 | 0.09 |
| 19.1 | 0.4 | 0.16 |
| 18.9 | 0.2 | 0.04 |
| 18.5 | −0.2 | 0.04 |
| 93.5 | 0 | 0.34 |

$\mu = 19.0$      Two-tailed

$H_0 : \bar{X} = \mu$     $H_1 : \bar{X} \neq \mu$
(Null)      (Alternate)

$$s^2 = \frac{\Sigma(X - \bar{X})^2}{n - 1} = \frac{0.34}{4} = 0.085$$

$\bar{X} = 93.5/5 = 18.7$

$$s_{\bar{X}} = \frac{s}{\sqrt{n}} = \sqrt{\frac{s^2}{n}} = \sqrt{\frac{0.085}{5}} = 0.13$$

From Table 4-3, $t_{0.05}(4 \text{ df}) = t_{4,0.05} = 2.776$

95% confidence limits for $\mu$: $\bar{X} + t_{4,0.05}\, s_{\bar{X}} = 18.7 + 27776(0.13) = 18.34$
                                                       19.06

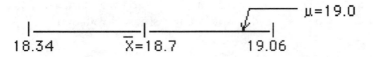

$\mu = 19.0$

18.34         $\bar{X} = 18.7$         19.06

The sample is not significantly different than the standard. There are two other possible solutions to this problem.

(a) Find the region within which $\bar{X}$ is not significantly different from $\mu$ at the 5% level of significance.

$$\pm\, t = \frac{\bar{X} - \mu}{s_X}$$

$$X = \mu \pm t_{0.05}s_{\bar{X}} \quad t_{4,0.05} = 2.776$$

$$= 19.0 \pm 2.776(0.13)$$

$$= 19.0 \pm 0.36 \sim 18.64 \text{ and } 19.36$$

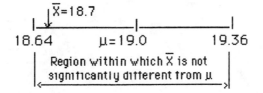

$\bar{X} = 18.7$

18.64      $\mu = 19.0$        19.36

Region within which $\bar{X}$ is not
significantly different from $\mu$

(b) From the sample mean, find the region within which the population may be expected to fall 95% of the time.

$$\mu = \overline{X} \pm t_{0.05}s_{\overline{X}}$$
$$= 18.7 \pm 0.36 \sim 18.34 \text{ and } 19.06$$

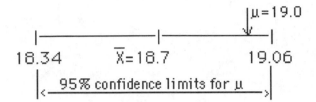

## DISTRIBUTION OF THE DIFFERENCE BETWEEN MEANS (UNPAIRED OBSERVATIONS)

If $X$ and $Y$ are normally and independently distributed populations with means $\mu_x$ and $\mu_y$, and standard deviations $\sigma_x$ and $\sigma_y$ respectively, then the difference between sample means $\overline{X}$ and $\overline{Y}$ is normally distributed with mean $\mu_x - \mu_y$ and standard deviation

$$\sigma_{\overline{x} - \overline{y}} = \sqrt{\sigma_{\overline{x}}^2 + \sigma_{\overline{y}}^2}$$

where

$$\sigma_{\overline{x}} = \frac{\sigma_x}{\sqrt{n_x}} \text{ and } \sigma_{\overline{y}} = \frac{\sigma_y}{\sqrt{n_y}}$$

Then

$$Z = \frac{\overline{X} - \overline{Y} - (\mu_x - \mu_y)}{\sigma_{\overline{x} - \overline{y}}}$$

will satisfy the normal probability curve.

Usually, $\sigma_x$ and $\sigma_y$ are unknown and must be estimated from the two samples. Since their estimates $s_x$ and $s_y$ will in general not be the same, and since we assume that the sample comes from populations having equal standard deviations, a pooled estimate $s$ is used in the calculations. The pooled estimate is calculated as follows:

$$s = \sqrt{\frac{\sum(X - \overline{X})^2 + \sum(Y - \overline{Y})^2}{n_x + n_y - 2}} = \sqrt{\frac{\sum X^2 - \frac{(\sum X)^2}{n_x} + \sum Y^2 - \frac{(\sum Y)^2}{n_y}}{n_x + n_y - 2}}$$

Then the $t$-distribution is appropriate where

$$t = \frac{\overline{X} - \overline{Y} - (\mu_x - \mu_y)}{s_{\overline{x} - \overline{y}}} \text{ with } n_x + n_y - 2 \text{ degrees of freedom,}$$

where

$$s_{\bar{x}-\bar{y}} = \sqrt{s_{\bar{x}}^2 + s_{\bar{y}}^2}, \quad s_{\bar{x}} = \frac{s}{\sqrt{n_x}}, \quad \text{and} \quad s_{\bar{y}} = \frac{s}{\sqrt{n_y}}$$

To test the calculated value of $t$ for significance at some level $\alpha$, it is compared with the tabular value $t_a$ with $(n_x + n_y - 2)$ degrees of freedom (from Table A-3). If the calculated value exceeds the tabular value, the means are said to be significantly different at the $a$ level of significance. The usual null hypothesis is $H_0: \mu_x = \mu_y$ resulting in the tests

$$Z = \frac{\bar{X} - \bar{Y}}{\sigma_{\bar{x}-\bar{y}}} (\sigma \text{ known}) \quad \text{and} \quad t = \frac{\bar{X} - \bar{Y}}{s_{\bar{x}-\bar{y}}} (\sigma \text{ estimated by } s)$$

## Example 1

To determine the differences in the butter consumption between rural and urban consumers, data were obtained from 12 farm communities and 14 metropolitan areas, as indicated below. Determine if there is a significant difference in the per capita consumption of butter between the two groups.

Since the standard deviation must be estimated from the samples, the $t$-distribution is appropriate. The hypotheses are $H_0: \mu_x = \mu_y$ and $H_1: \mu_x \neq$ (two-tailed).

| Pounds | | $\Sigma X = 125.6$ | $\Sigma Y = 122.1$ |
|---|---|---|---|
| | | $n_x = 12$ | $n_y = 14$ |

| Rural X | Urban Y | | |
|---|---|---|---|
| | | $X = 10.47$ | $Y\,5\,8.72$ |
| 12.1 | 8.3 | $\Sigma X^2 = 1345.38$ | $\Sigma Y^2 = 1103.37$ |
| 6.8 | 9.3 | $(\Sigma X)^2 = 15{,}775.36$ | $(\Sigma Y) = 14{,}908.41$ |
| 9.1 | 9.2 | | |
| 11.1 | 11.1 | $\frac{(\Sigma X)^2}{n_x} = 1314.61$ | $\frac{(\Sigma Y)^2}{n_y} = 1064.89$ |
| 11.4 | 10.7 | | |
| 13.3 | 4.6 | | |
| 9.8 | 9.9 | | |
| 11.3 | 7.9 | $s^2 = \dfrac{1345.38 - 1314.61 + 1103.37 - 1064.89}{12 + 14 - 2} = 2.89$ | |
| 9.4 | 9.8 | | |
| 10.2 | 7.9 | | |
| 11.3 | 8.5 | | |
| 9.8 | 9.1 | | |
| | 9.7 | | |
| | 6.2 | | |
| 125.6 | 122.1 | | |

$$s_{\bar{x}-\bar{y}} = \sqrt{\frac{2.89}{12} + \frac{2.89}{14}} \quad \text{and} \quad t = \frac{10.47 - 8.72}{0.67}$$

$$= 0.67 \qquad\qquad\qquad = 2.61 \ (24f)$$

Tabular $t$-values (two-tailed from $t$-tables—A-3 in the Appendix)

$$t_{0.05(24\ df)} = 2.064$$
$$t_{0.01(24\ df)} = 2.797$$

Since $t = 2.61$ is larger than 2.064 but less than 2.797, there is a significant difference in butter consumption between the two groups of consumers at the 5% level, but not at the 1% level.

If the question had been, "Is the per capita consumption of the rural group greater than that of the urban group?" the hypothesis would have been $H_0$: $\mu_x = \mu_y$ and $H_1$: $\mu_x > \mu_y$, and a one-tailed test would have been appropriate. Then

Tabular $t$-values (one-tailed from $t$-tables—A-3 in the Appendix)

$$t_{0.05(24\ df)} = 1.711$$
$$t_{0.01(24\ df)} = 2.492$$

Since $t = 2.61$ is larger than both 1.711 and 2.492, the per capita consumption of the rural group is significantly greater than that of the urban group at both levels of significance.

## Example 2

Find 95% confidence limits for the mean difference between the performance of two filling machines. Random samples from the two machines A and B gave the following results:

| A | B |
|---|---|
| $\bar{X} = 350$ g | $\bar{Y} = 325$ g |
| $n_x = 10$ | $n_y = 14$ |
| $\Sigma(X-\bar{X})^2 = 900$ | $\Sigma(Y-\bar{Y})^2 = 832$ |

$$s^2 = \frac{\Sigma(X - \bar{X})^2 + \Sigma(Y - \bar{Y})^2}{n_x + n_y - 2}$$

$$s^2 = \frac{900 + 832}{22} = \frac{1732}{22} = 78.7$$

then

$$s_{\bar{x}-\bar{y}} = \sqrt{\frac{s^2}{n_x} + \frac{s^2}{n_y}} = \sqrt{\frac{78.7}{10} + \frac{78.7}{14}} = 3.67$$

95% confidence limits for $\mu_x - \mu_y = (X - \bar{Y}) \pm (t_{22,0.05})s_{\bar{x}-\bar{y}}$

$$= 25 \pm (2.074)(3.67) = 25 \pm 7.61$$
$$= 17.39 \text{ and } 32.61$$

$$\frac{-7.61 \qquad\qquad +7.61}{17.39 \qquad X - Y = 25 \qquad 32.61}$$

Therefore, we are 95% confident that 17 and 33 grams are limits of the two confidence intervals which contain the difference in mean filling weights of the two machines. We then compare our specification to see if adjustments are necessary.

If instead of $\sum(X - \overline{X})^2$ and $\sum(Y - \overline{Y})^2$ being given, $s_x = 9.5$ and $s_y = 7.7$ had been given, then

$$s_{\bar{x} - \bar{y}} = \sqrt{\frac{s_x^2}{n_x} + \frac{s_y^2}{n_y}} \quad \text{with } n_x + n_y - 2 \text{ degrees of freedom}$$

$$= \sqrt{9.02 + 4.24} = 3.64$$

## PAIRED OBSERVATIONS

Many experiments are performed in such a way that each item of one sample is logically paired with a particular item of a second sample. For example, if members of a panel of tasters score a sample of each of two similar products, a set of paired scores results. With $n$ paired scores of $X$ and $Y$, the differences $D = X - Y$ are analyzed.

The mean difference $\overline{D} = \sum D/n = \overline{X} - \overline{Y}$. If the samples are the same, $\overline{D} = 0$. The question to be answered is, "Is the average difference $\overline{D}$ too large for the samples to be considered as having the same average scores?" To answer this question, one calculates

$$t = \frac{\overline{D}}{s_{\bar{d}}} \quad \text{with } n - 1 \text{ degrees of freedom,}$$

where

$$s_{\bar{d}} = \frac{s}{\sqrt{n}} \quad \text{and} \quad s = \sqrt{\frac{\sum(D - \overline{D})^2}{n - 1}} = \sqrt{\frac{\sum D^2 - \dfrac{(\sum D)^2}{n}}{n - 1}}$$

The significance of the result is obtained by comparing the calculated value of $t$ with the value from Table A-3 in the Appendix with the appropriate degrees of freedom.

### Example

Using a 12-point intensity scale, 10 trained judges scored the bitterness intensity of two samples of beer as shown below. Do the beers differ significantly in bitterness at the 5% and the 1% levels?

$$H_0 : \mu_d = 0; \quad H_1 : \mu_d \neq 0 \text{ (two-tailed)}$$

| Judge | Beer $X$ | Beer $Y$ | $D = X - Y$ | $D - \bar{D}$ | $(D - \bar{D})^2$ |
|-------|----------|----------|-------------|---------------|-------------------|
| 1     | 10       | 7        | 3           | 1             | 1                 |
| 2     | 6        | 3        | 3           | 1             | 1                 |
| 3     | 5        | 6        | −1          | −3            | 9                 |
| 4     | 7        | 7        | 0           | −2            | 4                 |
| 5     | 10       | 7        | 3           | 1             | 1                 |
| 6     | 6        | 4        | 2           | 0             | 0                 |
| 7     | 7        | 5        | 2           | 0             | 0                 |
| 8     | 8        | 6        | 2           | 0             | 0                 |
| 9     | 6        | 3        | 3           | 1             | 1                 |
| 10    | 5        | 2        | 3           | 1             | 1                 |
| Total | 70       | 50       | 20          | 0             | 18                |

$$\bar{D} = 2.0$$

$$s^2 = \frac{18}{9} = 2$$

$$s_{\bar{d}} = \frac{s}{\sqrt{n}} = \frac{1.414}{\sqrt{10}} = 0.447$$

$$t = \frac{2.0}{0.447} = 4.47$$

$$t_{0.05,9} = 2.263; \quad t_{0.01,9} = 3.250$$

Therefore, there is a significant difference at both levels.

If the question had been, "Is Beer X significantly more bitter than Beer Y?", a one-tailed test would have been appropriate, with

$$H_0: \mu d = 0; \quad \text{and} \quad H_1: \mu_d > 0 \quad \text{where } \mu_d = \mu_x - \mu_y$$

From Table A-3 in the Appendix,

$$t_{0.05,9} = 1.883, \quad \text{and} \quad t_{0.01,9} = 2.821$$

Since 3.16 is greater than both of these values, Beer X is significantly more bitter than Beer Y at both levels of significance.

## F-DISTRIBUTION

In applying the $t$-test for the comparison of two means, a pooled estimate may be made of the population standard deviation on the assumption that $s_x = s_y$. This assumption usually poses no problems. However, there are cases where the equality of the sample standard deviations is questionable. In such cases, the $F$-distribution provides a test of equality.

To test the hypothesis that two variances (or standard deviations) are equal, a form of the $F$-distribution is defined as:

$$F = \frac{\text{Larger variance}}{\text{Smaller variance}} = \frac{s_x^2}{s_y^2}$$

where

$$s_x^2 = \frac{\sum(X - \bar{X})^2}{n_x - 1}, \quad s_y^2 = \frac{\sum(Y - \bar{Y})^2}{n_y - 1} \quad \text{and} \quad s_x^2 > s_y^2$$

In this form, we can compare values of $F$ with critical $F$-values given in the $F$-distribution tables. If the calculated value is less than the tabular value, the variances may be considered homogeneous (equal). If the calculated value exceeds the tabular value, the numerator variance is considered to be significantly greater than the denominator variance at the indicated $\alpha$ level. The tabular values of $F$ depend on the degrees of freedom of both the numerator and the denominator. The $F$-distribution plays an important part in the analysis of variance.

## ANALYSIS OF VARIANCE

### One-way Classification

So far, we have found that the $t$-test provides a procedure for testing the equality of two population means when the data are composed of a random sample from each population. In many cases, more than two population means are under study and require a method of testing their equality from an evaluation of an independent random sample from each population. For example, we may wish to evaluate the performance of different can-closing machine heads, or the cost of running a process with several different raw materials, or determining the shipping quality of cases from several different suppliers.

It would appear that these problems could be easily solved by taking all of the possible pairs of samples and testing them individually by the $t$-test for significant differences between means. This could turn out to be laborious and, worse still, would have a high probability of leading to false conclusions. In a case where seven samples were to be taken from populations of identical means, if we were to test all of the possible pairs of samples for significant differences among their means, we would have to apply the test to 21 pairs, with a 66% chance of arriving at one or more incorrect conclusions. We have already discussed the $\alpha$ risk of making a Type I error in each individual $t$-test.

A more efficient procedure for making such comparisons is the analysis of variance (ANOVA), which examines all of the sample means together, and has a single $\alpha$ risk. Basically, it is a simple arithmetical method of sorting out the components of variation in a given set of data, and of providing tests of significance.

It is based on two principles:

- the partitioning of the sums of the squares
- the estimating of the variance of the population by different methods, and comparing these estimates.

## Partitioning the Sum of Squares—Equal Sample Sizes

Consider $k$ samples, all of which have $n$ variates (measurements). The notation is summarized in Table 4-5 in which double subscript notation is introduced. The first subscript identifies the sample to which the variate belongs, and the second subscript identifies the particular variate within the sample. Thus $X_{21}$ represents the first variate in the second sample, and $X_{ij}$ represents the $j$th variate in the $i$th sample.

1. The sums of squares of deviations from the respective means, and the corresponding degrees of freedom are defined as follows:

$$\text{SS} = \sum_{i=1}^{k} \sum_{j=1}^{n} (X_{ij} - \overline{X})^2 = \sum_{i=1}^{k} \sum_{j=1}^{n} X_{ij}^2 - \frac{T^2}{kn} \qquad df = kn - 1$$

   and is a measure of the dispersion of all of the variates about the grand mean.
2. The among sample means sums of squares SST (or sum of squares for samples) follows from the fact that

$$\sigma^2 = n\sigma_{\overline{x}}^2 \quad \text{or} \quad \sigma_{\overline{x}} = \frac{\sigma}{\sqrt{n}}$$

$$\text{SST} = n\sum_{i=1}^{k} (X_i - \overline{X})^2 = \sum_{i=1}^{k} \frac{T_j^2}{n} - \frac{T^2}{kn} \qquad df = k - 1$$

**Table 4-5.   One Criterion of Classification**

|        | Samples |          |          |      |                   |
|--------|---------|----------|----------|------|-------------------|
|        | 1       | 2        | 3        |      | $k$               |
| | $X_{11}$ | $X_{21}$ | $X_{31}$ | ---- | $X_{k1}$ |
| | $X_{12}$ | $X_{22}$ | $X_{32}$ | ---- | $X_{k2}$ |
| | $X_{13}$ | $X_{23}$ | $X_{33}$ | ---- | $X_{k3}$ |
| | .        | .        | .        |      | .                 |
| | .        | .        | .        |      | .                 |
| | .        | .        | .        |      | .                 |
| | $X_{1n}$ | $X_{2n}$ | $X_{3n}$ |      | $X_{kn}$ |
| Total | $T_1$ | $T_2$ | $T_3$ |      | $T_k$ |
| |          |          |          |      | Grand total $= T$ |
| Mean | $\overline{X}_1$ | $\overline{X}_2$ | $\overline{X}_3$ |      | $\overline{X}_k$ |
| |          |          |          |      | Grand mean $=\overline{X}$ |

3. The within-sample sum of squares SSE (or sum of squares for error) is the pooled sum of squares obtained within each sample, and the pooled value is unaffected by any differences among the means.

$$SSE = \left[ \sum_{j=1}^{n} X_{1j}^2 - \frac{T_1^2}{n} \right] + \left[ \sum_{j=1}^{n} X_{2j}^2 - \frac{T_2^2}{n} \right] + \cdots + \left[ \sum_{j=1}^{n} X_{kj}^2 - \frac{T_k^2}{n} \right]$$

$$= \sum_{i=1}^{k} \sum_{j=1}^{n} x_{1j}^2 - \sum_{i=1}^{k} \frac{T_i^2}{n} \quad df = k(n-1)$$

This quantity is known as the *experimental error* and is a measure of the variation which exists among observations on experimental units treated alike. From the above expressions, it is seen that

$$SS = SST + SSE$$

and the same relationships hold for the corresponding degrees of freedom. Because of this relationship, it is customary to obtain SSE as the difference SS − SST, and the same with the degrees of freedom. These relationships show that we can always partition the total dispersion in such data into two components, one due to the dispersion existing among the sample means and the other due to the dispersion existing within the samples.

From the sums of the squares and degrees of freedom, we can obtain two estimates of the variance $\sigma^2$. They are

$$s_{\bar{x}}^2 = \frac{SST}{(k-1)} \quad \text{and} \quad s_e^2 = \frac{SSE}{k(n-1)}$$

The first of these is influenced by differences among means, whereas the second is independent of any differences; therefore any difference existing between these two estimates of $\sigma^2$ is the result of differences among sample means. The two estimates may be tested for a significant difference by calculating

$$F = \frac{s_{\bar{x}}^2}{s_e^2}$$

If $s_{\bar{x}}^2$ is significantly greater than $s_e^2$, significant differences among means are indicated.

The *correction term* for the set of variates is defined as

$$C = T^2/kn$$

which occurs in the calculations of both SS and SST. The calculations in the analysis of variance are shown in the following example.

**Example**

Four scores of each of five products are shown in the following table. Are there significant differences among these mean scores?

|       | Products |      |      |      |      |
|-------|------|------|------|------|------|
|       | A    | B    | C    | D    | E    |
|       | 10   | 8    | 5    | 8    | 7    |
|       | 8    | 7    | 7    | 7    | 6    |
|       | 9    | 9    | 6    | 7    | 5    |
|       | 9    | 8    | 5    | 5    | 7    |
| Total | 36   | 32   | 23   | 27   | 25   |
| Mean  | 9.00 | 8.00 | 5.75 | 6.75 | 6.25 |

$$C = (143)^2/20 = 1022.45$$

$$SS = (10)^2 + (8)^2 + \cdots + (5)^2 + (7)^2 - C = 1065 - 1022.45 = 42.55 \ (19 \ df)$$

$$SST = \frac{(36)^2 + (32)^2 + (23)^2 + (27)^2 + (25)^2}{4} - C = 1050.75 - 1022.45 = 28.30 \ (4 \ df)$$

$$SSE = 42.55 - 28.30 = 14.25$$

The calculations are usually summarized as shown below.

Analysis of variance

| Source of variation | SS | df | Mean square | F | $F_{0.05}$ | $F_{0.01}$ |
|------|------|----|------|------|------|------|
| Total | 42.55 | 19 | | | | |
| Products | 28.30 | 4 | 7.075 | 7.45 | 3.06 | 4.89 |
| Error | 14.25 | 15 | 0.95 | F values at 4 and 15 degrees of freedom | | |

Since the calculated value of $F = 7.45$ is larger than $F_{0.01} = 4.89$, significant differences among product mean scores are indicated at both the 5% and the 1% levels. Significance at the 5% level is usually indicated by one asterisk, and at the 1% level by two asterisks.

If the calculated $F$-value is not significant, further testing is not required. If, however, the $F$-value indicates significant differences among means, we need to know where the differences exist. Two procedures in common use are the Least Significant Difference (LSD) test and Duncan's Multiple Range test.

## Difference Testing

**Least Significant Difference (LSD).** The difference between the means $\bar{X}$ and $\bar{Y}$ of two independent samples of the same size $n$, from a normal population with variance $v = s^2$, is tested for significance by calculating

$$t = \frac{\overline{X} - \overline{Y}}{s_{\overline{x} - \overline{y}}} = \frac{\overline{X} - \overline{Y}}{\sqrt{\dfrac{y}{n} + \dfrac{y}{n}}} = \frac{\overline{X} - \overline{Y}}{\sqrt{2\dfrac{v}{n}}}$$

where $v = s^2$.

If, in this equation, we replace $t$ by $t_\alpha$ and solve for $\overline{X} - \overline{Y}$, we have

$$\overline{X} - \overline{Y} = t_\alpha \sqrt{2\frac{v}{n}}$$

which is the boundary line between significance and nonsignificance at the $\alpha$ level, and is called the least significant difference between means, that is,

$$LSD = t_\alpha \sqrt{\frac{2 \text{ error variance}}{n}} \; n = t_\alpha \sqrt{\frac{2s_e^2}{n}}$$

where $n$ is the sample size and $t_\sigma$ is based on the number of degrees of freedom for error.

**Duncan's Multiple Range Test.** This is a newer approach to difference testing and provides a series of shortest significant ranges with which to compare differences between means. The shortest significant range $R_p$ for comparing the largest and the smallest of $p$ means, arranged in order of magnitude, is given by $R_p = Q_p s_{\overline{x}}$, where the values of $Q_p$ can be obtained from Duncan's Multiple Ranges Tables (Table A-7 in the Appendix). The number of degrees of freedom for the error variance, and the standard error (standard deviation) of any mean determined from $n$ individual variates, is

$$s_{\overline{x}} = \sqrt{\frac{\text{error variance}}{n}} = \sqrt{n}$$

**Example**

Use (a) the LSD and (b) the multiple range tests to establish significant differences among means at the 1% level for the data of the analysis of variance example.

In that example, the calculated $F = 7.45$ was larger than $F_{0.01} = 4.89$, so further testing is justified. The error variance $= 0.95$.

(a) $t_{0.01,15} = 2.947$, and

$$LSD = 2.947 - \sqrt{\frac{2(0.95)}{4}} = 2.03$$

| A | B | D | E | C |
|---|---|---|---|---|
| 9.00 | 8.00 | 6.75 | 6.25 | 5.75 |

Any two means not underscored by the same line are significantly different, and any two means underscored by the same line are not significantly different. At the 1% level, means A and B are not signifcantly different; B, D, and E are not significantly different; and D, E, and C are not significantly different.

(b) $v = 0.95$; $p = $ number of means being compared; $Q_p$ with degrees of freedom for error from Duncan's Multiple Ranges Table (1% level, 15 degrees of freedom); and

$$R_p = Q_p s_{\bar{x}} = \text{shortest significant range for } p \text{ means;} \quad \text{and}$$

$$s_{\bar{x}} = \sqrt{v/n} = \sqrt{0.95/4} = 0.49$$

Shortest significant ranges:

| $p$: | 2 | 3 | 4 | 5 |
|------|------|------|------|------|
| $Q_p$: | 4.17 | 4.35 | 4.46 | 4.55 |
| $R_p$: | 2.04 | 2.13 | 2.19 | 2.23 |

| A | B | D | E | C |
|------|------|------|------|------|
| 5.00 | 8.00 | 6.75 | 6.25 | 5.75 |

| Difference of | Used to compare |
|---------------|-----------------|
| 2.04 | A & B, B & D, D & E, and E & C |
| 2.13 | A & D, B & E, and D & C |
| 2.19 | A & E, and B & C |
| 2.23 | A & C |

The results of both methods were the same.

**Unequal Sample Sizes.**   If unequal numbers of measurements are involved in the samples, only minor modifications are required in the analysis. The formula for SST is replaced by

$$\text{SST} = \frac{T_1^2}{n_1} + \frac{T_2^2}{n} + \frac{T_3^2}{n_3} + \cdots + \frac{T_k^2}{n_k} - C$$

The procedure will be illustrated by the following example.

| A | B | C | D | |
|------|------|------|------|------|
| 9 | 10 | 10 | 11 | |
| 12 | 7 | 8 | 10 | |
| 16 | 6 | 8 | 7 | |
| 15 | 12 | 12 | 7 | |
| | | | 7 | |
| | | | 9 | |
| | | | 14 | |
| 52 | 35 | 38 | 65 | $T = 190$ |

$$C = (190)^2/19 = 1900$$

$$SS = (9)^2 + (12^2) + \cdots + (9)^2 + (14)^2 - 1900 = 152$$

$$SST = \frac{(52)^2}{4} + \frac{(42)^2}{5} + \frac{(38)^2}{4} + \frac{(58)^2}{6} - 1900 = 50.5$$

$$SSE = 152 - 50.5 = 101.5$$

| | Analysis of variance | | | | |
|---|---|---|---|---|---|
| Source of variation | SS | df | Mean square | $F$ | $F_{0.05}$ |
| Total | 152 | 18 | | | |
| Samples | 50.5 | 3 | 16.83 | 2.49 | 3.29 |
| Error | 101.5 | 15 | 6.77 | | |

Since the calculated $F$-value is less than the tabular value for 3 and 15 degrees of freedom, no significant difference among the samples is indicated.

Whenever unequal numbers of measurements occur in the samples and the $F$-value indicates significance, there is no simple way to decide which pairs of means differ significantly. All possible pairs must be tested individually by means of the $t$-test.

## TWO CRITERIA OF CLASSIFICATION
## (RANDOMIZED COMPLETE BLOCK DESIGN)

In the case of one criterion of classification, no allowance is made for the effects which are common to specific variates in all samples. Where such effects exist, a second criterion of classification is present. This occurs, for example, in a taste-testing experiment in which each judge scores each sample. This design permits another measure of variability, that due to the judges. Then, as before, the total sum of the squares can be subdivided with the addition of a term (SSB) for the judges. Then

$$SS = SST + SSB + SSE \quad \text{and} \quad SSB = \frac{B_1^2 + B_2^2 + B_3^2 + \cdots + B_n^2}{k}$$

where the $B$'s are total scores for the $n$ judges. The pattern is shown in Table 4-6.

**Table 4-6.  Two Criteria of Classification**

| Blocks | Samples (Treatments) | | | | Total |
|---|---|---|---|---|---|
| | 1 | 2 | 3 | $k$ | |
| | Judges | | | | |
| 1 | $X_{11}$ | $X_{21}$ | $X_{31}$ | $\cdots$ | $X_{k1}$ | $B_1$ |
| 2 | $X_{12}$ | $X_{22}$ | $X_{32}$ | $\cdots$ | $X_{k2}$ | $B_2$ |
| 3 | $X_{13}$ | $X_{23}$ | $X_{33}$ | $\cdots$ | $X_{k3}$ | $B_3$ |
| $\vdots$ | $\vdots$ | $\vdots$ | $\vdots$ | | $\vdots$ | $\vdots$ |
| $n$ | $X_{1n}$ | $X_{2n}$ | $X_{3n}$ | | $X_{kn}$ | $B_n$ |
| Total | $T_1$ | $T_2$ | $T_3$ | | $T_k$ | Grand total $= T$ |
| Mean | $\overline{X}_1$ | $\overline{X}_2$ | $\overline{X}_3$ | | $\overline{X}_k$ | Grand mean $= \overline{X}$ |

Computations

(a) Correction term $C = T^2/kn$

(b) Total sum of squares $\quad$ SS $= \sum_i \sum_j x_{ij}^2 - C \;\; df = kn - 1$

(c) Sum of squares for samples SST $= \sum_{i=1}^{k} T_i^2/n - C \;\; df = k - 1$

(d) Sum of squares for judges SSB $= \sum_{j=1}^{n} B_j^2/k - C \;\; df = n - 1$

(e) Sum of squares for
error $=$ SS $-$ SST $-$ SSB df $= (kn - 1) - (k - 1) - (n - 1)$

## Example

To determine which of three Napa Valley Chenin Blanc white table wines had the best quality, five expert wine judges evaluated each wine using a 20-point evaluation scale. Did the three wines differ in quality? Did the five judges agree in their evaluations?

$$C = (237)^2/15 = 3744.60$$
$$SS = (16)^2 + (15)^2 + \cdots + (16)^2 + (16)^2 - C$$
$$= 3757 - 3744.60 = 12.40 \qquad (14 \; df)$$

| Judges | Wines | | | |
|---|---|---|---|---|
| | X | Y | Z | Total |
| A | 16 | 17 | 15 | 48 |
| B | 15 | 16 | 14 | 45 |
| C | 15 | 16 | 15 | 46 |
| D | 16 | 18 | 16 | 50 |
| E | 16 | 16 | 16 | 48 |
| Total | 78 | 83 | 76 | 237 |

$$SST = [(78)^2 + (83)^2 + (76)^2]/5 - C$$
$$= 3749.80 - 3744.60 = 5.20 \qquad (2\ df)$$

$$SSB = [(48)^2 + (45)^2 + (46)^2 + (50)^2 + (48)^2]/3 - C$$
$$= 3749.67 - 3744.60 = 5.07 \qquad (4\ df)$$

$$SSE = 12.40 - 5.20 - 5.07 = 2.13$$

### Analysis of Variance Table

| Sources | SS | df | MS | F | $F_{0.05}$ | $F_{0.01}$ |
|---------|------|----|-------|---------|------|------|
| Total   | 12.40 | 14 |       |         |      |      |
| Wines   | 5.20 | 2  | 2.60  | 9.77**  | 4.46 | 8.65 |
| Judges  | 5.07 | 4  | 1.268 | 4.77*   | 3.84 | 7.01 |
| Error   | 2.13 | 8  | 0.266 |         |      |      |

*,** Significant at 5% and 1% levels, respectively.

Since the $F$-values for both wines and judges are significant, we can proceed to determine where differences exist. Suppose that we use the LSD procedure with the 5% level of significance.

$$\text{Wines: } LSD = t_{0.05,8}\sqrt{\frac{2V}{5}} = 2.306\sqrt{\frac{2(0.266)}{5}}$$

$$= 2.306(0.326) = 0.75$$

| | Y | X | Z |
|------|------|------|------|
| Mean | 16.6 | 15.6 | 15.2 |

Therefore, wine $Y$ is, in the opinion of the judges, signifcantly better than the other two wines.

$$\text{Judges: } LSD = t_{0.05,8}\sqrt{\frac{2V}{3}} = 2.306\sqrt{\frac{2(0.266)}{3}} = 2.306(0.421) = 0.97$$

| | D | A | E | C | B |
|------|------|------|------|------|------|
| Mean | 16.7 | 16.0 | 16.0 | 15.3 | 15.0 |

There is no significant difference between the judging of $D$, $A$, and $E$. There is no significant difference between the mean scores of Judges $A$, $E$, and $C$; and none between judges $C$ and $B$.

The analysis of variance procedure is applicable to many different and complicated experimental designs.

# 5    Sampling

## SAMPLING PLANS

In order to measure the quality characteristics of a population, be it a process, a product, or a lot, it is usually desirable to select a sample from the population and examine it. Rarely is it either desirable or even possible to examine 100% of a lot to determine its quality, hence some assumptions must be made so that a rational sample can be selected which represents the quality of the underlying population. Needless to say, if the sample selected does not represent the underlying population, then results of examining that sample are meaningless, and can produce some incorrect and perhaps costly conclusions. If a sample of cherries is selected from trees on the southern side of an orchard, a study of the degree of maturity of the sample may result in an erroneous conclusion that the orchard is ready for harvesting, whereas the conclusion should have been that only the southern side of the orchard was mature.

The study of sampling is enormously complex, and yet the results of analysis of possible sampling procedures ultimately lead to only four simply defined parameters:

1. Where is the optimum location of the sampling point?
2. How should the sample be taken from the population?
3. With what frequency should the sample be selected?
4. What is the optimum sample size?

Methods for analyzing the samples once the system has been established, and decisions to be made regarding the population quality based on the results of these analyses, will be discussed later. First, let us consider the factors inherent in populations which must be carefully considered so that a reliable sampling plan can be created.

The cherry orchard example above exhibits the need to ensure that the sample represents the population, and not merely a portion of it. Unfortunately, the variables which exist differ from industry to industry, and from product to product. For example, selecting a completely representative and random sample of corn from the first truckload in the morning, and analyzing the sample for sugar–starch ratio, may provide the plant operations manager with the information required to determine whether to pack whole kernel or cream style—*but only for that truckload!* Experience might show that the ratio could change drastically in a very few hours, and that subsequent truckloads might no longer be suitable for the process initially selected. This leads us to the first decision to be made.

## WHY SAMPLE?

In the case of the corn, the need for sampling is to establish the sugar–starch characteristic of the incoming raw materials so that the process can be fixed. It is essential that a sample be selected in such a way as to ensure that it represents some defined population. If experience has shown that the ratio changes slowly during the morning for product from a given acreage, it may be possible to sample from the first, third, and fifth truck from that acreage to establish the ratio, and to observe process drift for the conditions existing on that particular day (wind velocity, temperature, humidity). Thus, a sample can be used to reliably measure this one corn characteristic for this morning period. The answer to the "why sample?" question is obvious in this example: it is not possible to determine the ratio of the entire truckload, and so a few ears are selected at random in such a manner as to represent the entire population's composition. The difficulties inherent in this example are apparent because of the wide variations possible from farm to farm, from row to row, and even from ear to ear on a single plant. Yet, it is unlikely that a controlled quality canned or frozen corn product could be produced without this critical measurement of the raw material. Many years ago, the tests for suitability conducted in small plants may have consisted of an experienced foreman's selecting several ears from each truckload, cutting a few kernels and chewing them to rate the color, texture, and flavor. Although these tests were inexact and primitive, the conclusions drawn from the sample were as dependent on the relation of the sample to the lot as the conclusions drawn from today's more sophisticated testing methods are dependent on the need for a representative sample.

Sampling a tank truck of liquid sugar or corn syrup for concentration is a far simpler task. It may be shown, for example, that a composite created from 500 cc samples drawn from the top, middle, and bottom of a tank truck constitutes a consistently representative sample of the entire contents. The mere fact that it is possible to obtain a reliable sample with relative ease, however, is not a valid reason to set up a sampling system. The test results should have some meaningful function in the overall product quality system.

If, for example, the tank truck arrives with a certified Brix, and if the process requires that every batch of product using that syrup must be adjusted to some unique concentration, which is carefully monitored during the process, there may be no advantage to knowing that each shipment is at the concentration expected. This certainly does not preclude the necessity for periodic monitoring, but large volumes of data that are not required to control the quality are of no use, and merely add to the cost of quality operations.

There is a universal definition of sampling: a procedure which is used to draw inferences about a parent population from results obtained in the sample. To the uninformed, sampling is generally considered to be a simple procedure to discover defects in products, and it does indeed fit that definition on occasion. However, in the examples above, the cherry samples, the corn samples, and the sugar samples were drawn to measure some aspect of the quality of the population from which they were selected; they were not selected to find defects.

## SAMPLES FROM DIFFERENT DISTRIBUTIONS

Randomly selecting samples from a population to determine facts about its composition can produce meaningless data if the distribution of data within the entire population is unknown, or if the purpose for which the data is collected is not clear to the technician obtaining the information. For example, if the heights of a sample of 100 tomato plants are measured (Figure 5-1), and there are ten deformed plants only 3-in. high, along with 90 plants which vary from 2 to 4 feet in height, an uninformed technician might report the average height of the tomato plants in the field at 2.7 feet. The average 3-foot height of *normal* plants would be of more significance if the harvester equipment were to be set according to the average height of normal plants.

Representative samples taken from any distribution can be expected also to exhibit characteristics of that distribution. For example, representative samples from a bimodal distribution should also have bimodal measurements. Let us consider deliveries of shipping cases received simultaneously from two different plants, with different average bursting strengths, piled together in a plant warehouse. If truly random samples are selected, bursting strength tests of the samples should show two peaks (Figure 5-2).

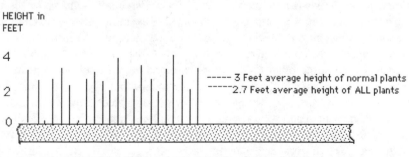

A portion of the 100 tomato plant sample, showing the effect of including deformed plants in the average height calculation.

**Figure 5-1.** Average height of tomato plants.

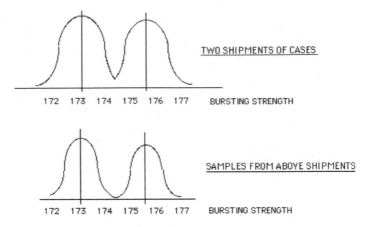

**Figure 5-2.** Distribution of samples from two shipments.

Similarly, samples taken from skewed distributions can be expected to exhibit skewed measurements mirroring the parent population. It should be noted, however, that compositing samples before analyzing, or averaging the results of groups of samples, will tend to create normally distributed sample characteristics, thus masking the true character of the population.

## SAMPLE SIZE

There are at least five common policies in use to determine the size of sample selected, and there is a place for each of them. Each policy is intended to satisfy two goals: the sample should accurately describe the population, and the sample size should be the most economical to reach this goal.

### Sample Size of Zero

There are occasions where a sample need not be drawn at all. Some materials are so incidental to the process that there is no need to examine them. Silicones used to lubricate conveyor rails might be an example. Or there may be instances where an uncomplicated raw material received from a reliable supplier will immediately be converted to some other form before use, and which may not constitute a particularly critical step in the process. An example of this might be common salt which is to be converted to various brines for grading green peas. Other classes of materials used in quantity but requiring no sampling might be industrial cleaning solutions and some maintenance supplies. As part of a "just-in-time" production system, the goal of zero raw material inventories is usually accomplished by shifting the quality control back to the supplier. In this case, examination of the supplier's quality control records eliminates the need for sampling, except on a monitoring basis. By far, the greatest use of a zero sample size quality control program is found within operations where a company depends on luck, supplier reputation, or ignorance. The policy statement: "We have never had trouble with this before, so we need not sample it" makes as much sense as the remark made by the farmer whose horse suddenly dropped dead: "That's strange; this never happened before."

So far we have considered only raw materials. Sampling along the production line is a step which cannot be avoided if costs are to be minimized. Yet, there are managers unable to see the value in avoiding the costs of scrap and rework by spending a far smaller amount of money in controlling the process quality to avoid making mistakes. It is difficult to conceive of a situation in which it would not be more expensive to correct a faulty finished product than it would be to avoid making the mistake of using an incorrect raw material, or preventing errors made on the production line. A zero sampling program of the finished product is unthinkable; yet, it would make more economic sense to reduce the sampling at the end of the line where mistakes are most costly, and increase the sampling on the production line and on the raw materials, where errors are less costly. It would seem almost axiomatic that a high production record is nowhere as economically important to a company as a high *acceptable quality* production record.

## 100% Sampling

At the other end of the spectrum is a sample size of 100%. In a previous chapter, we have learned that 100% sampling is not effective. Repetitive sampling is boring, and leads to inattention. To demonstrate this point, the author counted all of the "*es*" in the above paragraph. The first check produced 70. Rechecks counted 75, 71, and 87! It is suggested that the reader take on the role of inspector and attempt to decide which figure (if any) is correct.

There are many instances where 100% inspection is considered necessary because of the importance of some attribute to the consumer. Packages of multiple units can be the source of innumerable consumer complaints of missing product unless some method is devised to perform reliable 100% inspection for a correct count. Tea bags are produced at rates varying from 30 to over 400 bags per minute, and are either machine or hand packaged into cartons of many different counts (usually 8, 16, 48, and 100). Even at the lower machine speeds, counting individual bags for hand-filling operations is tedious work, and makes accurate counting over an 8-hr shift extremely difficult. Tea bag machine manufacturers have devised counting mechanisms which have greatly reduced the chances for a miscounted package. The machines may be equipped with automatic spacers which can be adjusted to nest stacks of 8 or 16 tea bags so that the operator need not count each bag. The operator places one 8-count stack in an 8-count package, or two 8-count stacks in a 16-count package, eliminating the human error factor. Machines can also be equipped with signal devices to indicate missing bags in the stacks, thus alerting the operator to a possible miscount. As a matter of interest, the 100-count packages are generally designed for three rows of tea bags. In hand-packaging operations, each row is loaded with 32 bags (from either 4 stacks of 8 bags, or 2 stacks of 16), and the operator distributes a total of 4 individual bags at the end of the rows to complete the 100-count. The 100% inspection in these instances is performed by the machine and the operator together, with the most difficult part left to the machine.

Where the units have a substantial weight, 100% inspection for count is simplified. Consider a case of 50 packages of 4-ounce trail-mix bags. By determining the tare weight of the empty case plus 50 empty trail-mix bags, the correct count can be verified by weighing each case on a 25-lb scale, and comparing the weight to the combined tare plus 12.5 lb of product. Two precautions are to be considered. First, the variation in filler weight deliveries must be controlled; otherwise, 49 packets may be overfilled to such an extent as to produce a case with the proper gross weight, but with a missing package. Second, a history of the tare weight must be acquired to enable the line supervisor to determine how often the scale setpoint needs to be adjusted for tare variations.

Many other methods have been devised to 100% inspect for counts, either direct or indirect. The sensing device may be an electric eye to produce an electric pulse each time a package trips the light beam; or timing devices which automatically speed up a line to separate products; or scaling mechanisms attached to, for example, condenser plates, thus permitting a change in electric output as a fixed number of products passes over a point in the production line. Signals to initiate electronic counting may also be in the form of proximity switches, or contact micro switches. Each of these, and similar systems, may be tied to counters or dials to verify counts.

Automatic checkweighers are available to 100% inspect the gross weights of filled packages on the production line. These may be equipped with various degrees of sophisticated readouts or mechanical devices which can calculate quality control charts for averages and ranges, and which can remove substandard weight packages from the line. Others can be electrically or pneumatically tied back to the weighing devices so that automatic weight adjustment can be performed by the checkweigher. As above, precautions must be taken to assure that the tare variations do not interfere with the accuracy of the measuring devices.

Among other 100% inspection devices commonly used are various optical scanning devices such as the electric eye or the television camera. Infrared or visible light beams are used to test for adequate fill height in rigid containers, and to detect cocked or missing package labels. There is no limit to the list of ingenious devices which may be applied to the 100% inspection concept to dramatically improve the reliability of testing compared to that of the human inspector.

It must be emphasized that no mechanical or electrical system for 100% testing can be installed and allowed to operate without periodic monitoring. The laws of variability which these mechanisms are investigating also apply to the devices themselves. Gears wear, mechanisms stick, metal fatigues, tubes burn out, connections loosen, electric current varies, levers expand and contract with temperature, and pivot points wear.

In a method analogous to the one used to determine sampling frequency, a periodic audit system should be developed by which the testing device can be evaluated before it would normally be expected to allow measurements to drift. The audit should consist of some reliable secondary standard which has proven reliability.

## Spot Checking

A quality control system using only spot checking is doomed to failure. Similarly, a quality control system set up without the additional use of occasional spot checking is on shaky ground. A spot check is an infrequent, nonscheduled extra examination of a process or product to reassure the inspector that no mistakes are made, and to reassure management that the system is working.

Spot checking can also uncover defects in the quality system. In a plant producing beverage syrups, one of the periodic tests was determination of pH. The quality control supervisor made an informal spot check every other day or so, using a hand-held pH meter. After dozens of uneventful spot checks, the supervisor found a major discrepancy between his reading and that of the line technician. A detailed review of the technician's procedures revealed that the sample had been correctly removed from the line, and that the steps required to place the sample in the meter and to take a reading had also been according to instructions. The problem was solved when it was discovered that the buffering agent used to calibrate the meter had somehow become contaminated, with resulting incorrect readings. Fortunately, the plant operated under a batch system, and the few affected batches were isolated and adjusted. As a result of this spot check, the procedure for pH determination was revised to include an extra step for affirming the accuracy of the buffer against a secondary standard.

Spot checks should not take the place of quality audits. A quality audit is a rigorous procedure to insure that all aspects of a quality system are working satisfactorily. A spot check may be conducted by anyone at any time, using any type of examination as an informal test of a fraction of the quality system, but has no statistical significance.

## Constant Percentage, Square Root

One might expect that replacing the haphazard approach of the spot sample with some sort of systematized sample would be an improvement and it frequently is. Without a statistical approach, it is unlikely that a rational system can be devised to provide—with a high level of confidence—a sample that describes the quality level of a population. Short cuts to statistical determination of sample size are convenient and are easy to explain, but they are rarely reliable.

One such short cut is the use of a constant percentage method. There is no common rule to establishing the percentage. A small percentage is selected if the population (lot size) is large, if quality variation is expected to be minimal, if the samples are to be examined by destructive tests, if the testing is time-consuming or costly, or if the population is made up of expensive units. Conversely, a large percentage is selected if the lot is small, quality is probably nonuniform, testing is simple or inexpensive, or the units are not costly. The errors inherent in such a system should be apparent.

Consider the size of sample required to determine the quantity of oxygen in a tank of liquid nitrogen to be used for gas-packaging peanuts.

- The sample will be discarded after testing
- The population (cubic feet of gas) is large
- The oxygen is uniformly dispersed through the tank
- Sampling and testing a large number of samples would be costly.

Even though the last condition (expensive units) is not met, certainly all of the other conditions would dictate taking a single or perhaps duplicate samples of a very small quantity of gas from the tank. A one-cubic-foot sample might typically represent 0.001% of a small tank's contents. Because of the uniformity of the population, the results of testing this small sample would provide a reliable picture of the gas quality level.

"Small" means something quite different for different products. A 4,000-lb truckload of pineapples might seem to satisfy the conditions for small percentage sampling, but 0.001% sampling would amount to perhaps one pineapple per truckload, certainly not enough to determine the number of moldy fruits present. If there are 1,000 fruits on the truck, a reasonable number to select might be 10. In this case, a "small" sample would be 1%.

There remain two gnawing doubts about the selection of percentage for sampling: (1) Would a 1% sampling method detect moldy fruit if it were present at 2 per 100; or at 6 per 100? and (2) Should trucks from ranches with a history of lower quality be subjected to a higher percentage sampling plan than those with higher quality?

Consider question 1. If 2% of the fruit were moldy, there would be 0.2 moldy pineapples in the sample of 10; and there would still be less than a whole moldy fruit (0.6) in the sample if the quality level were three times as poor (6 moldy fruits per 100). In fact, if the sample of 10 disclosed one moldy pineapple, it would indicate that the truck contained 10% mold, and this might be far too high to be acceptable. Perhaps a sample of 1% is not the correct number.

Intuitively, one might answer question 2 affirmatively. On the other hand, a larger sample might not be required for poorer quality since the defects would more easily be found. The problems with the use of an arbitrary sampling plan should now be apparent.

In an effort to improve on the percentage sampling techniques, some advocate the use of the square-root method (Table 5-1). Again, this is a simple method, easily explained to management, which requires smaller sample sizes for large lots than does the percentage sampling scheme. The following is an example of the sample sizes required on two production lines running the same product in different sizes. It assumes that samples will be drawn every hour.

In spite of the convenience of the square-root system, there is little to be said for its selection. It has no relationship to the number of defects in the lot, and has

Table 5-1.   Square Sampling Plan

| Package size | Production rate | Cans per hour | Samples required | |
| --- | --- | --- | --- | --- |
| | | | 3% Plan | Square root |
| 1 gallon | 12/min | 720 | 22 | 27 |
| 1 quart | 80/min | 4,800 | 144 | 69 |
| 1 pint | 200/min | 12,000 | 360 | 109 |

no statistical basis for establishing the confidence level of the results. It assumes that a sample size determined solely by the square root of the lot size has equal ability to distinguish between lots with 0.25% defectives and lots with 6% defectives. Unfortunately this is not true. There is no rational expression for the results of testing lots by this method other than to state that there were $x$ defectives in the samples, or that the values for $x$ in the samples tested ranged from $y$ to $z$. It is not possible to state the level of quality using this method except to state that the lot passes or fails some arbitrary values.

## Statistical Samples

So far we have discussed some of the difficulties associated with types of sampling procedures, and have implied that there is no way to be sure of the quality level of the population from analyzing samples obtained by using these procedures. However, by the use of statistical sampling techniques, it is possible to describe the quality of the population from which the samples were drawn, and to define the probabilities associated with the correctness of that evaluation. The techniques used in statistical sampling procedures will be described in detail in the discussion of operating characteristic (O.C.) curves later in this chapter. Statistical sampling procedures answer two very difficult quality questions:

1. How big a sample is required to define the quality of a population?
2. Based on the sample characteristics, is the quality of the parent population satisfactory?

## HOW TO TAKE SAMPLES

Sampling procedures must be tailored to the nature of the population being evaluated. The concept of a random sample seems clear enough: a random sample is one which has been chosen by a process designed to give every item in the population an equal chance of being chosen. If the population has satisfactory and unsatisfactory product scattered uniformly, then *simple random* sampling is effective. If the population is divided into small subgroups each containing product of uniform quality, then *clustered random* samples are indicated. If the population consists of layers of good quality and layers containing defects, then *stratified random* sampling is required. Establishing the type of population is usually accomplished by observation of the process just prior to the sampling point. If the type is not easily determined by observation, analysis of groups of samples taken throughout the population will readily define it.

## Simple Random Sample

To examine the differences between these types of sampling techniques, let us consider a lot of 5,000 cans of green peas from which we wish to draw a sample of 50 cans in order to determine the drained weight. A possible method of creating a simple random sample might be as follows: number each of 5,000 balls with a different integer from 1 to 5,000. Each ball now represents one of the cans of green peas. Place the balls in a drum, mix, and remove 50 of them. Next, number each of the 5,000 cans from 1 to 5,000. The numbers on the 50 balls would represent a *simple random sample* of the 5,000 can lot. This procedure would satisfy the two requirements for a random sample: (a) every subset of size 50 *can* be selected; and (b) every subset of size 50 has an *equal chance* of being selected. Obviously this is an impractical procedure, and only serves to illustrate the principle of simple random sampling. A more useful method would be to remove a total of 50 cans from the line, one at a time, intermittently, and spread out over the period required to produce a lot of size 5,000. Some bias is bound to be introduced by this compromise procedure: the technician might subconsciously avoid ever selecting a sample from the first 25 cans; or might sample more slowly at the beginning, faster in the middle, and not at all near the end of the 5,000-can run.

The mathematical notation commonly used when referring to the simple random sample is as follows:

$$
\begin{aligned}
&\text{Population size:} \quad N \\
&\text{Sample size:} \quad n \\
&\text{Sample subset:} \quad \binom{N}{n}
\end{aligned}
$$

A variation of the simple random sample is the *systematic random sample*. This is generally known as a $1/k$ random sample, and is obtained by selecting every $k$th unit from the population, beginning with a unit selected at random from the first $k$ units. Referring to the 5,000-can lot of peas in the example above, we again would number them consecutively as produced, and separate them into sublots. If a 1% systematic sample were desired, the 50-can sample would have to be selected in such a way that each of 50 sublots was represented. That means that one and only one can should be removed from each sublot of 100 cans. In order to be a *random* sample, the can selected from the first sublot of 100 should be randomly chosen. Let us assume that can number 7 were chosen from the first sublot. To be a *systematic* plan, can number 7 should then be selected from each of the remaining 49 sublots. In other words, once a starting point is selected, every 100th can shall be selected, thus generating a total sample of 50 cans.

Note this difference between simple and systematic random sampling: in systematic 1% sampling of a 5,000-can stream of production, there are only 100 different sets of 50 cans which may be drawn; in random 1% sampling, *any* combination of 50 cans may be drawn. Systematic sampling provides a far more meager selection of the population. Furthermore, if only the center of each subgroup is to be systematically sampled, then there is only *one* possible set of 50 cans which can be selected. On the other hand, a simple random sampling procedure could conceivably miss the first 200 or 300 cans entirely because of the true random nature of the selection; the systematic random sample guarantees that each 100 sublots shall be included in the 50-can sample.

Choosing between these two types of sampling techniques might be shown in Figure 5-3.

For uniform batch production, use simple random sampling; for continuous production, use systematic random sampling. As with any generality such as this, there are bound to be exceptions, and selection of a system should be tailored to the characteristics of the production line. If, for example, the cans are filled from 6 heads, the diagram above would suggest that only head number 3 will be sampled; this may or may not be of importance to the analysis of the subsequent examination of the samples. If head delivery is to be evaluated, the system should be modified to include each head: select head number 1 for the first sample; head number 2 from the next subset; head number 3 from the next subset, etc. Alternatively, all six heads might be sampled for every 800 cans produced—or some similar system might be randomly selected.

Note in the above example that if one of the six heads were defective, the systematic random sampling techniques in which each head is examined will quickly detect the problem. On the contrary, it is possible that a *simple* random sample system might never select a can from the defective head. But the systematic random sample system is far from perfect. If the samples are taken starting with every 800th can, and a defect due to a malfunction in the feed hopper causes

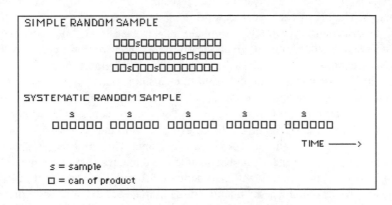

**Figure 5-3.** Simple random sample and systematic random sample plans.

the peas to stop flowing momentarily in between the regular sampling cycle, then the systematic random system might never select a can from this period. This suggests that before a sampling procedure is established, the possible causes for the process going out of control should be listed, and a sampling procedure selected which will detect them.

We have looked at relatively simple examples to illustrate the two types of random sampling techniques, and have suggested that there are hidden weaknesses in each. This should not imply that the methods are unreliable; instead, it suggests that no matter which method is used, it should occasionally be audited by use of a procedure which is likely to reveal weaknesses if they exist. One final caution: since processes are always subject to improvement or accidental changes, the sampling system should be reviewed periodically to verify its effectiveness in assessing the process quality.

## Stratified Random Sample

In its classical form, a stratified population is one which is made up of distinct and nonoverlapping groups with differing characteristics. An example might be a hopper (the population) containing a layer (the strata) of mature (the characteristic) blackberries (the group), covered by a layer of slightly reddish immature blackberries, which in turn are covered with a layer of overripe berries. If samples are drawn at random, a representative sample of the entire hopper would likely determine that the maturity was about average. If samples were drawn from each strata, it is likely that the conclusion would be that the bin contained three maturities. The sampling technique to be selected would depend upon the process which followed the bin storage. If the entire contents were to be mixed into a single batch, a random sample would suffice; if the contents of the bin were to be used continuously, a stratified sample would indicate that the end product would vary as the bin was gradually emptied.

Figures 5-4 and 5-5 illustrate the principle of stratified populations of tomatoes. The pH of each truckload of tomatoes is measured as received at the dock by selecting seven random samples. In this case, each truckload may be considered as a strata, and the graph clearly shows the variation as received both within each strata and between each strata. In the second graph, the tomatoes contained in five truckloads are considered to be a single batch, and the 35 random samples show only an approximate range of variation between strata. As with the blackberry bin in the previous example, if the truckloads are used in sequence, the pH of the processed tomatoes can be expected to fluctuate from 4.2 to 4.1, to 4.3, etc. as shown in the stratified sample graph. The information from the random sample graph merely indicates that the tomatoes should average about pH 4.2, and that the process will drift between 4.1 and 4.3. The effect of pH on the quality of the product being manufactured from these raw tomatoes will determine which data (and thus which sampling technique) is required to control the process.

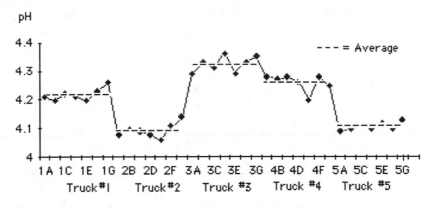

**Figure 5-4.**   Stratified samples (seven per truck).

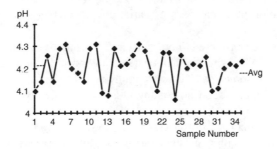

**Figure 5-5.**   Random samples (35 per lot).

## Cluster Random Sample

Cluster random sampling is a technique used when the population is naturally divided into a large number of clusters, where the units within the clusters are similar to units between clusters. Under these conditions, a cluster random sample will yield an estimate of the population mean which is likely to be much more accurate than an estimate based on a simple random sample of the same size.

Consider a production line filling a lot of asparagus into 4,800 cans which are then packed in sequence into 200 cases of 24 cans per case. The cases are then stacked in the warehouse. If we wish to take a sample of size 50 (1 can per 96 produced), we are faced with a number of decisions. Are the major quality differences to be considered due to variation from can to can, from case to case, from quarter-hour to quarter-hour (or other) period of production?

A random sample could be taken to establish the overall quality of the lot by randomly selecting 50 cans from the lot. This is far simpler said than done: the cases are relatively inaccessible in their stacks, the cases are probably sealed and would have to be destroyed to obtain the samples, the samples removed would

**Table 5-2.  Examples of Choices for Cluster Sample Selection**

| Cluster size | Subsample size | Total sample |
|---|---|---|
| 50 cases | 1 can from each | 50 |
| 25 cases | 2 cans from each | 50 |
| 10 cases | 5 cans from each | 50 |
| 5 cases | 10 cans from each | 50 |
| etc. | etc. | etc. |

have to be replaced with cans bearing the same code numbers, and all 12 cans in each case opened would then have to be repacked and sealed in a new case which would have to be properly coded and/or stenciled. After all of this effort, the sample would represent the average quality of the lot, and would not indicate possible variation between cases or variations occurring over the time interval produced.

If the cases are considered to be "clusters," a group of random cluster samples could be drawn. All of the problems listed above would still be present, but they might be reduced somewhat. There are many possibilities of arriving at a sample size of 50, depending upon how many clusters are selected, as is illustrated in Table 5-2.

The variation between clusters is best shown by the use of 50-case-cluster size; the variation within clusters becomes more apparent as the subsample is increased. As with all sampling plans, the goal of the sampling procedure must be established before a rational selection can be made.

It should be emphasized that none of the cluster sampling techniques is the same as a simple random sample from the 4,800 cans. The cluster technique includes selection of many subsets, which means that there are many subsets which are not included in the sample. It is also important to realize that one sampling plan does not necessarily provide representative samples for more than one variable. If the variable under consideration is the satisfactory glue application of the label, then the sample should consist of one can from each of many cases; if the variable is the performance of the sealing rollers, then a representative sample might be obtained by examining half the cans in each of a few cases.

## TYPES OF SAMPLES

One of the few generalities found in quality control systems is that samples fall into two classes: those in which variables are measured, and those in which attributes are counted. Decisions regarding the quality are based on the degree to which the variables depart from some normal or desired figure, or the extent to which the attributes exceed some specified limit.

Variables are product or process characteristics which can be measured. There are thousands of variables which affect the quality of food products. As an example, consider the variables commonly analyzed to establish the quality of wheat flour:

| Variable | Measurement units |
|---|---|
| Solids | Percent solids |
| Moisture | Percent water |
| Cold-water soluble | Percent |
| Ash | Percent |
| Added inorganic phosphate | Percent calcium phosphate |
| Iron | mg Fe per pound |
| Calcium | mg Ca per pound |
| Phosphorus | Percent P |
| Carbon dioxide | Percent carbon dioxide |
| Fat | Percent fat |
| Crude fiber | Percent crude fiber |
| Acidity of fat | mg KOH per 100 g |
| Hydrogen ion concentration | pH units |
| Sugars | Percent sucrose |
| Protein | Percent protein |
| Lipids | Percent lipids |
| Unsaponifiable residue | Percent unsaponifiable |
| Starch | Percent starch |
| Chlorine in flour fat | mg per gram of fat |
| Nitrite | ppm N |
| Benzoic acid | ppm benzoic acid |
| Bromates | ppm bromates |
| Iodates | ppm iodates |
| Pigment | ppm carotene |
| Diastatic activity | mg maltose per 10 g |
| Alpha-amylase | Falling number, seconds |
| Proteolytic activity | Hemoglobin units per gram |

The table above refers to product variables, and applies specifically to wheat flour. In addition to the product variables, there are thousands of process variables which may affect the quality of the end product. The following table is far from complete, and is intended only to illustrate the array of variables which may be selected for process control:

| | | |
|---|---|---|
| Temperature | Pressure | Density |
| Specific gravity | Color | Weight |
| Flow | Texture | Oxygen |

| Sediment | Solids | Particle size |
|---|---|---|
| Vacuum | Volume | Headspace |
| Bacterial count | Mold count | Yeast count |
| Hardness | Solubility | pH |
| Vitamin content | Brix | Contaminants |
| Nitrogen | Viscosity | Flavor |
| Char | Stone weight | Shell weight |
| Husks | Additives | Humidity |
| Carbon dioxide | Chlorine | Level |

The one common characteristic of all of the variables in both of the above tables is that each item can be measured and expressed in increments of inches, milligrams, degrees, pounds per square inch, percent, etc. Variables can be measured by use of some type of scale which can theoretically be divided into infinite subunits.

When it is not considered feasible (or possible) to conduct scalar measurements, some other type of quality rating must be used. Frequently, the most practical method of checking such a process or a characteristic of the product is to classify it as acceptable or unacceptable. This is the method of classifying attributes. Attributes, as contrasted with variables, are quality characteristics of either a product or a process, which may be tabulated as either "present or absent"; "satisfactory or unsatisfactory"; "go or no-go"; "within limits or outside of limits"; "right or wrong." Note that attributes are *tabulated* whereas variables are *measured*. These are the key descriptors.

Tabulating attributes as the basis for quality control is generally less expensive than measuring variables, but attributes require many more observations to obtain the same information than can be obtained from a small number of measurements of variables. One of the reasons for this is that it generally is not possible to tell how close an item comes to meeting the desired quality level when it is rejected on an attribute basis.

One of the complications of attribute counting is the necessity to distinguish between defects and defectives. A defect is a characteristic of either a process or a product which is not in conformance with requirements. A defective is a product containing more defects than allowed. In some instances, this may be as few as one defect.

It may be convenient to reclassify a variable as an attribute. For example, a package of peanuts may be considered to be a defect if it exceeds 4.075 in. in length. In this case, a go-no-go gage exactly 4.075 in. long would be used to count the package defects in a lot. This type of sampling and testing might be used if the purpose of the test were to assure that all of the peanut packages would fit into a dispensing machine without jamming. If, on the other hand, the goal were to control the amount of packaging material used, then exact measurements of package length might be required.

# SAMPLING PLANS

The most common procedure used to determine the acceptability of the quality level of a lot is to select a random sample, examine it for one or more quality characteristics, tally the number of defects found, and then decide on acceptance or rejection of the lot. The sample size may vary from one unit to many, depending upon the size of the lot and the probability of acceptance or rejection. Sampling plans for acceptability of lots are discussed at length at the end of this chapter.

When the decision to accept or reject is based on the test results from one sample, the plan is referred to as a *single sampling plan*. In some cases, rejecting a lot on the basis of a single sample may raise doubts in the mind of management, the customer, the vendor, or production personnel. In other cases, a single sampling plan requires destructive testing which might be costly. Because of these considerations, it might be desirable to consider multiple sampling procedures which start with smaller size samples.

In double or multiple sampling plans, a failure of the first sample to contain an acceptable number of defects automatically permits drawing a second sample (double plans) or in the event that fails, additional samples may be drawn (multiple plans).

There would appear to be justification for double sampling plans where the costs of running the tests are high, since a lot can be considered satisfactory on the first small-size sample if it proves to be defect free. On the other hand, if the purpose of a double sampling plan is to "give the lot another chance," then the principle behind sampling plans is misunderstood.

When a lot is rejected, the news is often received by the production personnel with little enthusiasm. The first reaction might be "you pulled a bad sample. Take another sample and let's see how that one comes out." Depending upon the circumstances, perhaps the best answer the quality manager can use to respond to such a request is "perhaps we should also pull another sample from each of the last three lots which were found to be satisfactory."

# TYPES OF INSPECTION

"Sampling Procedures and Tables for Inspection by Attributes" was originally issued as a Military Standard Number 105 and is available as MIL STD 105E. It has been slightly modified and issued as American National Standard ANSI/ASQC Z1.4 1981. These standards provide guidance for initiating inspection procedures, as well as instructions on switching to tightened or reduced inspection.

The suggestion that flexible procedures be allowed has been subject to much debate. The logic is fairly clear, but the rules for switching from one procedure to another are arbitrary, and open to considerable question.

The principle seems to be that if a producer has a history of acceptable quality, lots from that producer should be examined by a less stringent procedure. The standards state that normal inspection will be used at the start of inspection "unless otherwise directed by the responsible authority." Reduced inspection shall be instituted providing that all of the following conditions are satisfied:

1. The preceding 10 lots have all been on normal inspection, and have all been accepted. (The reason for 10, rather than some other number, is not explained.)
2. The total number of defects found in the samples from the 10 lots does not exceed designated values presented in a table of the standard.
3. Production is at a steady rate. ("Steady" is not defined.)
4. Reduced inspection is considered desirable by the responsible authority. ("Responsible authority" is not defined.)

The standards have provided rules to follow in switching from reduced back to normal, normal to tightened, and tightened to normal. In spite of the arbitrary nature of these rules, the standard is accepted worldwide, and has been very effective in the field of quality control.

## CLASSES OF DEFECTS

Many subjects are taught where specific areas are classified to make it easier to learn. In the case of defect classification, this is certainly true, but there are other advantages. For one thing, the interrelationships between classes of defects and their respective costs to the company dictate where cost reduction efforts are most likely to have the greatest effect. A second need for classification of defects is to determine which types of defects cause the more severe problems with quality control, thus dictating which areas need the most attention.

Quality defects are classified into three broad areas: critical, major, and minor. *Critical defects* are those which are certain to cause failure of the product to function as designed. The most serious of critical defects are the types which can endanger the health of the consumer—contamination by toxic chemicals, salmonella, botulinum, metal or ceramic fragments, etc. Other defects which may or may not directly endanger health may be classified as critical if they cause the product to become inedible or distasteful—severely discolored, infested with insects, moldy, etc. A third type of critical defect is one which is in violation of the law, and may not be marketed without risk of severe penalties—unapproved additives, deceptive labeling, ingredients exceeding regulatory limits, etc.

The second classification, *major defects*, is a difficult one to define. It generally includes those defects which are likely to have important adverse effects on the appearance or function of the product. Dented cans might appear in this classification, but the severity of denting could determine the importance of this defect. A dent causing a fracture of the metal would be critical; a dent which is barely detectable might not be considered a defect at all.

A useful criterion for classifying major defects is the probable evaluation that a consumer might be expected to make when observing the defect. If it is likely that a consumer would hesitate to buy a product containing a defect which appeared to affect adversely the product quality, then that defect would be considered major. An example of this situation might be a chipped olive jar. One would expect that most consumers might not purchase a chipped jar because of the possibility of fragments of glass inside of the container. It might be expected that the consumer would choose an adjacent undamaged jar instead. On the other hand, if one of a group of jars of olives on a store shelf were severely discolored or black-ened, the consumer might be expected to select another brand of olives—not another jar. The severely discolored jar would then be classified as a critical defect.

The third classification of defects is *minor*. Among minor defects are those which would be expected to have unimportant effects on either the performance or the shelf life of the product. A slightly scuffed label is a minor defect which pres-ents a somewhat undesirable appearance to a package, but otherwise has no effect on the product itself. (A torn or heavily scuffed label, on the other hand, might indicate to the consumer that the package has been handled roughly and could somehow have damaged the product within; this could be a major defect.) Other common minor defects include: illegible code, slightly soiled exterior, minor blemishes or imperfections, minimal net weight or headspace variation, a few broken or dusty items in a package of friable product, a trace of weeping of a gel, some indication of freezer burn, somewhat stronger or weaker aroma than usual, color a trace lighter or darker than normal, product slightly thick or thin, etc.

Each industry tends to subclassify these three classes of defects into a lan-guage unique to the industry, and frequently unique to the company within that industry. Over the years, this practice tends to get out of hand, and needless sub-classes are collected and analyzed to no particularly useful purpose. There are occasions when detailed analyses are quite valuable. One such case would be the study of a particular defect with the goal of eliminating it. By classifying major defect dents into numerical demerits, and tracking these over time and location on

| Line location | Dent count (500-can sample) | | |
| --- | --- | --- | --- |
| | 0–5 mm | 6–10 mm | 11+ mm |
| Can cleaner | 3 | 0 | 0 |
| Filler head | | | |
| 1 | 2 | 5 | 0 |
| 2 | 1 | 1 | 0 |
| 3 | 0 | 0 | 17 |
| 4 | 2 | 4 | 1 |
| Line transfer | 5 | 22 | 0 |
| Seamer head 1 | 1 | 0 | 0 |
| etc. | — | — | — |

the production line, it may be possible to identify the various causes of the defect, thus leading to elimination of those which are most troublesome.

In this simplified, partial study, it should be apparent that there is considerable damage to the cans at filler head No. 3 and at the line transfer station. The need for adjustment, maintenance, or possibly redesign is indicated.

It has been suggested that converting the three classifications to numerical terms will permit analysis by statistical means. One such scheme is to apply values such as 100, 30, and 10 to critical, major, and minor defects. Another suggestion is to apply probable scrap or rework costs to each defect as accumulated. It is difficult to perceive of this type of quality control technique as being particularly useful over a long term, but perhaps the use of the dollar analysis for a short-term study might highlight the significance of a problem. It might also provide operations management with data to justify a capital expenditure to improve the product quality.

Two other important systems for classification of defects are the grading procedure for raw fruits, vegetables, grains and meats (U.S. Department of Agriculture [USDA]), and the standards and contaminant regulations for specific foods (Food and Drug Administration [FDA]). These procedures and standards are rigorously defined, and rarely (if ever) require statistical analysis for quality control.

Perhaps a philosophical note would be in order at this point. There will always be variability in the raw material and the processing of food products. There will always be defects in food products. Some of these defects are preventable. Since quality managers and their corporate leaders are humans (for the most part), it can be expected that here, too, there will be variability in the selection of projects to eliminate defects. Look at some of the choices in selection:

- some will save more money than others
- some will not save any money, but will add to costs
- some might improve the image of the quality department
- some are very interesting projects
- some are extremely difficult
- some are very important to upper management
- some might provide solutions detrimental to others' reputation
- some might apply to short-lived products

One of the more powerful statistical tools used to evaluate the list of priorities is the Pareto Curve. It suggests that a "critical few" defects produce the bulk of the consumer complaints, and that the "trivial many" defects are responsible for a relatively small proportion of complaints. Logically then, the quality department can provide the greatest complaint reduction by eliminating the causes for the "critical few" defects. Often this is the case. But bear in mind that among the "trivial many" defects that cause little concern, there are some defects which could cause ruin to the company if they were not controlled rigorously. One cannot ignore the constant threat of salmonella even though it is among the "trivial many" defects which rarely, if ever, appear as a complaint.

If the powerful logic of the Pareto analysis does not always work, how does one select the projects to eliminate all of the defects? The answer here is a great deal simpler than most are willing to admit. Yes, Pareto is powerful, and it is useful. Yes, it is possible to select a project from the above partial list of eight choices. But it is not possible to provide a system which will magically solve all of the defect problems in one major effort. The control of quality cannot be established by formula since each company has unique variables of suppliers, processes, equipment, marketing goals, and (most important) people with unique needs, personalities, and motivation. Projects should be selected in accordance with the needs as perceived by the company (not by the quality department), and solutions will be created and implemented one at a time. Quality control and quality improvement do not arrive in a bolt of lightning as so many quality programs promise; quality control is established and reestablished one step at a time. Likewise, quality improvement is established one step at a time.

## SAMPLING RISKS

The risks associated with acceptance sampling have been classified into four groups, two for the producer, and two for the consumer. These can be tabulated as follows:

Probability of acceptance
   Producer (Type I risk); Consumer (Type II risk)
Probability of rejection
   Producer (Alpha risk); Consumer (Beta risk)

The first reaction to this classification is that it is unnecessarily complicated, but as we shall see in the discussion of O.C. curves (the technique by which an unlimited number of sampling plans can be derived), it is essential that these risks be understood.

Both the producer and the consumer would like to accept only perfect lots: zero defects. In some instances, this is an attainable goal, but only at considerable expense to the producer, and at prices which might be unrealistic to the consumer. Each party has to be willing to reach a decision as to the number of defective items in a lot which would be acceptable. There are no universal rules for determining the acceptable quality level. It is generally arrived at over a long period of time during which general management, production, quality control, purchasing, accounting, research, and sales all provide inputs until a general consensus is reached. This compromising process is a continuing one. As new needs arise, or as processes, costs and raw materials change, the acceptable quality level may be either narrowed or expanded.

Having established the acceptable quality level (AQL) both the producer and the consumer are now faced with an additional decision: how precise a figure is this? Is there any margin for error? This is where the factor of risk appears. Of course, the producer would like to manufacture and ship products containing defects at the AQL level, but there is the risk that the sampling plan will reject some of the lots which are really satisfactory. This is known as the *producer's alpha risk*. The producer runs the additional risk that the sampling plan will erroneously label lots as acceptable when they contain a number of defects which is outside of the AQL. This is referred to as the *producer's Type I risk*. The consumer is faced with similar risks of acceptance and rejection.

## SELECTION OF POPULATION TO BE SAMPLED

Determining the sample size is based heavily on statistical concepts. Determining the lot size is of equal importance, but is far more difficult to accomplish. There are no rules and few principles to follow. Let us examine a situation in which a processor receives about 20 truckloads of 100-lb bags of beans per month.

Assume that we are interested in assessing the quality level of the following:

- bag weight
- insect damage
- mold
- uniformity
- moisture
- bean size
- foreign material
- color
- flavor.

### Truckload Lot

If each truck contains approximately 400 bags, then each truck could be considered as the population to be evaluated, and a simple constant sampling procedure could be established. If the loads should vary from truck to truck, then the population size would require different sampling procedures for each truck, and perhaps this would unfairly penalize some by requiring more intensive inspection of smaller truckloads. There are more serious considerations. Considering a truckload as a population would be satisfactory only if the truckload were uniform. The load could consist of produce from several farms or several warehouses with varying quality levels. A sample from such a truckload would average the quality of all, and perhaps fail to uncover a group of bags with particularly poor quality.

### 100 Bag (or Other Number) Lot

Arbitrarily dividing each shipment into 100 bag lots would be likely to uncover the presence of unusual quality portions of a truckload. Such a plan would provide average quality information in a continuous stream, thus assisting production personnel in regulating process variables. Variations in quality would probably not be detected by a population definition of this type if the variations were due to such factors as location of the bags in the truck or in the warehouse preceding delivery, wherein the product may have been exposed to heat, moisture, light or contamination.

## Pallet-Load Lot

If there is no particular advantage to isolating portions of truckload deliveries, or if the variations between truckloads do not have an important effect on further processing in the plant, then the trucks might be unloaded and restacked on pallets in the plant warehouse, and each pallet load might then be considered as the population. In fact, if the variations are generally unimportant, the population might be defined as 2 pallets, or 5 pallets, or a larger number. In this case, composite samples may be obtained from several pallets, thus reducing the cost of sampling. Pallets are a convenient population (or subpopulation), but in some instances, it might be advisable to dump truckloads directly into bins, in which case the bin contents become the population.

There appears to be no limit to the size of the population which might be selected to establish the quality level of incoming truckloads of beans. There are, however, a number of guidelines which might be considered. First and foremost is an understanding of which quality factors have to be measured—which quality factors are important to further steps in the process. If moisture is critical, a series of studies will assist in determining the logical size of the population to be used routinely. The first study would be a detailed examination of the variation of moisture within a single bag. If it is found that there is no consistent difference in the moisture levels found at the top, bottom, center, or edges of a single bag, then the next series of tests would be detailed examination of the variation of moisture between bags. Similarly, if it is found that the moisture variation between individual bags taken from a single truckload is minimal, then the third series of tests would be to establish differences in moisture level between truckloads. This type of investigative testing is continued to establish the points at which significant variations begin to appear. Other possible studies might include differences resulting from farm-to-farm variation, seasonal variation, temperature and humidity variations, year-to-year comparisons, etc. Once the level at which significant variations begin to appear is determined, the population size required for quality control sampling for that variable becomes apparent.

There is a likelihood that conditions will eventually change—new trucks, different farms, varietal improvements, different process requirements, improved packaging of raw materials. Consequently, the procedure of selecting population size must be reviewed periodically to confirm its integrity.

## SELECTION OF SAMPLE FREQUENCY AND LOCATION

How often should a sample be selected for quality evaluation? The answer to this question is simple: if the quality variable or attribute to be examined normally remains within acceptable limits for a period of $P$ units, then the frequency of sampling should be no longer than $P$ units. The value of $P$ should be determined by test, and should be reviewed periodically to insure that conditions have not changed. Examples of $P$ are the following:

| | |
|---|---|
| 3 | truckloads of beans |
| 3 | each 250-gallon batches of jam |
| 1 | month of warehouse storage |
| each | supplier |
| 200 | jars from sealer |
| each | label change, at start-up |
| every | third pallet load |

Selecting the location of sampling can be more complicated. Before the advent of statistical quality control methods, attempts to control quality of food production were based on two principles: (1) strict adherence to tried-and-true procedures; and (2) inspection at key points. These procedures have not been totally replaced by statistical quality control. In fact, adherence to procedures is one of the principle tenets of SQC. But the inspection at key points has been transformed into a far more powerful tool by the introduction of rational means of evaluating the results of those inspections and tests. Perhaps another change has been de-emphasizing the importance of finished-case inspection at the end of the line, and concentrating on in-process test and inspection. End-of-line inspection in the pre-SQC days worked fairly well since labor was relatively inexpensive, and repair and rework cost was not as critical as it is today. For the most part, companies employing SQC techniques continue to use end-of-line inspection as an audit procedure to provide a score card to rate the effectiveness of the quality control system.

## HAZARD ANALYSIS CRITICAL CONTROL POINT (HACCP)

HACCP is a food safety system acronym for Hazard Analysis Critical Control Points, and has been used by several companies in the United States for over 30 years. The USDA and the FDA, have been incorporating this preventative tool into their food regulations. For example, the Code of Federal Regulations for the FDA 21 CFR part 113, which deals with thermally processed low-acid foods in hermetically sealed containers, was first adopted in 1974, and applies the principles of HACCP. Critical control points (CCPs) are covered in great detail: personnel, equipment, procedures, containers, closures, preparation, processes, thermal operations, deviations of critical control factors, records and reports. In 1981, the FDA published suggestions for a formalized HACCP program in which they recommended preparing a detailed process flow diagram to help locate key sampling points (process hazards and CCPs). It is expected that, ultimately, HACCP will be applied universally.

Seven principles of HACCP, adopted by the USDA and the FDA in 1997, are:

1. Conduct a hazard analysis
2. Identify critical control points (CCPs)
3. Establish critical limits for each CCP
4. Establish monitoring procedures for CCPs
5. Corrective action
6. Verification procedures
7. Record keeping and documentation.
   - by 1997: Mandated for meat, poultry and seafood plants
   - Under way: juice, egg

HACCP is recommended by Industry Associations:

International Dairy Foods Assn: milk, ice cream, cheese, yogurt, butter
Refrigerated Foods Assn: cole slaw and others
American Spice Trade Assn: whole spices, blended seasonings
American Institute of Baking: bread, buns, cake

Most of HACCP principles were already FDA-enforced in the low-acid canned food industry when the first of the regulations were enacted for meat and poultry by the USDA (1998). The FDA issued HACCP regulations for the seafood industry (1995), the bottled drinking water and the juice industry (2001). The FDA has conducted pilot programs for cheese, frozen dough, flour, bread, cereals, salad dressing and other products. It is just a matter of time before all food products are included under HACCP regulations.

## The Statistical Application to HACCP

HACCP principle No. 3 (establish limits for each CCP) and HACCP principle No. 4 (establish monitoring procedures) requires control chart analysis. As explained elsewhere in this book, either variable or attribute control charts should be used to evaluate the acceptance at each CCP. Dependence on antiquated procedures of periodic inspection will not suffice.

The task of the FDA as food industry regulator is to oversee the food industry to ensure that it meets its legal requirements. For years, the ability for the FDA to assess a food company's performance in meeting its legal responsibilities was based on observations of a plant's practices at the moment of regulatory inspection. The assumption of the inspector was that the plant operations on the day of the inspection were typical. HACCP thus became a more efficient regulatory tool.

FDA regulations for low-acid canned foods contain many of the HACCP principles. HACCP regulations for the seafood industry were introduced by the FDA in 1995 under 21 CFR 123, and for the juice industry in 2001, effective in 2002. The USDA established HACCP regulations for meat and poultry products in

1999. It can be expected that before long all food products will require HACCP regulations. The FDA continues to develop HACCP regulations as the food industry safety standard for domestic and imported food products. The Food Safety and Inspection Service of the USDA defined HACCP as a systematic approach to food safety. "The primary goal is to assess hazards and risks associated with growing, harvesting, raw material and ingredients, processing, manufacturing, distribution, marketing, preparation and consumption of food."

The FDA issued seven principles involved in HACCP:

1. Analyze hazards. Examples: biological, physical contamination, chemical.
2. Identify the CCPs. These include processing hazards such as foreign material contamination, heating, packaging.
3. Establish preventative measures with critical limits for each control point. An example would be cooking time and temperature.
4. Establish procedures to monitor the CCPs.
5. Establish corrective actions when monitoring indicates failure at a critical control point.
6. Establish procedures to verify that the processing operation is proceeding according to standards.
7. Establish effective record keeping to document the HACCP system. This should include records of hazards, control procedures, and monitoring.

Some of the advantages of a HACCP system, as expressed by the FDA in their *Backgrounder* publication October 2001 apply both to the food processor and to the FDA inspectors. It provides focus on identifying hazards from contaminating food and is based on sound science. The record-keeping requirements provide more efficient government oversight, and help the food processor to compete more effectively in the world market. The need for further regulation is indicated by the growth in the food industry and research projects which continue to uncover possible chemical and biological hazards in food production.

It must be emphasized that the HACCP procedures regulate critical health risks—not quality control of the food product. Critical Control Point is defined by the USDA as: "Any point or procedure in a specific food system where loss of control may result in an unacceptable health risk." Other serious quality characteristics such as underweights, wrong ingredient, over or under mixing time, incorrect color, strange texture, unusual odor, etc. are not a factor of the HACCP system. A suggested procedure for overcoming this potentially serious omission is discussed later.

First, a process flow diagram (Figure 5-6) is constructed for each product, showing the individual steps in sequence from receipt of packaging and raw materials, through plant processing steps, to final shipment or storage of the finished product. Then each step is analyzed for possible hazards. The FDA defines a process hazard as "a possible source of trouble along the processing chain which might be defined as a failure to identify: (1) critical materials, (2) critical processing points, (3) adverse environmental conditions, and (4) human malpractices."

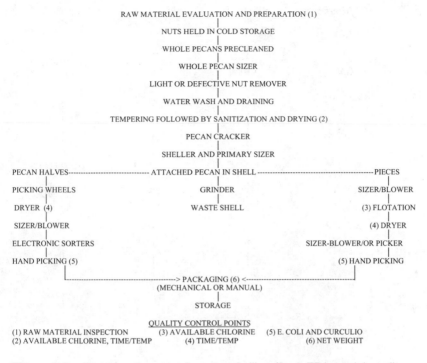

RAW MATERIAL EVALUATION AND PREPARATION (1)
|
NUTS HELD IN COLD STORAGE
|
WHOLE PECANS PRECLEANED
|
WHOLE PECAN SIZER
|
LIGHT OR DEFECTIVE NUT REMOVER
|
WATER WASH AND DRAINING
|
TEMPERING FOLLOWED BY SANITIZATION AND DRYING (2)
|
PECAN CRACKER
|
SHELLER AND PRIMARY SIZER
|
PECAN HALVES------------------------------ ATTACHED PECAN IN SHELL --------------------------------------------PIECES
|                                                                  |                                                               |
PICKING WHEELS                                       GRINDER                                          SIZER/BLOWER
|                                                                  |                                                               |
DRYER  (4)                                             WASTE SHELL                                     (3) FLOTATION
|                                                                                                                         |
SIZER/BLOWER                                                                                              (4) DRYER
|                                                                                                                         |
ELECTRONIC SORTERS                                                                          SIZER-BLOWER/OR PICKER
|                                                                                                                         |
HAND PICKING (5)                                                                                  (5) HAND PICKING
|------------------------------------------> PACKAGING (6) <------------------------------------------|
(MECHANICAL OR MANUAL)
|
STORAGE

QUALITY CONTROL POINTS
(1) RAW MATERIAL INSPECTION          (3) AVAILABLE CHLORINE     (5) E. COLI AND CURCULIO
(2) AVAILABLE CHLORINE, TIME/TEMP     (4) TIME/TEMP                  (6) NET WEIGHT

**Figure 5-6.**  Process diagram of pecan shelling (from FDA model quality
assurance plan).

Once these CCPs have been determined, a sampling system including sample
location, size, frequency, methods of selection, testing, analyzing and reporting is
established. From management's point of view, inadequate control of CCPs is
undesirable (or even fatal), since it leaves the company exposed to the dangers of
producing defective products which might, if shipped to the public, adversely
affect the reputation or the even the financial structure of the company. At the
very least, defective products could result in costly scrap or rework. At the other
extreme are the possibilities of product recall or adverse consumer action.

In addition to these immediate concerns, there are legal considerations as well.
Failure to recognize CCPs might result in unacceptable risks of adulteration, or failure to comply with federal, state, or local regulations with respect to net content,
food standards, good manufacturing practices, sanitation, or storage conditions.

Figure 5-7 is another example of a HACCP chart. This one is for quality control of empty cans received on pallets, can ends, insert coupons, and plastic
resealers. In this application, cases, case glue, coder ink and other case supplies
are handled on a separate HACCP chart, along with other cardboard and flexible
packaging supplies, and will not be considered here.

The numbers shown after critical equipment refer to standard quality control
procedures as described in the quality control manual. 1.01 refers to laboratory

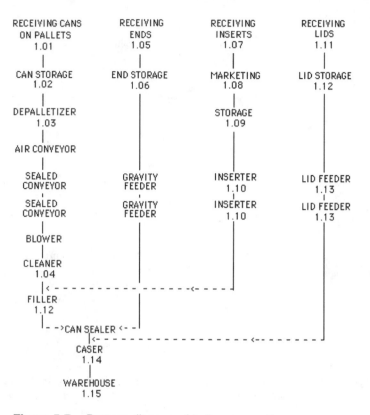

**Figure 5-7.** Process diagram of bulk can supplies.

tests for approval of can lithography (color and content), dents, leaks, side seam and throwback solder, leaks, foreign material, scuffs, etc. For those companies with standardized written operating procedures, another set of numbers might also appear on the HACCP chart, referenced to applicable operations to be taken by plant personnel. This might include cursory physical examination, count and tally with receiving documents, checkoff tickets with loads and incoming paperwork, and similar operations documents.

In any business endeavor, the management must be aware of the possible hazards in the conduct of the business which will adversely affect the outcome; and they also realize that some of these hazards are critical to satisfactory performance. To this extent, HACCP is not a unique concept, but it differs from informal procedures for avoiding mistakes and failures in that it is a detailed system which incorporates seven distinct areas:

1. *Identification of severity of a likely hazard.*
   The FDA and the USDA state that a hazard may be a biological, chemical or physical property that may cause a food to be unsafe for consumption.

Additionally, a commercial hazard is any condition which would make the product dangerous or unsalable.

2. *Locating the CCPs in which a failure would likely result in a hazard being created or allowed to persist.*

   A CCP is a point or procedure in a food system where loss of control may result in an unacceptable product (or in the case of FDA and USDA, an unacceptable health risk). Controls applied at these process steps shall prevent, eliminate, or reduce to acceptable levels the hazards identified.

3. *Documenting control limits for each point.*

   Tolerances of safety shall be established for each CCP to control a quality hazard (defined by FDA and USDA as a health hazard, or multiple health hazards).

4. *Creating a system for monitoring these CCPs.*

   Each CCP shall be controlled through a planned sequence of measurements or observations which are recorded for subsequent audit procedures. Continuous (100%) monitoring is recommended, but where it is not possible, the frequency of observation should be determined by statistically designed data collection systems. A system of calibrating the equipment, reagents and instruments used in controlling and monitoring should be included. This also applies to validation procedures of software for computer control systems.

5. *Establishment of corrective actions to be taken.*

   When a critical control limit has been exceeded, the cause of the deviation must be corrected and eliminated. The materials involved shall be tagged, or otherwise identified, and removed from the process for subsequent disposition. Each CCP will require specific corrective action in the HACCP plan. Precise records of each deviation and corrective action taken shall be maintained for the shelf life of the product.

6. *Preparing and maintaining records of these observations.*

   A written HACCP plan (manual, procedure) should be prepared, and continually updated. Records generated during operation of the plan should be current and complete. A check sheet consisting of "pass/fail" is not considered satisfactory for recording data at CCPs. Instead, actual values obtained during monitoring should be recorded, along with product description, product code, time and date of data collection. Consumer complaints relating to CCP should be included in the HACCP records.

7. *Providing routine audits of the system to verify its adequacy.*

   Audit requirements of the government may differ from those of a processor. Audits may include non-scheduled spot check sampling; review of CCP records for completeness; updating the HACCP plan to conform to changes in materials, process, product, packaging, storage or distribution.

Acceptable control limits for these tests and analyses, along with preventative measures for each control point, should be documented. Corrective action and record keeping are required for an effective system. Finally, a system of audit procedures will insure that the HACCP system is working satisfactorily.

Figures 5-6 and 5-7 illustrate relatively uncomplicated flow charts. Most processes are far more complex, and require considerable effort for use in designing a HACCP system. One of the most effective methods of creating a HACCP process flow chart is the use of team effort. A preliminary flow chart describing a manufacturing process from raw materials to the feed bin might start out as shown in Figure 5-8.

After a team of personnel from several departments have worked on this simplified flow chart, it might well result in the document shown as Figure 5-9.

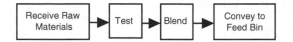

**Figure 5-8.**    Simple flow chart.

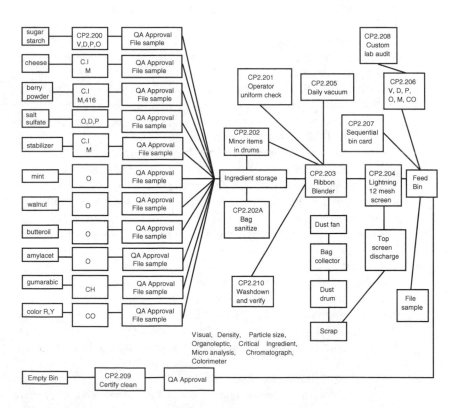

**Figure 5-9.**    Detailed flow chart, showing control points.

Note that Figure 5-9 shows several *control points*, but no *CCPs*. This distinction is important. For example, control point CP2.200 covers all of the raw materials for the product. At the bottom of the list is Color R,Y, with the test CO (or colorimeter) listed in the adjoining block. From the FDA viewpoint, approved color is not a CCP since it is not a health hazard to the consumer. On the other hand, an incorrect, sickly yellow color additive (instead of, e.g., a specified deep orange color required for the product) can certainly be considered critical to the acceptance of the product quality. In addition, there are several raw materials which are critical to the manufacturer and require organoleptic tests to assure that the product quality will be acceptable; but these materials are otherwise *safe* to use, and would not be considered critical by the FDA or USDA.

There are two solutions to this dilemma. First, for a company with sufficient personnel to handle the extra work involved, two separate systems can be designed. One for the FDA/USDA, in which only CCPs are shown, and another in which all points necessary to maintain quality control are shown. This would also require two separate methods manuals, and two separate data files.

For smaller companies, it would be more efficient to design a single flow chart with *all* of the control points shown. To distinguish the CCPs from the other quality control points, they might be differentiated by coloring the control point boxes or, as shown in Figure 5-9, by using extra heavy borders on the boxes considered critical. Again, two sets of records need to be kept so that the HACCP records are readily available for audit at any time.

Each HACCP flowchart requires an accompanying written explanation of each of the CCPs. For the chart illustrated in Figures 5-9 and 5-10, a format is suggested which identifies each CCP, those responsible for testing and approval, the control procedures, and documentation required (Figure 5-11). There may be several other control points which are not of a critical nature, but it is suggested that these *not* be included in the HACCP manual.

Note that although references are shown for the control procedures, sample size, sample frequency, test methods, specifications, tolerances, and report forms, they are not detailed on this HACCP document. Since these techniques tend to be fairly complex and voluminous, including them here would unnecessarily complicate the HACCP manual. Furthermore, these procedures should be readily available in manuals prepared by Quality Assurance, Production, Engineering, Purchasing, and other departments.

Of necessity, the above explanation of HACCP programs is rather generalized. It is expected that governmental regulations will become very detailed and increasingly specific as they are refined over the years. For example, the seafood regulations will not be applicable to the bakery industry or the canned foods industry except in a general way. The CCP problems tend to be peculiar to each industry: stones in raisin processing, grain toxins in the baking industry, cooling water contamination in aseptic canning plants, proteolytic decomposition in meats, pesticide residues in vegetables, frozen food refrigeration failure, etc.

The identification of CCPs is a relatively straightforward procedure. In the food industry, each step of the process can be examined and the CCPs identified: raw

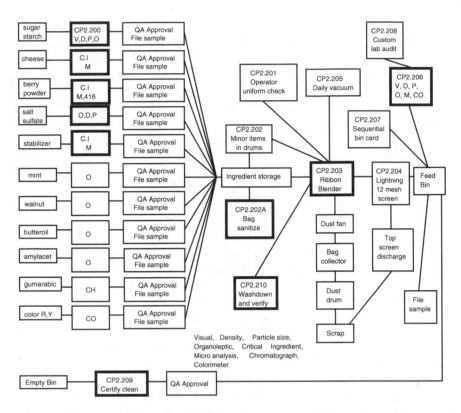

**Figure 5-10.**  Combination flow chart of control and critical control points.

materials (chemical, physical or biological fabrication, harvesting, and growing); receiving and storage; production steps (processing and packaging), distribution (including warehousing); and final consumption. Hazard analysis, on the other hand, contains some judgmental characteristics which are more difficult to evaluate. Some of the systems offered for analyzing microbiological hazards, for example, categorize the hazards according to the risk severity. *Clostridium botulinum* is classified as a severe hazard; *Staphylococcus aureus* is classified as a moderate hazard; and yet, neither one can be permitted in a food product.

Perhaps the same problem exists with chemical hazards. All chemicals can be toxic at some concentration, even table salt. Some cannot be permitted at any level in food products, while safe limits have been established for others. Safe levels of physical contaminants are even more difficult to evaluate. How much is "an acceptable level of a little dust"? How coarse is dust which is classified as "unacceptable grit"? 21CFR Part 109 attempts to specify definitions and tolerances for unavoidable contaminants in food and food packaging materials. Unfortunately, over the years, these tolerances have been misunderstood by the general public.

HAZARD ANALYSIS AND CRITICAL CONTROL POINT MANUAL
CONTROL POINT CODES

PRODUCT FAMILY 2.2
    PRODUCT: ROMAN HOLIDAY PUDDING FLAVOR BASE MIX
    PROCESS: RAW MATERIAL THROUGH FEED BIN

| CP Code | 2.200 | 2.201 | 2.202 |
|---|---|---|---|
| Identification | Raw materials | Personnel | Drums |
| Prime Responsibility | Quality assurance | Process supervisor | etc. |
| Secondary Responsibility | Purchasing | Personnel | etc. |
| Control Procedures* | QA  V #10.97<br>QA  D  #10.88<br>QA  P  #10.60<br>Taste Test  O #6.95<br>QA #416<br>R&D M  #3.22<br>QA  CO #10.86<br>R&D CH #3.15 | House Rules No.<br>  4, 6, 15, 22<br>  (Uniforms<br>  and<br>  sanitation) | etc. |
| Documentation | Lot number form and<br>  ticket.<br>Hold/release form | Checklist San#1 | etc. |

* The numbers refer to standard test procedures in various departments.

**Figure 5-11.**   Suggested format for HACCP manual.

Some of the hazards, referred to as "unavoidable contaminants in food" have been listed in detail; others may eventually be included in "Reserved Subparts." Other hazards have very clearly listed. For example, the tolerances for polychlorinated biphenyls (PCB) are:

1.5 ppm in dairy products
3 ppm in poultry
0.5 ppm in eggs
2 ppm in fish
0.3 ppm in infant and junior food
10 ppm in paper used for food packaging

Reference is also mentioned regarding microbiological contaminants: yeasts, molds, bacteria, and viruses. Other contaminants are also mentioned, but with no specific tolerances. These are parasites, chemical contaminants, chemical residues, unlawful pesticide residue, decomposition, natural toxins, unapproved additives, undeclared ingredients that might be allergens, and physical hazards. Other regulations are specific in the coverage of records, training, corrective actions, verification, and validation.

In the 2001 Edition of the Code of Federal Regulations CFR 21 Food and Drug, the HACCP rulings define many of the principles cited above in greater detail. For example, in 110.3(J), "Quality Control Operation means a planned and systematic procedure for taking all actions necessary to prevent food from becoming adulterated." In Part 120 (and in Part 110 as well), "Good Manufacturing Practice (GMP) refers to ... facilities, methods, practices and controls used in the manufacture ... administered in a manner adequate for the public health." Part 120 lists four generalized HACCP requirements:

1. Identify food hazards.
2. Evaluate an assessment of the severity of the illness or injury if the food hazard occurs.
3. Identify control measures.
4. Review process to determine whether modifications are necessary.

A weakness of the HACCP system is its pass/fail approach. A more effective procedure for quality control would be the use of a control chart showing the three-sigma limits of the process, in addition to the maximum/minimum levels permitted by HACCP. Sampling on a continuous basis for substance "A" and plotting the measurements on a control chart would provide far more information than pass/fail reporting. For example, if the permitted limit for "A" were 12 ppm, a control chart would show progressive concentration determinations, and might indicate not only the safe levels, but also the trends—if any. This type of continuous reporting could signal a need to adjust a process before a reject level was reached.

An additional advantage to continuous testing is the possibility of using the measurements as a tool for process improvement—thus lowering product costs and perhaps improving quality. An example of this effect might be found in testing outside manufacturers' components which might occasionally be produced near the HACCP reject limits in some respect. Using the statistical quality control procedures could alert the supplier to the dangerous trend, thus averting possible penalty from FDA inspection.

And a final emphasis: the governmental safety regulations, detailed though they may be, do not cover the CCPs which govern market acceptability of the food product. Net weight control, product color, flavor, texture, aroma, granulation, flowability, solubility, and dozens of other product characteristics may have no bearing on safety, but could mean the difference between an acceptable product and a market failure. A half-gallon container of vanilla ice cream may comply with every safety requirement of the HACCP regulation, but if it has a sickly green color and tastes like overcooked cereal, it is obviously nonmarketable. The problems should have been detected at several steps of a well-designed statistical quality control program. Any one of these characteristics may be considered a CCP in the manufacture of a specific product. Most companies have been well aware of these CCPs for years, and as pointed out above, they must continue to be monitored, either separately or as part of the overall HACCP system.

This examination of the many aspects of sampling might at first appear to take up a disproportionate part of this text. But it is likely that a quality control program assembled without a thorough understanding of the nature of sampling will accomplish little, if anything. Precise tests followed by accurate statistical analysis and detailed reports may be without any value if these efforts are based on samples which do not represent the population under study.

In order to obtain a truly representative sample as a basis for establishing a useful quality control system, there are probably only four absolute requirements:

- Determine the location of sampling points which are critical to the population under consideration.
- Establish a method of sampling which will represent the population characteristics.
- Select the sample size which will produce results with the probability needed, using statistical methods.
- Specify the frequency of sampling, based on the expected cycle of population quality variation.

## ATTRIBUTE SAMPLING PLANS

References have been made above to sample size, O.C. curves, sampling plans, types of inspections, and sampling risks. All of these subjects are contained in the Military Standard: "Sampling Procedures and Tables for Inspection by Attributes." This standard has been known for years as MIL-STD-105E, and has more recently been adopted in slightly revised form by the American National Standards Committee of ANSI, and issued as ANSI/ASQC Z1.4 with the same title as the MIL-STD-105E.

Either of these standards provides an extremely flexible series of sampling plans which encompass 15 lot size classifications, seven inspection levels, 26 acceptable quality grades, three degrees of inspection severity (normal, tightened, reduced), as well as double and multiple sampling. The O.C. curves, showing the producers and consumers risks for each plan, are also included in the standard. With all of this flexibility, the standard appears to be difficult to use, but it is actually quite simple. Once the quality and inspection parameters are agreed upon, the selection of the specific sampling procedure is easily found. The standard has long been accepted internationally for use in purchasing contracts by both seller and buyer. Its most serious drawback is the difficulty in explaining to nontechnical people how the tables are constructed, and why they work.

To arrive at a sampling plan, the following steps are taken:

1. Decide on the lot size to be examined, and select the corresponding code letter from the Code Letter table. Inspection level II is normally used. (For example, lot size 1000 has code letter J.)

2. Find this letter in the single sampling plan master table for normal inspection, and note the corresponding sample size. (For example, Code letter G requires a sample size 32.)
3. Decide on the acceptable quality level (e.g., let us select 1.5%) and find the accept/reject level. (For example, for letter Q, and 32 samples, with acceptable quality level 1.5%, the lot of 1000 would be accepted if 1 or less defects were found, and would be rejected if 2 or more rejects were found.)

Detailed instructions and definitions are found in the first several pages of the standard, including techniques for special sampling plans.

It is suggested that these special plans be avoided, or used with great care. They were probably introduced to limit the costs of testing when either the test itself was costly or time consuming, or when destructive testing of several expensive products was prohibitive. (For example, destructive testing of an engine, a parachute, or complex electronic assemblies.) For most food products, these are not particularly important considerations. There is a psychological approach to double sampling plans: they erroneously seem to present a "second chance" to the testing of the acceptability of a lot. (If it fails on the first test, draw additional samples.) To the statistically informed, this is nonsense. The multiple sampling plans become quite complex in their usage, and there is a tendency for the uninformed to attempt to use short cuts, which makes the results meaningless.

With these precautions in mind, the standard provides a useful attribute sampling plan for industry. In the introductory remarks, it suggests several inspection applications: end items, components and raw materials, operations or services, materials in process, supplies in storage, maintenance operations, data or records, and administrative procedures. The user is cautioned to consult the O.C. curves contained in the standard to find a plan which yields the desired protection (risk).

# 6    Test Methods

An attempt to cover the subject of test methodology in a single chapter would be close to impossible. There are hundreds of books describing thousands of test methods for a myriad of food products and their components. What we should be able to accomplish is a brief discussion of some of the relationships between test methods and quality control, along with a bibliography for guidance.

Selection of tests might be classified in three groups, those that provide information relative to:

- legal requirements
- process, product, packaging specifications
- special guides.

Legal requirements may be found for nearly every product: net contents, microbiological purity, nutritional claims, absence or presence of food components or of non-food components. The following are a few FDA requirements selected at random from the Code of Federal Regulations. If vitamin A is added to milk, each quart must contain not less than 2000 International Units. Egg rolls must contain over 2.56% by weight of whole egg solids. Each part of the contents of a package of margarine must bear the word "margarine" in type or lettering not smaller than 20-point type. Ninety-nine percent disodium ethylenediaminetetraacetate may be used up to 500 parts per million to promote color retention in canned strawberry pie filling. When processing bottled drinking water, cleaning and sanitizing solutions used by the plant shall be sampled and tested by the plant as often as is necessary to assure adequate performance in the cleaning and sanitizing operations. In these few examples, there is no question as to the tests to be conducted. Certainly, the first step in selecting tests is to find the legal requirements at all levels of government. Some city or state codes exceed the requirements of the federal laws.

When selecting test procedures not legally prescribed for controlling the quality of a process, product or package, methods used may be rigorously standardized, or they may be developed in-house along scientific but less stringent techniques. There are also many arbitrary methods which are developed in-house which are not recognized by industry or government, but which serve a specific purpose in controlling quality. The most important test of all is the user test. If a product meets all of the known legal requirements and company specifications, but doesn't work, the company is in serious trouble. Obviously, the specifications are incomplete. A painful example happened in a company who had successfully manufactured tea bags for years. The bag specifications appeared to be complete: bag dimensions, 35 grains of product per bag, 3 in. string length, in $\frac{3}{4}$ in. square tag, uniform seam width, seal complete on three sides, etc. Suddenly, an unexpected torrent of consumer complaints poured in, all describing the string tearing loose from the bag. The problem was traced to insufficient wet strength in a shipment of tea bag paper—for which there was no specification. The quality laboratory test was scientifically designed to evaluate all of the dimensions listed above, plus a "final product test" in the cup. Unfortunately, the scientific test did not include lifting the brewed tea bag from the cup by its string. The problem would have been revealed the day the first roll of defective paper was received, if only the lab had duplicated the consumer's method of preparing a cup of tea.

Selecting the formal test methods for legal requirements is usually not a problem, since the requirements generally refer to clear test requirements.

To set up a library for test methods, several books are clearly needed. A list of the most useful follows:

*Standard Methods for the Examination of Water and Wastewater*. American Public Health Association.

*Bacteriological Analytical Manual*. Food and Drug Administration.

Gruenwedel, D. and J. Whitaker. *Food Analysis*. ISBN 0-8247-7181-8. Dekker.

Harrigan, W. *Laboratory Methods in Food and Dairy Microbiology*. ISBN 0-12-326040-X. Academic Press.

Joslyn, M. *Methods in Food Analysis*. Academic Press.

*Microscopic Analytical Methods in Food and Drug Control*. Food and Drug Administration.

National Research Council. *Food Chemicals Codex*. ISBN 0-209-02090-0.

National Academy Press.

Pomeranz, Y. and C. Meloan. *Food Analysis Theory and Practice*. ISBN 0-442-28316-4. Van Nostrand Reinhold.

Richardson, G. *Standard Methods for the Examination of Dairy Products*. American Public Health Association.

Speck, M. *Compendium of Methods for the Microbiological Examination of Foods*. ISBN 0-87553-117-2. American Public Health Association.

*Training Manual for Analytical Entomology in the Food Industry*. Food and Drug Administration.

Williams, S. *Official Methods of Analysis*. ISBN 0-935584-24-2. Association of Official Analytical Chemists.

Zweig, G. and J. Sherman. *Analytical Methods* (12 Volumes) ISBN 0-12-784312-4. Academic Press.

Of course, many of these references will not be applicable to all food products, but it might be well to review each for possible adoption of modified test methods. The next list contains references fairly commonly used by the food industry, and are arranged by type of test.

## GENERAL ANALYSIS

Aurand, L. *et al. Food Composition and Analysis*. Van Nostrand Reinhold.
Birch, G. *Analysis of Food Carbohydrate*. ISBN 0-85334-354-3. Elsevier.
Chaplin, M. and J. Kennedy. *Carbohydrate Analysis: A Practical Approach*. ISBN 0-947946-68-3. Oxford.
Fennema, O. *Food Chemistry*. ISBN 0-8247-7271-7. Dekker.
Fresenius, W. *Water Analysis*. ISBN 0-387-17723-X. Berlin.
Gilbert, J. *Analysis of Food Contaminants*. ISBN 0-85334-255-5. Elsevier.
Kurtz, O. *Micro-Analytical Methods for Food Sanitation Control*. Association of Official Agricultural Chemists.

## SPECIAL INSTRUMENTATION

Gilbert, J. *Applications of Mass Spectrometry in Food Science*. ISBN 1-85166-801-X. Elsevier.
Lawrence, J. *Food Constituents and Food Residues: Chromatographic Determination*. ISBN 0-8247-7076-5. Dekker.
MacLeod, A. *Instrumental Methods of Food Analysis* (1973). Halstead.

## MICROBIOLOGY

*Recommended Methods for the Microbiological Examination of Foods*. American Public Health Association.
Beuchat, L. *Food and Beverage Mycology*. ISBN 0-442-21084-1. Van Nostrand Reinhold.
Corry, J. *et al. Isolation and Identification Methods for Food Poisoning Organisms*. ISBN 0-12-189950-0. Society for Applied Bacteriology.
Post, F. *Laboratory Manual for Food Microbiology*. Star.
Sharpe, A. *Membrane Filter Food Microbiology*. ISBN 0-86380-065-3. Wiley.

## SENSORY

Bourne, M. *Food Texture and Viscosity: Concept and Measurement*. ISBN 0-12-119060-9. Academic.
Moskowitz, H. *Food Texture: Instrumental and Sensory Measurement*. ISBN 0-8247-7585-6. Dekker.
O'Mahoney, M. *Sensory Evaluation of Foods: Statistical Methods and Practices*. ISBN 0-85334-272-5. Elsevier.
Piggott, J. *Sensory Analysis of Foods*. ISBN 0-85334-272-5. Elsevier.

Next, a list of fairly recent books covering many aspects of test methods follows.

Aurand, L. and A. Woods. *Food Composition and Analysis*. ISBN 0-442-20816-2. Van Nostrand Reinhold.

Banwart, G. *Basic Food Microbiology*. ISBN 0-87055-322-4. AVI.

Charalambous, G. *Analysis of Food and Beverages*. ISBN 0-12-16916-8. Academic.

DiLiello, L. *Methods in Food and Dairy Microbiology*. ISBN 0-87055-411-5. AVI.

Gorman, J. *Principles of Food Analysis for Filth, Decomposition and Foreign Matter*. Food and Drug Administration.

Harrigan, W. and M. McLance. *Laboratory Methods in Food and Dairy Microbiology*. ISBN 0-12-326040-X. Academic.

Jay, J. *Modern Food Microbiology*. ISBN 0-442-24445-2. Van Nostrand Reinhold.

Jellinek, G. *Sensory Evaluation of Food*. ISBN 0-89573-401-X. VCH Pubs.

Jowitt, R. *Physical Properties of Food*. ISBN 0-85334-213-X. Elsevier.

King, A. *Methods for Mycological Examination of Food*. ISBN 0-306-424479-7. Plenum.

Knorr. *Food Biotechnology*. ISBN 0-82477578-3. Dekker.

Lee, F. *Basic Food Chemistry*. ISBN 0-87055-4. AVI.

Macral, R. *HPLC in Food Analysis*. ISBN 0-12-46780-4. Academic.

Morris, B. and M. Clifford. *Immunoassays in Food Analysis*. ISBN 0-85334-321-7. Elsevier.

Moskowitz, H. *Applied Sensory Analysis of Foods*. ISBN 0-8493-6705-0. CRC.

Mountney, G. *Practical Food Microbiological Technique*. ISBN 0-442-22688-8. Van Nostrand Reinhold.

Okus, M. *Physical and Chemical Properties of Food*. ISBN 0-916150-82-8. American Society of Agricultural Engineering.

Osborne, B. and T. Fearn. *Near IR Spectrographic Food Analysis*. ISBN 0-470-2675-6. Wiley.

Post, F. *Laboratory for Food Microbiology and Bacteriology*. ISBN 0-89863-127-0. Star.

Rockland, L. and G. Stewart. *Water Activity*. ISBN 0-12-591-350-8. Academic.

Stewart, K. and J. Whitaker. *Modern Methods of Food Analysis*. ISBN 0-87055-462-X. AVI.

Vaughn, J. *Food Microscopy*. ISBN 0-12-715350-0. Academic.

Finally, we present a short list of industry methods. By contacting suppliers, it is possible to obtain other industry test methods which have been developed by private industry and by trade associations.

*Evaluating a Double Seam*. Dewey and Almy Chemical Co., Divn. W. R. Grace Co., Cambridge, MA.

*FPA Technical Manual of Specifications and Test Procedures*. Published by the National Flexible Packaging Association, Cleveland, Ohio.

*Index of CCTI Recommended Industry Standards and Testing Procedures*. Published by The Composite Can and Tube Institute, Washington, DC.

*Official Microbiological Methods of the American Spice Trade Association*. Published by the ASTA, Englewood Cliffs, NJ.

*Specifications and Methods of Analysis*. N.V. Cacaofabriek De Zaan, Holland.

Hundreds of test methods developed specifically for food analysis are available from instrument manufacturers. When contemplating additions to laboratory equipment, check with the manufacturer's technical department to obtain lists of applications. Contact the manufacturer of presently owned equipment for new applications developed since the original purchase. Although all of these industry tests may not be accepted as official analytical methods, they may provide useful rapid analyses on such equipment as infrared analyzers, HPLC, ultraviolet and visible spectrophotometers, mass spectrophotometers, ion chromatographs, polarographs, and others.

# 7    Product Specifications

Assembling specifications to describe a product would seem to be the simplest part of a quality control program. A bread bakery might decide to develop a cake to round out the product line. Once the kind of cake has been decided, a brainstorming session with the bakery staff can produce a list of attributes which need to be defined. Consider Figure 7-1. In order to convert the attributes listed into specifications, there are three basic questions to be answered:

- What do you want to make?
- What are you capable of making?
- How much can you afford to spend?

Converting descriptive quality terms into concrete specification numbers can be accomplished logically by the use of statistical tools. A statistical evaluation of process capability will define the limitations of cakes which the plant can consider manufacturing. Capability studies will also lead to decisions regarding the numerical values to be applied to each of the variables under consideration: the weight, moisture, height, color, acidity, shelf life, and cost. The range of attributes can be determined by the formulation of experimental designs, and analysis of the results: texture, grain, rigidity, sweetness, flavor, etc. Compromises regarding shelf life, packaging, ingredients, dimensions, and product characteristics suggested by the experiments can then be effected through discussions with marketing, accounting, distribution, engineering, production, and other departments. Finally, the specifications can be presented to management for approval.

Now that the cake and frosting specifications are spelled out, the details of the ingredients, packaging, processing and storage must also be specified in order to bake and market cakes with the consistent quality level required. Although these specifications usually require fewer department meetings, using the same statistical design and analysis of experiments is the surest path to reliable data.

If any of the above steps to converting an idea into a product specification is omitted, disappointment on the production floor and in the marketplace becomes

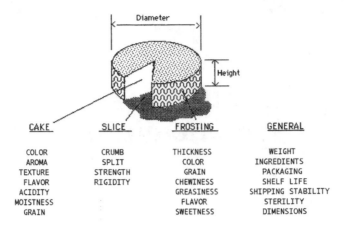

**Figure 7-1.** Quality attributes of a frosted cake.

likely. If specifications are either overlooked or unrealistic, quality failure on the production floor is bound to cause high scrap and rework costs; and quality failure in the marketplace is bound to cause loss of sales.

Listing the specification data for an existing product is a simpler task than writing specifications for raw materials or ingredients produced by others. The selection of the ingredients from the vast maze of materials available in the marketplace can appear overwhelming. Is it possible to select and then specify an ingredient which has performance, flavor, uniformity, price, stability, and purity characteristics which will in turn provide specific performance, flavor, uniformity, price, stability, and purity characteristics to the finished product? And assuming that such a selection exists, is it then possible to assign to that ingredient the required specification data which the suppliers can meet consistently? The answer to both questions is "yes." For the sake of expediency, it may be practical to establish greatly simplified and fairly broad specifications at the outset, but by the continued use of a program of statistical experimental design and analysis, the specifications can be fine-tuned to assure uniform quality.

The techniques of experimental design and analysis, process capability study, and sensory testing will be discussed later. For the present, let us examine the content of specifications. To start with, an ingredient specification for cocoa is presented in Figure 7-2. The numbers shown are illustrative only, and represent values which might be required for a frosting mix ingredient. In order to arrive at a specification shown in Figure 7-2, the Boston Biscuit Bakery will have completed several series of tests in which each of the data listed was found to be required for the quality cake desired, or were required by various laws and regulations. Much of the information is available from cocoa manufacturers, but not all of the available analyses are required in the ingredient specification. For example, composition data is available for 14 chemical elements, 7 vitamins, oxalic acid, theobromine, starch, sugars, sulfates, and organic acids all of which may be found routinely in cocoa products. There is no point in listing this data in the specification merely

**BOSTON BISCUIT BAKERY**

*Chocolate Cake #CC071 Ingredient Specification*

INGREDIENT: Cocoa powder
          #FCP 071

APPROVED SUPPLIERS:
    Andrus Cocoa Co. Supl#C3
    Filbert Specialties Supl#C1

DATE APPROVED: Sept. 20, 19–    APPROVED BY: *ACG*

DESCRIPTION: Medium fat cocoa, conforming to the standard of identity as described in 21CFR 163.112 and 163.113, and produced in accordance with Good Manufacturing Practices detailed in 21CFR Part 118, Cacao Products. Gras Clearance 21CFR 182.1.

| | ATTRIBUTE | MEASUREMENT | TEST METHOD |
|---|---|---|---|
| PHYSICAL: | Color | B2–B4 | Agtron A#76 Phototron P#3 |
| | Flavor | Betw.Std A and Std B | BBB Method #0449 |
| | Fineness | 99% minimum | Wet, thru 200 Mesh sieve (0.075 mm) |
| | Odor | Free from off odors | BBB Method #0006 |
| | Shell Content | 5% maximum | BBB Method #0720 |
| CHEMICAL: | Fat Content | 10.0–12.0% | AOAC 13.036 |
| | Lecithin | 4.0–5.75% | BBB Method #0724 |
| | Moisture | 5.0% maximum | AOAC 13.003 |
| | pH | 7.0–7.4 | AOAC 13.008 |
| | Lipase activity | Negative | BBB Method #0726 |
| MICROBIOLOGICAL: | | | |
| | Std Plate Count | 5,000 maximum | FDA/APHA |
| | Molds per g | 50 maximum | BBB Method #0728 |
| | Yeasts per g | 50 maximum | BBB Method #0729 |
| | Enterobacteriaceae | Negative in 1 g | BBB Method #0730 |
| | *E. coli* per g | Negative | FDA/APHA |
| | Salmonellae | Negative | FDA Bacty Manual |

PACKAGING:
50-pound multiwall paper bags, polyethylene-lined. Each bag labeled with product name, manufacturer, production lot.

STORAGE, SHIPPING:
Shipped with no more than 2 sublots per delivery. Transported according to 21CFR 110.80. Pallets to be stored in cool, dry (R.H. less than 50%) area, isolated from spices and flavorings. Shall comply with storage conditions in accordance with FDA good manufacturing practices.

SAMPLING PROCEDURE:
Samples shall be taken according to BBB Procedures specified in sampling manual #BBB Manual #4.

**Figure 7-2.** Cocoa powder specification.

because the figures are available. The specification requirements should be concerned only with those variables which might affect the quality of the final product.

Similar specifications are prepared for each ingredient used in the standardized BBB cake formulation. This insures that the starting conditions for the product manufacture will be correct. This is the second of the three classes of specifications required for a quality control system. In addition to the ingredient specifications and the finished product specification discussed above, the need for a process specification should be obvious.

Process specifications are proprietary, and are usually in a continuous process of evolution as new machinery, new raw ingredients, new process tests, and new formulations are discovered. In the case of the BBB cake, process controls are evolved for each of the unit operations:

- Ingredient weighing
- Blending
- Mixing
- Depositing
- Baking
- Cooling
- Packaging.

Process specifications for each operation are developed from process capability studies, and test methods are improvised as necessary to assist in measuring the attributes considered critical to the process. As discussed under HAACP, the control points are discovered from analysis of the process steps and their effects on the product. Both the product and the process equipment are included in the process specification network. For example, time–temperature relationships during the baking cycle must be measured and recorded for both the oven equipment and for the cake itself during the baking cycle.

Process specifications are the most complex of the three types. Even starting with raw materials which fall within the ingredient specifications, small variations in the flour composition, coupled with trace variations in the leavening agents and minor differences in atmospheric pressure can combine to upset the normal leavening process and thus create an in-process oddity which cannot be readily corrected by unbending process controls.

Each industry seems to collect its own group of such oddities. The density of a dehydrated food product was found to shift markedly each afternoon, even though no known changes in either the ingredients or the process were occurring. The cause was finally found to be an afternoon wind blowing against the dehydrator air inlet which upset the drying air flow. In a french fried potato plant, the raw potatoes tested satisfactorily when received, but turned darker than process specifications permitted when processed. It was discovered that rotation of stock was not being carefully followed and, as a result, some of the raw potatoes had started to increase their sugar content in storage, thus frying darker. Another mystery went unexplained for years in a tea packaging operation where, periodically, packages would appear with unusually high (and undesirable) dust content.

All ingredients were within specification, and all process controls were within specification; yet, the problem persisted. Ultimately it was discovered that dust was accumulating in the ends of a lengthy feed hopper above the scales, and then abruptly emptying into the production line packages. It is this type of exceptional line or product performance which makes process specifications so difficult to compile.

---

**BOSTON BISCUIT BAKERY**

*Chocolate Cake #CC071 Process Specification*

DATE APPROVED: Sept. 20, 19–                                    APPROVED BY: *ACG*

INGREDIENT WEIGHING

> *Small Ingredients*            Scale #S2007 Precision 0.25 ounce

|                     |       |         |
|---------------------|-------|---------|
| Butter flavor B03   | 3 lbs | 7.75 oz |
| Albumen powder      | 74    | 8.00    |
| Vanillin            | 1     | 4.25    |
| Salt REF17 S        | 20    | 2.00    |

Mix 4.5 minutes on Ingredient Blendor #BI02. Temperature 65–80 deg.

|                          |   |      |
|--------------------------|---|------|
| Sodium tartrate ST04     | 5 | 0.00 |
| Calcium phosphate CP04   | 2 | 0.25 |

Add to Ingredient Blendor #BI02. Blend 1.75 minutes. Maximum 90 deg.

Store in sealed drums at 65–90 deg. for no more than 8 hours.

> *Major Ingredients*                                        etc. etc.

---

This portion of the process specification may read more like a recipe than a process control. Subsequent data concerned with air flows, purge times, temperatures, oven speeds, and interactions of these variables with the product variables are often more descriptive. Since moisture content of the cake is a critical variable, a process specification would have to include a table of expected values of moisture levels for various combinations of oven conveyor speed and temperatures in both primary and secondary sections of the oven. Similarly, a process specification covering the cooling cycle would need to list either the cooling time or the cooling conveyor speeds at various ambient temperatures in order to provide uniform product quality.

To summarize this discussion of specifications:

1. Start with a description of the ideal product.
2. Through studies of available ingredients and process experimentation, define attainable product specifications.
3. Prepare optimum ingredient specifications.
4. Prepare optimum process specifications.

# 8     Process Capability

Before examining the mechanics of process capability analysis, it would be wise to define the subject. With only two words to define, this would appear to be a trivial task. According to the dictionary, "process" refers to steps (or operations, or changes) leading to a particular result. "Capability" is defined as the capacity for an indicated use. Let us explore a few of the definitions proposed by various authorities in the field.

> "Process capability is the measured, inherent reproducibility of production turned out by a process."
> Juran, *Quality Control Handbook*

> "The minimum spread of a specific measurement variation which will include 99.7% of the measurements from a given process."
> B. Hansen, *Quality Control Theory and Applications*

> "The minimum range of values of the process output variable which includes at least 99% of the observations under normal operating conditions."
> D. Braverman, *Fundamentals of Statistical Quality Control*

> "Process capability is quality-performance capability of the process with given factors under normal, in-control conditions."
> A. Fiegenbaum, *Total Quality Control, Engineering and Management*

> "Process Capability—the level of uniformity of product which a process is capable of yielding."
> DataMyte Corporation, *DataMyte Handbook*

> "Process capability is determined by the total variation that comes from common causes—the minimum variation that can be achieved after all special causes have been eliminated."

W. Edwards Deming, "On some statistical aids toward economic production," *Interfaces*, Vol. 5 No. 4, August 1975. The Institute of Management Sciences, Providence, Rhode Island

No two of these definitions are in complete agreement. Nor do they clarify whether we are concerned with a complete product, a single step in the process, a vendor's product quality (raw material), or if a machine can be considered to be a "process." One definition refers to the "spread of a measurement;" another, the "reproducibility of production;" a third, "range of values," and "quality performance," "99.7% of the measurements;" "minimum variation," etc. Perhaps the simplest definition above is the most useful: "the level of uniformity of product which a process is capable of yielding." We shall start working with this one.

Note that this definition does not concern itself with how well the process is currently working, but aims at what the process is *capable* of yielding. The implication here is that at times the process works more uniformly than at other times, and more important, there is a possibility that the process might be further improved. Herein lies the value of process capability studies: they will establish the best quality the process is currently capable of producing so that realistic control limits and specifications can be drawn up. That's the immediate goal. In addition, the studies required for capability analysis will also suggest areas where modifications might be looked at for further improvement.

Deming refers to variability in manufacturing processes as consisting of two types: common and assignable. The common causes are those which are inherent in the process, remaining there until management sees fit to change the process to modify or eliminate them. Assignable causes are external factors which can be eliminated or controlled, and although they affect the process, are not a part of it. Examples of assignable causes are incorrect packaging material, improper identification of raw material, wrong laboratory data. Examples of common causes: poor lighting, gradual machine wear, inadequately trained operator, inexact measuring equipment. *Process capability studies are conducted on processes which have all assignable causes eliminated, and insofar as possible, have all common causes reduced to a minimum.* By definition, these conditions exist when a process is in *statistical control*. Herein lies the problem: more often than not, a process capability study is required for new or proposed processes which have not had an opportunity to be freed from as yet unidentified assignable causes. This might explain why there are definition conflicts. Different quality control practioners have conducted process capability tests under unique conditions for which they have later evolved a definition which suited their purpose.

Having cleared the air of that semantic problem, let us examine a simplified example of a process capability study. The example is taken from a plant processing melba toast. One of the quality criteria is the color development in the final stages of oven browning. Individual pieces of toast are scanned by a spectrophotometer which has been calibrated to reflect yellow-brown light over a reduced spectrum. The readout is expressed in percent of a standard yellow-brown reference plate, and a value of 41 is considered to be optimal. Data is collected every 30 min starting at early morning and continuing over 15 hr. The results are tabulated and graphed as shown in Table 8-1 and Figures 8-1–8-3. Superimposed on the graph are the calculated values for the average and for the upper and lower three-sigma control limits (UCL and LCL). These limits are considered to be totally unsatisfactory; the upper limit is far too light, and the product tested at the lower limit is nearly burned. There are many variables at play over these 15 hrs which contribute to this unsatisfactory range.

**Table 8-1.   Example of Process Color Capability Data**

| Sample no. | Color units every 30 min | Color units consecutive | Color units improved proc |
|---|---|---|---|
| 1 | 34 | 42 | 41 |
| 2 | 45 | 44 | 41 |
| 3 | 62 | 41 | 41 |
| 4 | 35 | 44 | 42 |
| 5 | 44 | 45 | 41 |
| 6 | 33 | 43 | 45 |
| 7 | 42 | 44 | 42 |
| 8 | 50 | 44 | 43 |
| 9 | 44 | 41 | 41 |
| 10 | 53 | 41 | 40 |
| 11 | 33 | 38 | 39 |
| 12 | 32 | 39 | 38 |
| 13 | 57 | 40 | 39 |
| 14 | 45 | 41 | 40 |
| 15 | 40 | 41 | 41 |
| 16 | 33 | 40 | 41 |
| 17 | 45 | 43 | 42 |
| 18 | 20 | 40 | 41 |
| 19 | 44 | 44 | 42 |
| 20 | 31 | 38 | 42 |
| 21 | 59 | 39 | 41 |
| 22 | 34 | 40 | 40 |
| 23 | 52 | 44 | 40 |
| 24 | 29 | 42 | 40 |
| 25 | 45 | 45 | 41 |
| 26 | 44 | 42 | 42 |
| 27 | 52 | 44 | 41 |
| 28 | 33 | 43 | 41 |
| 29 | 34 | 47 | 42 |
| 30 | 29 | 42 | 41 |
| Average | 41.1 | 42.0 | 41.0 |
| Standard deviation | 10.0 | 2.23 | 1.29 |
| $3 \times$ Standard deviation | 30.1 | 6.7 | 3.8 |
| Control Limits | 11.0–71.2 | 35.3–48.7 | 37.2–44.8 |

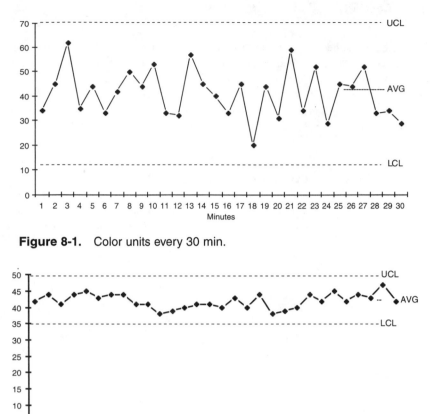

**Figure 8-1.** Color units every 30 min.

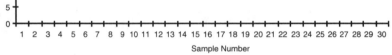

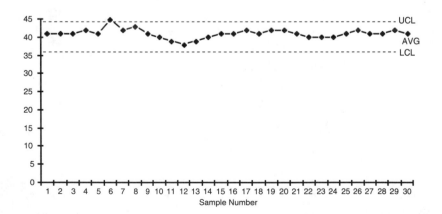

**Figure 8-2.** Color units consecutive.

**Figure 8-3.** Color units consecutive, improved process.

To limit the variables, another set of data is obtained in which the samples are taken continuously, 15 seconds apart, until 30 samples have been obtained. These are then examined, data recorded and graphed, and control limits applied as above (see Figure 8-2). Now we find control limits within 6.7 units of the average value of 42.0. Contrast this with the range of 30 units in either direction of the average observed in the 15-hr sampling period above. What we have discovered is the capability limits of the process when most of the variables are unchanging. To phrase this simply: the process under the best conditions now available will process toast with a range of 13.4 color units.

During the course of the study thus far, certain observations of the process may have been questioned with the possibility of reducing the variability. For example, it may have been observed that the widest swings in data over the 15-hr test period were the result of the coarse control setting on the trim burners. That is, the sensitivity to temperature control was set at the lowest range, causing the burners to alternately surge and then coast. This resulted in marked differences in the product color units. Let us further suppose that the range setting on the burner was moved up a notch so that the burners cycled more often as the temperature started to drift. With this improved process control, it was decided to take another set of data, (see last column Table 8-1 and Figure 8-3) as shown on the accompanying table under "improved process."

Here we find that the consecutive sample series obtained showed a reduced variability, and an improved process capability: control range reduced to 7.8 color units.

This example demonstrates how a process capability study may be made using simple statistical control charts. It also demonstrates how process improvements may be discovered, implemented, and verified.

Again, it should be emphasized that the above example is simplified. In the real world, there are at least three major changes to be considered:

1. 30 samples are not enough. Generally, at least 100 samples will be required to obtain a reasonable estimate of population mean and standard deviation.
2. Subgroup size of one is more difficult to calculate than larger sizes. Subgroups of three or five may be used, thus permitting use of tables to estimate standard deviation. Bear in mind that there we are dealing with means, not individuals, and the results should be interpreted with constant reference to the *means*.
3. After conducting the first series of determinations to establish process capability, there is rarely an opportunity to *immediately* run process improvement tests as shown in the example. These require time to plan adequately.

A word of caution regarding the second paragraph above. Not all quality control practitioners would agree that there is justification in labeling the control limits for means as synonymous with process capability. But there is certainly a practical use for determining means control limits to explore, at least initially, the capabilities of a process. Eventually, it may be necessary to measure the variability and control limits of individual products. After all, the customer examines the product purchased, not a subgroup of products. In fact, the customer doesn't care what the average quality is—only the quality of the individual product purchased. This is not necessarily true for intermediate products, however. In a corn cannery, there is no need to know ear-to-ear variability; bushel-to-bushel quality data may readily suffice.

A more complete determination of process capability may be arrived at by investigating the time-to-time variability of the process. The "consecutive" data are perfectly valid, but they define the best possible control limits of the process as it presently operates. It would add to the value of these data if some measurement of process shift could also be measured. An examination of the data in the "every 30 min" graph shows no particular drift pattern, but clearly indicates that the process does not meet the control limits determined in the "consecutive" process capability study. Specifically, if the capability control limits of 35.3–48.7 were applied to the "every 30 min" data, 11 points would be out-of-control on the low side, and 7 points would be out-of-control on the high side. That means a total of 18 of the 30 points are out-of-control. This process is obviously in need of attention.

By contrast, the "consecutive" data show a rather tightly controlled process over the period of time required to obtain the data. Somewhere between that period and 30 min, the process has drifted several times. To locate that interval, it will be necessary to conduct a detailed test in which samples are taken at fairly short intervals for a total time of 30 min. To simplify calculations, it is suggested that the average color units of every five samples (in order of production) be plotted on a graph. A visual examination of the graph should reveal the interval of time during which the process is stable, and the approximate time after which the process tends to drift in one direction or another. Once this interval has been uncovered, the causes for drift can be studied so as to reduce variability. Without taking this step to determine the minimum drift interval, attempts at process improvement testing are likely to be disorganized, random, and frequently fruitless.

An example of a graph showing process drift is shown in Figure 8-4.

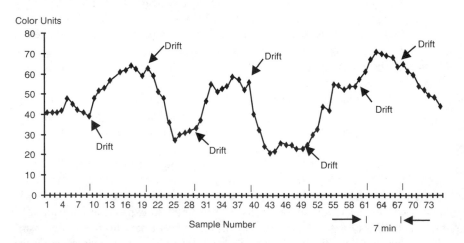

**Figure 8-4.**   Average of five successive samples taken at 12 s intervals.

It appears that the shortest interval between a stabilized process and the start of a drift is about 7 min. Therefore, the process should be evaluated every 7 min and adjustments made if required. If, in the case of melba toast processing, the key to the color unit drift is found to be oven temperature, then the evaluation and correction should be made by a temperature sensor and controller if available; in a manual process the evaluation would be performed by an operator who would be instructed to adjust the oven temperature if the reading is outside of the process control limits. If the key to color drift is found to be oven conveyor speed, then this would be the factor to be adjusted to maintain process control.

A number of sampling and adjusting schemes can be devised, depending upon the process, the equipment, and the operators. For example, it would be possible to establish a 14- min evaluation cycle, with the understanding that if the process drifts out of limits, as much as 7 min of production might be out of the color control limits.

Note that the process control limits determine the *obtainable* uniformity of product. In some instances, this might be far more uniform than is required by the customer. In such a fortunate situation, the process need not be adjusted until the specification limit is reached, possibly reducing the cost of production.

As we have already seen, there are many uses to which process capability studies can be applied. Here are some others:

1. *Process selection.* When several processes are under consideration, capability analysis will assist in selection of the most effective one.
2. *Tolerance compliance.* Process capability analysis (PCA) will predict the extent to which the process is capable of holding.
3. *Machine selection.* PCA can be used to assign machines to the classes of work for which they are most suited.
4. *Selection and training of inspectors.* When a PCA has been prepared, the control limits provide reliable benchmarks to evaluate inspectors' ability to adequately measure quality.
5. *Identification of troublesome steps.* In a series of sequential processing steps, PCA of each step may identify the one responsible for quality abnormalities. This reduces the research necessary to find a solution.
6. *Research criteria.* Theories of defect causes may be readily evaluated by PCA during quality improvement programs.
7. *Audit tool.* A series of PCAs performed on the same process will often disclose the frequency of process reviews required to insure adequate maintenance, inspector instruction, operator performance, personnel supervision, analytical instrument calibration, etc.
8. *Setting specifications.* If the product is to succeed, its quality specifications must lie beyond the process capability limits.

## CAPABILITY INDEX

One of the end products of a process capability study is the control chart showing the upper and lower control limits of a controlled process. The control limits
cover a range of six-sigma. Another end product is the *capability index*. The
capability index compares the product specification range with the six-sigma
range. The index may be useful to those who prefer to express test results in
numbers, rather than in words. It may also find value where a company has
a vast number of specifications and is constantly in need of comparing process
capabilities with control limits.

The calculations are simple:

$$\text{Process capability index} = C_p$$

$$\text{Capability index} = \frac{\text{Allowable process spread}}{\text{Actual process spread}}$$

$$= \frac{\text{Specification limits}}{\text{Process control limits}}$$

$$= \frac{\text{USL} - \text{LSL}}{\text{UCL} - \text{LCL}}$$

$$= \frac{\text{Tolerance}}{6\sigma}$$

All of these equations say the same thing: subtract the lower specification limit
from the upper specification limit, and divide the result by the difference between
the upper and lower control limits (or six standard deviations). The number found
by this simple calculation determines whether or not the process is capable of
meeting the specifications.

If the specification spread is the same as six-sigma ($C_p = 1$), then approximately 0.3% of the production can be expected to be out of limits. If six-sigma is
less than 75% of the specification spread, then the process is generally considered
acceptable and capable. (That is, $C_p = 4/3 = 1.33$.) In tabular form:

| Capability index | Conclusion |
| --- | --- |
| 1.33 or greater | Process is satisfactory |
| 1.00 to 1.33 | Process is capable, but marginally as the index approaches 1.00 |
| 1.00 or less | Process is unsatisfactory |

## Example

Let us assume the specification for a food product limits the ferric iron content between 14 and 22 parts per million. A process capability study shows control limits at:

$$\text{LCL} = 15 \text{ ppm}; \quad \text{UCL} = 21 \text{ ppm}.$$
$$C_p = (22 - 14)/(21 - 15)$$
$$= 8/6 = 1.33$$

Therefore, this process is capable of meeting the specification. A control chart illustrating this example is shown in Figure 8-5.

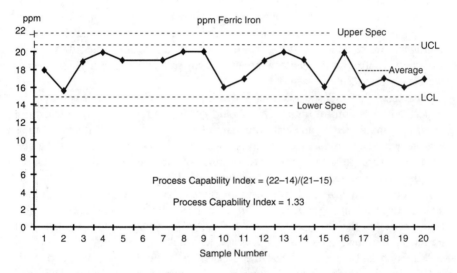

**Figure 8-5.** Control chart—ferric iron content (in ppm).

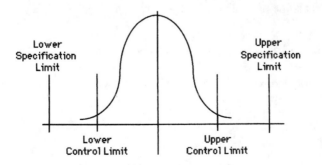

**Figure 8-6.** A capable process.

To review: if a process is in control, and if all of the measurements of that process which lie between the lower control limit (three standard deviations below the average) and the upper control limit (three standard deviations above the average) also lie between the lower and upper specification limits, then the process may be considered "capable." An illustration of this is shown in Figure 8-6.

A method to avoid calculating the standard deviation of individuals data: Calculate the average range ($\overline{R}$) of differences of each successive pair of data points, and find the control limits for individuals using the following formulas:

$$UCL = \overline{X} + 2.66\overline{R}$$
$$UCL = \overline{X} - 2.66\overline{R}$$

A short cut to finding the standard deviation for individuals when a previously calculated control chart for averages is available: multiply the UCL and LCL by the square root of n, where n is the sample size used in the control chart for averages. The square root of n converts average data to individual data. These individual control limits may now be used to calculate the capability index using the above formulas.

The capability index discussion above refers to $6\sigma$ as the area between the upper and lower control limits. This should not be confused with the $6\sigma$ quality program popularized by an electronic manufacturing company. This organization was determined to reduce defects to parts per million, rather than the generally accepted 3 parts per 1000. In order to accomplish this, they started by placing the *specification* limits at 6 standard deviations on *either side* of the mean, shifting the *process* mean 1.5 standard deviations higher than center, and improving the process until the upper *process control* limit was within the specification limit. (Note carefully the italicized *process* and *control* verbiage in this explanation.) Perhaps the following table and discussion will make this more clear.

Research had determined that the process mean normally shifted higher by $+1.5\sigma$ for the company which evolved this quality control system. The units out of specification under these conditions would have been expected to rise to 6.8 per million; however, it was theorized that since the lower control limit was also raised by $1.5\sigma$, there would be zero defects at the lower level, and 1/2 of 6.8 or 3.4 defects at the upper level. With this reasoning, the company identified the

**Table 8-2. Relationship Between Capability Index and Defects**

| Capability Index | Specification Limits | Units per Million out of Specification |
|---|---|---|
| 1.0 | $\pm 3\sigma$ or $6\sigma$ | 2700 |
| 1.33 | $\pm 4\sigma$ or $8\sigma$ | 63 |
| 1.48 | $\pm 4.5\sigma$ | 6.8 |
| 1.67 | $\pm 5\sigma$ or $10\sigma$ | 0.570 |
| 2.0 | $\pm 6\sigma$ or $12\sigma$ | 0.002 |

quality program as the Six-Sigma Quality Program, which produced 3.4 defects per million.

Since the program would be successful only if process improvements could be accomplished to fit within it, one might suggest other, perhaps simpler ways to accomplish the same defects per million goal. On the other hand, the numeric goal-oriented system has motivated employees, and positive results have been obtained.

## BENCHMARKING

The usual understanding of the benchmarking process is based on finding an external goal obtained from the best of industry practices. By investigating and incorporating these best practices into its operations, a company can expect to better serve its customers, and improve its processes and profitability. Definitions of external benchmarking suggest continuously measuring and comparing a company's products and processes with those of industry leaders in similar products or services, and adopting the best of them with whatever modifications are necessary.

Direct competitors are unlikely to embrace this policy of sharing best practices, depending partly on the market share of each company. However, companies with different products or uncommon market niches may both benefit from frank discussions of each other's best practices. Other sources of information are trade journals, the internet, business consultants, market research organizations, business and professional association meetings and publications, customers, and equipment vendors.

There are many differing benchmarking plans offered in the literature. Rather than recommending a single series of steps and rules for conducting a benchmarking program, it might be of greater value to consider some of the many questions which must be resolved before starting.

What management level should interview which host management level
How is the benchmark target selected
How can host's information be tactfully verified
Should the first project be a surefire success or a major goal
Must contacts be performed with written questionnaires, or verbally
Do host's measurements have the same meaning
Are personnel qualified to perform, or are consultant's required
Should informal contacts be made in addition to office visits
May an undefined goal be used as a starting point
Is a project schedule desirable—or possible
Does the host's differing corporate culture account for success
Are the units of measurement the same for the two companies
Would information sharing violate any law
How will use of new information be processed in the company
Should benchmarking be conducted by a team or a department

The techniques involved in benchmarking have been formalized by several companies and authors, and they all seem to have a central theme: Plan / Do /

Check / Act (the Deming Cycle.) However, the terminology used in benchmarking steps varies: analysis, implementation, integration, measurement, identification, data collection, functional goals, establishment of performance levels, develop action plans, monitor.

Not to be overlooked is a perhaps more valuable approach to benchmarking: internal process and product improvement made possible through examining the most successful principles and techniques which exist in one's own company. The discussion of benchmarking has been included in this chapter on process capability for that reason.

Internal benchmarking is an extremely effective tool and is usually welcomed by both production personnel and management, since it represents the best efforts and previous attainable successes found within the company. In essence, it is based on a simple principle: "we did it before and we can do it again." Perhaps most companies have repeatedly used a form of this technique without formalizing it with a title.

Although internal benchmarking can be successfully handled by a single engineer, production manager, researcher, or department head, it is more likely to be successfully implemented if a team is formed from various company operations. The increased success observed may be due in part to the possibility of contributions of more diverse ideas; but since the team includes members whose departments are affected by the final decisions, they would likely be motivated to contribute willingly and to actively implement the findings.

The principle of internal benchmarking is to select an operation which worked better than normal one which exceeded the normally acceptable (and desirable) control limit for quality, cost, sales, personnel requirements, productivity, line changeover, absenteeism, consumer complaints, mistakes, scrap, production interruptions, accidents, billing, specifications (such as color, flavor, weight, dents, leakers, lumps, solubility, flow, granulation, printing) or other. In a well-structured quality control system, this selection process can be as simple as forming a team to identify favorable outliers on process control charts, and selecting the one which appears to have the greatest potential. In its simplest terms, benchmarking now proceeds to find out how the favorable outlier occurred, set up a test to duplicate those conditions, and if successful, change the methods of operation to include the new conditions.

How does internal benchmarking differ from the normal principles of quality control and process improvement which have been used for years? Not by very much. The major change has been the introduction of a formalized structured system which includes (usually) team members from several departments. Each company can best decide how to structure the benchmarking process so that it will fit in with their culture. Following is a suggested framework for such a structure.

Form a search team to select a project area
Establish a specific goal and major objective
Select a team consisting of members affected by the goal
Train team to follow specific procedures

Brainstorm goal to select specific project
Establish resources necessary to achieve goal
Determine data collection procedures
Gather and analyze existing data
Generate additional data as required
Redefine goal and target
Propose improvement plan (PLAN)
Educate and train those involved with plan execution
Implement plan (DO)
Determine effects of implementation (CHECK)
Apply corrections or modifications to implementation
Standardize procedures, specifications (ACT)
Prepare report on findings, action taken and benefits
Publicize successes
Review benefits and start a modified or new improvement project

The major advantages of internal over external benchmarking are the availability of readily interpreted data, some of which might be proprietary, and the knowledge that the culture, environment and operations are constant.

# 9     Process Control

Once the initial investigative procedures have been completed on a process (specifications, capability analysis), control charts can be constructed and evaluated continuously to guide in the production of acceptable quality products. From the company's viewpoint, this is the most visible and the most useful function of the quality control manager. When progressing from capability analysis to production quality control, use of single sample measurements of variables is generally superseded by subgroups so that control charts for averages and for ranges may be constructed. The procedures for establishing and maintaining these charts have been described elsewhere. Our concern here is the interpretation of the charts.

In order to interpret a chart, it is essential that a number of factors concerning its selection and construction be reviewed. The function of charts used by a long-established company frequently erodes from the original purpose into routines which lose considerable impact on the control of quality. Under these conditions, it would not be unusual to expect that the question "why do we sample at this point in the line?" might be answered with a shrug and the reply "we always have done it this way." Many things change over time. New equipment is added to the line which makes some older measurements less significant; changes in raw products or packaging materials might not react to obsolete controls; or new methods on the line which critically affect quality might be overlooked by the existing quality control procedures. Periodic review of the principles on which the charts were constructed will avoid these problems.

Such a review should consider:

1. Training of the people currently involved with preparing the charts.
2. Assuring that the process location where the chart is generated is still a critical control point.
3. Determining whether or not this chart has an appreciable bearing on the customer's needs.

4. Judging the adequacy of the measurement system used from a standpoint of calibration for precision and accuracy, and state-of-the-art techniques.
5. Considering the possible need for an updated process capability study and control point revision, as dictated by either process improvement, or specification change.

Control charts are a continuous "picture of the process," and when properly constructed, provide the means of continuously producing product of uniform quality. When a single point falls outside of the control limits, something is probably wrong. There are two key words in that sentence. "Something" might be the miscalculation or misplotting of the offending point, or the sampling may have been improper due to stratification or ignorance, or the data may have been edited. The other key word is "probably," since 0.3% of the observations of a controlled process can fall outside of the control limits by chance alone. Because of these two possible situations, a number of ground rules have been established by various quality control practitioners. Some say that a control chart need not cause alarm until two points are out of limits on the same side of the control limit. Others say three. Some will call for action when there seems to be a trend or a pattern in the data. And there are systems which call for two standard deviations as a warning limit. Or seven points all on the same side of the process average line. There are also some shortsighted (and probably shortlived) companies which use as a ground rule: "No matter how many chart dots are out of limits, we are so far behind in production we will not stop the line for anything short of a fire."

There are special cases where some ground rules might be applied, but it is unlikely that misinterpretation of a control chart will occur if *every out-of-limits point were investigated, and if all other combinations of points within the control limits were ignored*. This statement applies equally to $X$-bar, $R$, $p$, $n$, or $c$ charts. One of the greatest sins in attempting to control quality, and one of the major contributors to production costs, is attempting to control quality by adjusting or modifying when the process is already within the control limits.

With the understanding of the italicized statement in the paragraph above, we will shortly examine some of the special cases where peculiar chart patterns might be handled as exceptions to the rule. But first, there are some general principles relating to data points which appear outside of the control limits, and a brief discussion of these follows.

Data on variables charts may first exceed the control limits in one of three ways:

1. If the average is out of limits, but the range is not, chances are that there is a serious problem with the process. When this does occur, chances are that all of the observations in the subgroup are out of control, or nearly so.
2. If the range is out of limits but the average is not, the process may be on the verge of developing an oscillation. It can be expected to soon drive the average out-of-limits as well. It is likely that individual items are already close to or outside of the specification limit.

3. If both the average and the range are out of limits, something drastic has occurred, and it can be expected to become worse unless immediate action is taken.

When attribute charts ($p$, $n$, or $c$) exceed control limits, it may be that a substantial quantity of questionable quality products has been produced, and corrective action should be swift. If not, then charts become useless scorecards, and reject/rework loads will build up.

When any chart shows a reading below the lower control limit, immediate attention is needed, but frequently this may not require shutting down the line. In fact, exceeding the lower limit may be advantageous: fewer can leaks than permitted, lower bacteria count, less than the minimum moisture vapor transfer rate, unusually low bitter flavor characteristic, a total absence of contamination, or zero sediment. These and many other similar variables and attributes may all be very desirable in the product, but are generally unattainable. When this type of outlier data occurs, the process obviously may not need to be stopped, but the causes for the unexpected quality improvement may form the basis for an improved product, and should be explored immediately before they disappear. Once uncovered, permanent changes may be instituted, and new process control limits adopted.

Of course, this might also be true for charts of some variables where exceeding the upper control limit may produce a more desirable quality product. Here too, the causes should be explored with the goal of product improvement.

## CHART PATTERNS

### Normal, in Control

The pattern is random and within the control limits (Figure 9-1). No action is either needed or desirable.

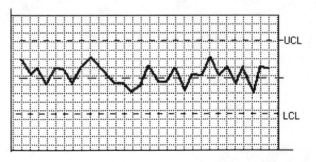

SAMPLE NUMBER

**Figure 9-1.** Normal, in control.

## Random Cycles

When found in the $X$-bar chart, the causes may be due to rotation of operators or machines, merging processes (such as feeding reworked material into the line, or alternating raw material and supplies), changes in environmental conditions (temperature, humidity), or worker fatigue (Figure 9-2). On the range chart, the causes may be due to inadequate maintenance of equipment, resulting in wear and intermittent operation. On $p$-charts, cycles may occur from merging processes as above. In addition, failure to use the correct sampling plan for the chart in use can produce cycles.

## Trend (Up or Down)

The chart for averages may trend in one direction (see Figure 9-3) as one raw material is used up and another is introduced. Gradual deterioration of equipment, continuous change in the environment, or change in the production rate will also

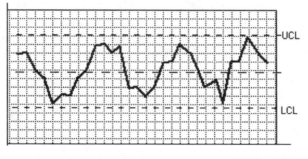

SAMPLE NUMBER

**Figure 9-2.**   Random cycles.

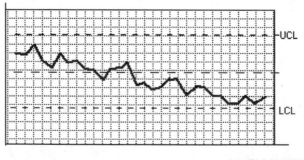

SAMPLE NUMBER

**Figure 9-3.**   Trend (down).

produce a trend. A trending range chart may be the result of gradual improvement or deterioration of a worker's skill, a production rate change, or a change in the number of components reaching the process. *P*-charts are similarly affected.

## Abnormal Uniformity

All charts may show a reduced scattering of observations (Figure 9-4) when a superior skilled employee is present, when the process has been changed (intentionally or not), when a key step or key ingredient has been omitted, when defect-free raw materials are introduced, when unusually favorable environmental conditions exist, or when the wrong control limits are used. Nonrandom samples selected by an overzealous or improperly trained employee can result in uniform results. Samples tested on a malfunctioning instrument (such as a sticking scale, electronically insensitive meter, worn vibrator cam, dusty colorimeter lens) or subjected to chemical or biological tests using outdated reagents may also show abnormally uniform quality. Range charts are particularly sensitive to conditions leading to abnormal uniformity.

## Recurring Cycles

Charts which show cycles may be related to activities which occur at repeated times (days, shifts, hours, break periods) when changes in the process or the environment occur regularly (Figure 9-5). The changes may be in workers, supervisors, inspectors, scheduled maintenance, ambient temperature or humidity, lighting, or production equipment rotation. Nonuniform cycles may be traced to occasional temporary equipment stoppages for adjustment or repair. These may produce new quality levels when restarting. Cycles may occur occasionally on range charts from these causes, but the *X*-bar and *p* charts are more likely to uncover cyclical quality problems.

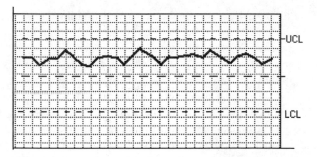

**Figure 9-4.** Abnormal uniformity.

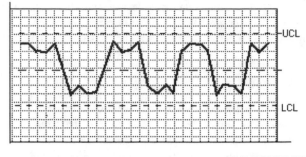

**Figure 9-5.** Recurring cycles.

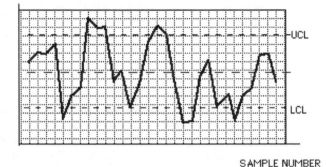

**Figure 9-6.** Excessive variability.

## Excessive Variability

When found on the chart for averages, excessive variability most frequently is caused by overadjustment of a process (Figure 9-6). When an operator adjusts the pH downwards simply because it has been above the average for a few readings, the chances are that the pH will shortly drift below the lower control limit. Blending production from two or more lines may drive the range chart out of limits. (Drawing samples of pouches for length measurements when produced on blended lines, where one line is at the minimum length and the other is at the maximum, would be an example.) Malfunctioning equipment and intermittent use of variable raw material (such as feeding into the line raw material which is of variable quality or nearly out of specification) may produce peaks and valleys on the range chart and the $p$-chart. A malfunctioning inspection device may also produce apparent wide variations.

## Wild Outliers

When an occasional control chart plot lies far outside the limits with no apparent cause, it is generally the result of a quality testing error (Figure 9-7). The list of errors include: mathematical calculation, plotting, sample size, sampling procedure, testing, sample contamination, etc. Problems associated with the production line are rare, and generally non-repetitive: an interruption of power or gas supply, a sticking solenoid or chattering relay, or possibly an omitted operation. Samples removed from the line and inadvertently returned to the wrong part of the line can also produce an otherwise unexplained outlier.

## High/Low

$X$-bar and $R$ charts will often show extreme values with few in between when two sources of raw materials are used, one of which is at the upper end of the quality specification, and the other at the lower end (Figure 9-8). As the variable supply reaches the production line, the process swings rapidly from one end to the other,

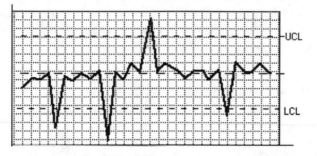

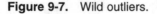

SAMPLE NUMBER

**Figure 9-7.**   Wild outliers.

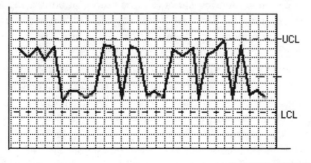

SAMPLE NUMBER

**Figure 9-8.**   High/low.

rarely pausing in the center. Since product defects would tend to average out, $p$-charts might not be affected by two source materials. However, nonrandom sample selection could show high/low $p$-chart values. The blending of product from two dissimilar lines or from two operators with varying techniques may be responsible for this unusual chart. The output of "twin-pack" packaging machines may produce units which are within the control limits, but at opposite ends. Depending on which side of the machine the sample is taken from, variables such as gas content, fill weight, package dimensions, dust content, seal strength and registration may appear at either end of the control chart, but rarely in the middle.

The above explanations for patterns in control charts are necessarily incomplete. Considerable detective work is required for explanations for most out-of-limits data. Most of the answers are found on the production floor (or at the raw materials supplier plant), and as the relationship between quality control personnel and operating personnel grows closer, the answers and solutions to production quality problems are found more readily. Sometimes the causes are strange and require a great deal of ingenuity to uncover: vibrations from the freight elevator upsetting the line checkweighers, burnt paper ashes strewn over the oven contents during the Kosher blessing, static buildup during thunderstorms causing blocking of plastic films on sealers, conscientious line inspectors who want the record to look good for the company by eliminating out-of-limits data, afternoon wind impinging on air-intake filters and upsetting airflows in dehydrators.

## USING THE CONTROL CHART AS A QUALITY MANAGEMENT TOOL

What does one do with the charts at the end of the shift—or day, or longer? Some benefit can be found in posting the charts on the production floor as they are prepared so that the operators can see how well they are accomplishing their tasks. A continuing interest is unlikely unless some kind of attention-getting technique is included: perhaps posting the name of an outstanding operator for the week; or preparing a score card comparing shifts (can be risky), or comparing the current month with the same month last year, or displaying a "thermometer of accomplishment." All of these ideas tend to lose effectiveness because of overexposure. Some are difficult to turn off after they have served their purpose, since a few employees remain attached to them. These plans should be considered carefully before obtaining approval of management since competitions can cause jealousy, animosity, or even dishonesty if improperly administered. However, the proper use of a carefully planned motivational quality tool is recommended if it is used infrequently and has a worthwhile message.

At one time or another, copies of control charts should be sent to everyone in the company who is even remotely concerned with product quality. Subsequent copies should not follow unless requested. If few requests are made, a way must be discovered to make the charts more colorful, more informative, and more interesting; then they should be sent out again. Notes and comments should be included showing solved problems, quality improvements, productivity gains, and cost reductions. Most important, credit should be shared (deserved or not) with production, engineering, accounting, purchasing and marketing departments whenever even remotely possible. Suggestions should be solicited.

For those few managers who do ask for daily charts, their interest should be rewarded with brief summaries, weekly and monthly reports and short notes about other industrys' quality problems and solutions. Charts should be informative, and they should be "spiced" and embellished with cartoons, or photographs of quality improvements, so that the recipients actually look forward to the next issue. As an additional courtesy, provide managers with informal tours of the production floor where they can watch a chart being constructed.

We have left the most important point for last: how does one actually use these charts to control quality effectively; that is, what physical steps do quality control personnel take to make the charts work? When an observation exceeds the control limits, the line should immediately be stopped, if permitted by management, and all production since the previous observation must be set aside and clearly labeled with a "HOLD" or "NOT TO BE SHIPPED" or "REWORK" tag. Once the cause for the out-of-limits condition has been identified and corrected, the line may be permitted to start up again, and paperwork is completed for isolating the defective material and providing instructions for its disposition.

The alert reader might have noticed "if permitted by management" in the paragraph above. For many, this may be difficult to understand, but there are companies which permit only production supervisors to stop production lines. It is still possible to operate a quality control system effectively under these conditions, for it is unlikely that any management would also forbid the quality control department from isolating and holding any defect, whether it be raw material or finished product. Consequently, the out-of-limits material can be isolated at the end of the production line, and then held for corrective action. This can be continued by quality control until the production department realizes that all production is headed for the rework or scrap area, and will continue to be until the line is stopped and the offending condition corrected. Needless to say, one or two such experiences will lead to a revision in management philosophy, and quality control personnel will eventually be authorized to stop defective production.

This procedure of eliminating out-of-limits materials as they are uncovered in the process by active use of control charts and immediate corrective action is at the heart of quality control, and is the means to reducing defect production, reducing scrap, increasing productivity, and—perhaps most important—reducing costs. Some quality leaders seem to wish to label this procedure as a newly discovered principle of quality control, and insist on providing it with a label: "statistical process control," or "zero defects production," or "process quality assurance," or other. So be it! Regardless of the title given to this procedure, it works well, and has for many years. Properly administered, it can work successfully for any food manufacturing process.

# 10      Sensory Testing

Many think of sensory testing primarily as a research tool. It is widely used in product development, to match products with competition, to determine consumer acceptance, to improve products, and to conduct shelf-life studies (storage stability). But there is little doubt that sensory testing and evaluations are used far more routinely for quality control functions such as raw material control, evaluation of new sources for raw material, process control (at every step of the process), evaluation of process change, product grading, and multi-plant quality coordination.

Ask any healthy person if he or she can compare food flavors. Chances are that the response will be "yes." The next question is, "Why are you qualified as a flavor judge?" The answer will be something like "because I have been tasting foods every day for my entire life." The only reasonable conclusion to be drawn from these answers is that this individual is capable of judging food flavors which will be personally pleasing or displeasing—but this does not necessarily speak for anyone else. Many questions remain unanswered, and we will consider some of them here.

For example, sensory evaluations of food are concerned with much more than flavor. Texture, color, size, density, aroma, and many other factors have sensory values which are often of vital importance to the quality acceptability of foods. The threshold level at which food characteristics can be detected, or the levels at which they can be differentiated, need to be understood. Then there are preference values dictated by cultures, ethnic practices, age, geographical area habits, etc. So we are concerned not merely with one person's evaluation of flavor when we refer to sensory testing; we have a vast field of sensory perceptions which require sorting out. Fortunately, a great deal of research has been devoted to this field, and sensory testing programs can be established at an almost infinite number of degrees of complexity.

The scientifically trained person finds it difficult to understand why a clear preference for one food item over another does not lead to instant success in the marketplace. The significance of preference values such as those listed above is clearly demonstrated by a test conducted many years ago by the U.S. Army's Subsistence Research and Development Laboratory during investigations of foods for the military. A consumer panel of nearly 300 subjects nearly unanimously rated a "new flavor" ice cream as "excellent." When the same golden-colored ice cream was presented to the same group three weeks later, but labeled as "carrot-flavored ice cream," the ratings were "unacceptable" or worse. The lowest rating was "revolting." The art of marketing may start with the logical findings of a sensory panel, but presenting a winning product to the market-place requires techniques beyond the scope of quality control, and will not be discussed here.

In 1975, the Institute of Food Technologists agreed on a definition of Sensory Evaluation: "a scientific discipline used to evoke, measure, analyze, and interpret reactions to those characteristics of foods and materials as they are perceived by the senses of sight, smell, taste, touch and hearing."

## THE SENSES

### Taste

By far, this is the most commonly evaluated characteristic of foods. Taste can be classified into four basic flavors: sweet, salt, sour, and bitter. (Note that the thousands of characteristics which we loosely call "flavors," such as onion, strawberry, mint, butter, etc. are better described as "odors." If one is blindfolded and fed a small piece of onion while squeezing the nose shut, it will most likely be identified as apple. The characteristic onion "flavor" is actually onion *odor*.)

### Odor

There are thousands of odors associated with foods. To date, a simple classification of odors acceptable to the food industry has not been devised. In a paper delivered to the 1982 Institute of Food Technologists Annual Meeting by Boelens and Haring of Naarden International (and others), the following 30 odor classifications were utilized to compare odors of a single group of products:

| | | | | |
|---|---|---|---|---|
| Fresh | Green | Sourish | Tart | Citrusy |
| Watery | Metallic | Floral | Fatty | Aldehyde |
| Vegetable | Lavender | Coniferous | Minty | Medicinal |
| Fruity | Honey | Buttery | Animal | Erogenic |
| Sweet | Aromatic | Anisic | Spicy | Powdery |
| Dusty | Earthy | Smoky | Woody | Balsamic |

Each of these classifications was described by standardized chemicals, natural flavorings, oils, or other substances, and may not be universally accepted as descriptive. For example, Minty was defined as "like peppermint oil"; buttery as "diacetyl"; aromatic as "vanillin." For purposes of the study however, these descriptors proved satisfactory.

## Combinations and Interactions of Taste and Odor

Here we find an infinite number of possibilities. No attempt has yet been made to categorize them for all applications, but simple definitions can be used successfully.

## Heat, Cold, Pain, Touch, Sight

At first, these terms might sound far-fetched, but it is difficult to find an expression more easily understood for the sensation of hot pepper than "heat." Perhaps there are some wintergreen mixtures which are "colder" than others. "Touch" might be interpreted as "texture" or "greasiness," or "sound." Yes, sound can play a significant part in the perceived quality of some products: the crackle of cold cereal in milk, the snap of chewing nuts, the hiss of carbonated drinks. Sight is so critical to flavor perceptions that some tests may be conducted blindfolded, and many tasting booths are equipped with special lighting to accentuate or deaden color effects. When meat is presented as an appetizing steak surrounded by its own juices, and with barbecue grating marks on the surface, the flavor perception is totally different from the same meat presented as a ground, dull-brown, shapeless mass.

The physiological aspects and the nomenclature of sensory perception is obviously quite complicated. The above examples are included here to show how the simple "yes" answer to whether or not one can qualify as a flavor judge is not the way to select a sensory panel. For a more detailed examination of stimuli and the physiological background to sensory evaluation, see "Basic Principles of Sensory Evaluation," American Society for Testing and Materials Special Publication No. 433.

## SENSORY TESTING METHODS

Selection of a sensory test method is dependent first on the type of information required, and second on the precision required of the results.

Some of the common questions to be answered are:

How do you like this product?
How do you like the color of this product?
Are these two products the same?
Are all of these samples identical?
Arrange these products in a column with the strongest at the top and the weakest at the bottom, with intermediate ones in order of their strength. Arrange these products in a column with the one you like the most at the top, decreasing until the one you dislike the most is at the bottom. Identify the one you neither like nor dislike.

Which samples have vanilla flavor?
Which two of these three samples are alike? Are they better flavored than the
     single sample?

Once the question to be answered is determined, the choice of test will usually
be found among the following list of methods.

**Monadic.**   Evaluate a single sample, based on past experience only. Useful
when an experimental product contains a major departure from similar products
on the market. An example might be lime-flavored apple juice, or hamburger
containing tiny julienne potatoes.

**Paired Comparison.**   Compare sample A with somewhat similar sample B.
Describe specific perceived differences. The results of the test are analyzed using
the two-tailed test, since either sample could be preferred. In contrast, the asser-
tion that one mean is greater or lesser than another (more salt, less sugar, stronger
flavor, etc.) requires a one-tailed test.

**Threshold.**   Identify which of several samples (usually five or more) do not
contain a given characteristic, and which do. The samples are prepared by adding
some characteristic to a series of "blanks" in small incremental quantities.

**Dilution.**   To determine the lowest concentration at which characteristics
can be detected. Differs from the threshold test by use of complete product at different strengths, whereas threshold tests generally are concerned with a single
component in a bland carrier.

**Triangle.**   Two samples are identical, and one is different. Identify the pair.
Is the pair preferred for some designated characteristic?

**Double Triangle.**   There are two sets of triangles, not necessarily with the
same pair in each. The pairs and their preferences are to be identified.

**Duo–Trio.**   Two samples are identical, and one is different. One of the
identical samples is presented as a "control." The tester is asked to pick the
unidentified sample which differs from the control.

**Rating Scale.**   Apply a rating to each of a coded group of samples. The scale
may be a line with "good" at one end, and "poor" at the other; adjectives such as
"poor," "fair," "OK," "good," "excellent," numerical scales representing successive levels of quality; or other scalar devices. This is also referred to as "optical
intensity scale." Results may be analyzed using Analysis of Variance.

**Magnitude Estimation.**   Similar to rating scale, except that here the tester is
asked to apply quantifying scales to some characteristic of a series of samples.

For example, the strength of some component might be assigned values from 0 to 100 by the panelist, the values being of any perceived interval, not necessarily uniform.

**Ranking.**  The tester is asked to place samples in order from "most" to "least," or "best" to "worst." Differs from Rating Scale in that no suitable scale may be available. Works best for preliminary screening and where the differences are fairly obvious, requiring little time for evaluation.

**Attribute Analysis.**  A specifically trained tester is presented with a single sample and is asked to list all of the attributes which define the quality of the sample. Example (a candy bar): chocolate flavor, sweetness, size of nuts, darkness of color, crunchiness, creaminess, grit, chocolate aroma, sour aroma, bitterness, aftertaste. Results from 6 to 12 panelists may be analyzed for significance by using Analysis of Variance.

**Hedonic Scale.**  Similar to Rating Scale, except the criterion here is for the extent to which the samples are liked. No attempt is made to direct the responses or to define the meaning of scale categories. The scale is presented in nine categories: like extremely, like very much, like moderately, like slightly, neither like nor dislike, dislike slightly, dislike moderately, dislike very much, dislike extremely. Where less experienced panelists are used, the "moderate" categories are eliminated. Where a panel consists of children, the following descriptors are suggested: love it, like it a lot, like it a little, it's just so-so, dislike a lot, hate it. Hedonic scale testing is most useful when conducted with 40 or 50 non-trained office personnel. If conducted outside the company, over 100 consumers should make up the panel. Results may be analyzed using Analysis of Variance.

**Flavor Profile.**  (Also adaptable for texture or other characteristics.) Five characteristics are identified by the subject: character, intensity, order of appearance, aftertaste, and fullness. "Order of appearance" refers to the time sequence at which the various aroma or flavor components appear. "Fullness" refers to the overall impressions of blending of quality components and the appropriateness of the factors to the product. This method is not subject to statistical analysis. After panelists independently evaluate one to three samples, they then report to a panel leader in open discussion.

**Texture Profile.**  Three classes of characteristics are identified and described: initial perception on first bite (hardness, viscosity, fracturability), masticatory perception during chewing (gumminess, chewiness, adhesiveness), and residual perception during mastication (rate of breakdown, type of breakdown, moisture absorption, and mouth coating). Statistical analysis is not appropriate to this method. Individual judgments are discussed to arrive at an average rating for each of the three classes.

**Routine Testing.**    When a large number of production samples is presented to a quality control technician for comparison to the quality standard, the usual procedure is to taste each rapidly, referring to the standard only when there is some doubt about the characteristics of an individual sample. Generally the products are smelled, tasted and chewed (if applicable), but not swallowed. If there are lingering characteristics, a rinse is recommended between samples. In the event of a suspect sensory characteristic, a second opinion is recommended. When the number of flavor variables is relatively small (such as lemonade, coffee, flavored pudding) as many as 30 samples can be evaluated at one time. When the number of flavor or texture variables is large (beef stew, mixed vegetables, fruit pie), it is unlikely that even an experienced technician can handle more than six samples at a time.

Many questions arise when looking at this formidable list of available tests. There are no universal answers, but guidelines such as the following might be considered.

**Fatigue and Frequency of Testing.**    In-house panels consist of office, plant and laboratory personnel who have responsibilities, interests and time constraints which may tend to pressure them to rush through the sensory tests. This obviously will affect their reliability. On the other hand, some subjects enjoy the break from the monotony of their routine. With the obvious exception of those whose duties require continuous sensory testing, such as quality control laboratory inspectors and technicians, it is wise to restrict the number of tests in which a subject participates to no more than one per day, and preferably only one every other week. At a single sitting, the product will determine the limitations as much as the testers' abilities. Weak- or mild-flavored products can be presented in as many as three simple tests without fatigue; but strong or harsh flavors can dull the senses very quickly, suggesting that a single test of, say, spiced olives will end the day's testing for any panelist. Alcoholic beverages present very obvious limitations to the amount of product which can be consumed before the test results become suspect.

**Performance Grading.**    Many panelists will hand in their score sheets with the question, "did I get them right?" For some tests there is no right or wrong, but where a participant selects a pair from a triangle test, or grades five aged products for staleness, there is no harm in disclosing the accuracy of the reponses. In fact, there are some advantages. There is a certain amount of pride in knowing that one has the ability to discern the correct answers. This might also motivate the tester to continue to concentrate on future tests, as well as look forward to participating again. Knowing that there are correct responses may help some people avoid guessing at the answers. For those who missed the correct answers, perhaps they will try harder on the next test.

**Blind versus Open Testing.**    Exploratory tests may be performed open. That is, the panelists may be told which sample is the standard, what the other samples are, and the purpose of the test. It instills a measure of confidence, and speeds up the testing time. In blind testing, where unbiased data is essential, some information can and perhaps should be made available to the tasters. For example, if they are presented with five oranges, and asked to evaluate them "for appearance," without further explanation, some panelists might look for size differences. The test might have been designed to evaluate the effectiveness of various levels of controlled atmosphere storage, or different temperature storage, or irradiation, and the testers might have totally missed rating the samples on a basis of color, absence of mold, shriveled skin, or odor—all of which could have been the major goal of the test.

**Order of Presentation.**   When samples are presented one at a time for evaluation, the order of presentation becomes significant. Each sample is subconciously related to decisions about the preceding sample. This "contrast effect" varies between subjects, and it is generally agreed that neither experience nor training will eliminate it. It can be reduced by changing the order of presentation to each subject, and if possible, changing the order on repeat testing as well. Even when samples are presented simultaneously as in triangle tests, "position error" can affect results. For that matter, the first sample selected in a triangle test will bias evaluation of the next two.

**Temperature Control.**   For difference testing, most products should be presented at room temperature. For preference testing, they should be at the temperature normally served. If unsure of the effect of temperature, tests may be repeated at various temperatures and the results compared.

**Reference Standards.**   If the company's product A is to be rated against similar competitive products X, Y, and Z, should A be identified as "our product" for comparison purposes? Generally, the answer is no. Most employees on in-house testing panels tend to have a certain loyalty to the company and tend to rate "our product" as superior whenever it is identified as such. After the test is completed, however, there probably is no harm in informing the tester. Where the development of off-flavors in storage are of concern, presenting the testers with an identified fresh standard is desirable. It allows the tester to concentrate on the possible degradation of product quality, rather than expending unnecessary time and effort in trying to establish which sample is the standard.

The techniques for preparing a true standard for a taste test are often difficult to establish. When comparing the browning reaction in sliced apples, the standard selection is fairly simple—nearly any fresh apple slice from the same variety can be used. But in storage tests where staleness development is the key characteristic to be tested, it may be extremely difficult to present an identical (but fresh) sample as a standard. Samples of potato chips selected off the production line cannot be used as a standard for test samples which have been stored under various gas mixtures for various lengths of time at various temperatures, since current production may represent potatoes of a different type with different flavor, texture and color characteristics. The same would be true in testing coffee for staleness development during storage: the current production might consist of a different blend with significant flavor differences. In general, the best standard would be one selected at the same time the storage test was prepared, hermetically sealed, and held frozen at $-15°F$ until required for the test. Obviously, this cannot work for all products because of physical changes at freezing temperatures. Where no satisfactory standard can be provided, it is best to offer none, rather than to provide a sample with instructions to ignore the mushy texture, the excess fluid, or the unusual appearance. This would lead to questionable test evaluations.

Where it can be demonstrated that the reference standard does change with time, it may be possible to provide a continuous series of updated reference samples. Let us assume as an example that a reference sample, stored frozen in a hermetically sealed container, changes imperceptively for the first four months, but has been proven to show a gradual change after that period. At the end of three months, an updated sample is prepared by selecting a series of samples directly from the production line, and comparing them to the thawed standard by use of triangle tests. When a new sample has been found to be indistinguishable from the thawed standard, it may then replace the present standard (which should be discarded).

In addition to the stored product standard and the fresh product standard, there are three other standards in general use:

1. *Written standard.*  A documented flavor profile may suffice as a flavor standard for relatively uncomplicated generic types of products. It may be made somewhat more definitive by including charts of chromatographic or spectrophotometric peaks. Although this may function satisfactorily for awhile, its effectiveness becomes questionable as sensory judges move on and are replaced by others who are less familiar with the sensory objectives of the original product defined by such a document.
2. *Photographic standard.*  These standards are quite effective for defining physical characteristics (colors, blemishes, defects, size, proportions, etc.) provided the photographs are suitably protected from changes.
3. *Verbal or mental.*  Some flavors and textures can only be described and taught to others through experience. Examples of these: wood, tallow, hide, wine, cardboard, nut-like, straw, acidic, gummy, etc. Training most healthy people to recognize these characteristics is not particularly difficult. The main problem is defining the extent or strength of these characteristics in a given sample. A secondary difficulty is training panel members to rate uniformly the degree or intensity of these characteristics. By using larger panels for these inexact measurements, the average reaction becomes more meaningful.

## TYPES OF PANELS

Experts in the sensory evaluation field tend to treat the subject as if it were directed to large companies with vast financial resources. This leaves the small company quality control supervisor in a dilemma: his sensory panel consists of himself,

Mary Smith in purchasing, and the plant manager. Occasionally the company owner will sit in with comments from his wife regarding the desirable flavor characteristics for the product in question. How can he provide separate panels for the many sensory applications which arise? He cannot. But armed with the information regarding the functions of specialized panels, he certainly should be able to provide at least the basics of sensory quality control to his company. And in many instances this might be all that is needed. Where specialized assistance is really required, it is available in outside consulting organizations, or even in noncompetitive food processing companies.

## Screening Panels

These are generally found in the research department where an informal decision is reached to continue development of an idea or to try again with a different direction. The panels should be staffed with members competent in the area under consideration, but since fine tuning is not required at this point, reasonable experience with the product class is generally satisfactory. Other screening functions may be carried out in the quality control laboratory for a variety of needs. In sorting rejected material, there is usually only one major factor to be evaluated (the one which caused the rejection) and it is customary to use semiskilled personnel to "screen out" the offending material. Another quality control use of the screening function is to discover unwanted sensory characteristics in the process stream at each critical step. This is accomplished by use of a highly trained individual (preferably with a backup trainee) who samples the process every 15 minutes (or other time interval, or batch) to identify any undesirable sensory characteristic at the earliest step possible and at all subsequent steps. This would include, but is not limited to, unusual color, texture, flavor, mold, foreign materials, etc.

## Expert Panels

These are groups of specially trained individuals whose function is to make fine sensory distinctions between products, or evaluations of single products. In addition to the research department, experts are often utilized routinely by raw material purchasing departments to evaluate both offerings and receipts. Frequently they will also check the finished product to determine if their selection of raw material withstood processing without showing undesirable characteristics. It is relatively unusual for a quality control panel to maintain a special expert panel. In the rare event that the routine quality control sensory panel finds an unidentifiable characteristic in the product, it can be sent to the expert panel elsewhere in the company for further analysis. If none exists, it can be sent out to an independent laboratory (or to the material supplier, in some cases).

## In-House Panels

The term "in-house" merely identifies the panel as consisting of company employees who may or not be specially trained, and who may or may not regularly participate in sensory testing. Frequently, an "in-house" panel is one which evaluates specific products (new developments, critical complaints, competitive comparisons) as an inexpensive way to determine if outside technical assistance is required. An in-house panel may occasionally be considered as a miniature consumer test to provide rapid and tentative answers to broad-scale consumer tests being conducted by outside marketing organizations at the same time.

## Informal Panels

In addition to the research and control panels which use relatively rigorous procedures in conducting their functions, companies may also have a number of informal groups who meet periodically to "see how things are going." An example might be the monthly sales or marketing meeting at which samples of the company's major products are compared to those of competition. The procedures are usually casual, with prepared samples being offered to each participant for comment. Results may or may not be formally reported to other departments. Under these conditions, major product differences uncovered are liable to be viewed as catastrophes, and either research or quality control departments quickly become involved. The same type of informal competitive sampling may be performed by each concerned department: research, quality control, purchasing, and others. Since these tests are similar to the actual treatment the product might receive by the consumer, occasionally some surprising characteristics are revealed. For example, a soup mix pouch may be found to be difficult to open because the tear strip is not clearly identified. Or the countersink may be excessive so that a hand can opener will not work at all on a vegetable can. Or reclosing a partially used frozen food package might be difficult. Or reheating a gravy product might cause curdling. Or the tea bag string tears loose in hot water.

## Acceptance Panel

We have here a rather loosely-defined term which describes the end result of any number of sensory panels: acceptance or rejection of a sample. In quality control, an acceptance panel is one which can be used in audit functions, in daily routine sampling, or sorting rejected material. In a manner of speaking however, screening panels and in-house panels are also acceptance panels.

## Interplant Panels

Where a company has a multiplant operation, some companies might wish to dignify the function of those panelists in quality control who are responsible for maintaining identical product quality characteristics from all plants with some special title: interplant panelists. At best, this might be a part-time duty of those in quality control responsible for routine sensory testing. This is not to belittle the function, since it is of prime importance, and requires special quality control efforts. The greatest contributing factors to multiplant or even international quality control is the development and maintenance of strictly enforced standard methodology, including a communication network. Interplant panels are an integral part of the methodology.

## Outside Panels

Assistance outside of the company exists in many forms. Marketing and advertising companies frequently maintain panel facilities for advertising copy, label design, flavor testing, and focus groups. A word of explanation about the latter. Focus groups are panels of varying sizes selected or maintained by market research organizations to explore existing or new concepts concerning products. The groups are carefully selected to represent the potential consumer of the product, and the evaluations conducted by the panels are carefully monitored and directed by a professional group leader. In addition to marketing and advertising organizations, outside assistance is available from consultants and commercial testing laboratories. Although the client company has the final word on how the studies are to be conducted and analyzed, it is wise to allow these outside facilities to provide the expertise which they have developed over the years.

## SELECTION AND TRAINING

In order to obtain meaningful and uniform evaluations from a panel, the members should be selected and trained with care. Most uninformed prospective panelists consider sensory panel assignment as a possible fun-and-games diversion from the routine of their regular jobs within the company. Although this attitude must be modified, there is little to be gained by treating the assignment as another chore. Participating in panels can be a pleasant experience, but must not be allowed to be treated casually. Occasional light humor in the panel room may be allowed, but by no means encouraged.

As a guide, an in-house panel of eight to ten members can be selected from a group of 15 to 20 volunteer employees. After two or three training sessions, the six most reliable and consistent candidates should be identified. Others might be held in reserve as needed. The training sessions should include such techniques in which the panelists follow a well-defined procedure:

1. Rinse mouth with room-temperature water.
2. Sip (or nibble) a sample.
3. Chew (if required) and roll over front, sides, rear of tongue.
4. Spit sample into cup provided (do not swallow).
5. Record observations as instructed.
6. Rinse mouth until flavor sensation has disappeared.
7. Repeat for next sample.

It is suggested that the tests used for training start with a ranking test in which a company's product is presented in four levels of either flavor strength, sweetness, saltiness, sourness, firmness, or other sensory characteristic appropriate to the product. The differences between the samples should be substantial for the early training sessions. The panel members selected in the screening process should then receive training in the other types of test methods, starting with triangle testing. It is generally preferable to conduct the training program on consecutive days for at least a week. During these training periods, terminology can be gradually introduced as examples appear. Frequently, panelists will suggest the use of descriptive terminology other than that commonly used. This should be encouraged, and adopted if it clarifies the perception. Silence should be stressed during the training period; subconscious grunts, groans or snickers may affect the judgment of other panelists.

The ability to duplicate results is as important as sensitivity to sensory stimuli, and a series of replicate tests must be conducted using the selected panel members. The record keeping becomes voluminous at this stage, and the use of a computer and some sort of grading system should be devised.

## PREFERENCE RATING

Name_____ Date_____

Booth_____

You will receive a reference and two samples of product. Taste the reference, but do not rate it. The Expert Panel ratings are listed under "REFERENCE" and shall be used as a guide for rating the unknowns. Taste and evaluate the other samples in the order given.

|  | RATINGS |  |  |
|---|---|---|---|
| APPEARANCE | REFERENCE | CODE__ | CODE__ |

Color-Yellowness

├----┼----┼----┼----┼----┼----┼----┤                3.0        _____        _____

  extremely     moderately    slightly     white,
   yellow        yellow      yellow   not yellow
    7     6     5     4     3    2    1

Hardness (Firmness)- Force required for initial bite

├----┼----┼----┼----┼----┼----┼----┤                3.1        _____        _____

  extremely     moderately    slightly     not
    hard         hard       hard     hard
    7   6     5     4     3    2    1

Crispness (Fracturability/Brittleness)

├----┼----┼----┼----┼----┼----┼----┤                3.6        _____        _____

  extremely     moderately   slightly     not
    crisp        crisp      crisp     crisp
    7   6     5     4     3    2    1

Flavor (Sweetness)

├----┼----┼----┼----┼----┼----┼----┤                3.7        _____        _____

  extremely     moderately   slightly     not
    sweet       sweet      sweet    sweet
    7   6     5     4     3    2    1

COMMENTS:_____

_____

**Figure 10-1.** Preference rating form.

Finally, some system of retraining should be considered. If tests are run frequently, panelists might drift away from standard procedures. An occasional refresher course in the form of a rigged test in which participants openly discuss their findings might reveal developing weaknesses in the system which need to be corrected.

The form shown in Figure 10-1 is a model for use in sensory panel testing for preference.

# 11　Net Content Control

Controlling net content of food packages has two goals: first, the food processing company must be able to evaluate performance to assure that production is within the governmental limits for net package content; and second, optimize (which really means "minimize") overfill.

Federal regulations for net content of foods are covered by the U.S. Department of Health and Human Services, Food and Drug Administration, and by the U.S. Department of Agriculture Food Safety and Inspection Service. Alcoholic beverages are controlled by the U.S. Department of the Treasury Bureau of Alcohol, Tobacco and Firearms. Nonfood consumer commodity net content regulations are covered by the Federal Trade Commission, and pesticides are under the jurisdiction of the Environmental Protection Agency.

The regulations dealing with net content control for food are less specific than one would desire. For example, the Code of Federal Regulations 21, Food and Drugs, Section 101.105 requires that "the declaration of net quantity of contents shall express an accurate statement of the quantity of contents of the package. Reasonable variations caused by loss or gain of moisture during the course of good distribution practice or by unavoidable deviations in good manufacturing practice will be recognized. Variations from stated quantity of contents shall not be unreasonably large."

The U.S. Department of Agriculture (CFR 7) Meat and Poultry Act (Section 317.2) requires that the label "shall express an accurate statement of the quantity of the contents of the container exclusive of wrappers and packing substances."

Neither of these regulations clarifies the word "accurate." Does it mean "close to," does it mean "at least equal to," or does it mean "average"? Nor, for that matter, is the meaning of "unreasonably large" variations clarified. As a general practice, the American food industry overpacks, but according to the letter of the law, an extra $\frac{1}{4}$ ounce of product over the label claim does not comply with the requirement that the label "express an accurate statement of the contents." To further complicate the issue, each state has the power to enforce compliance with its particular weights and measures laws, usually through the county bureaus of weights and measures.

Individual company policies vary in their attempt to comply. Some packers with "Zero Defects Programs" may state that no product shall be underweight (although overweights might also be considered defects). Some managers might have 5% underweight/95% overweight as the goal, or 10/90, or 20/80, or even 50/50. Others may have a more scientific approach. It is also likely that there are companies who have no policy at all regarding underweight/overweight ratios.

Fortunately, there is an answer to this confusion. The U.S. Department of Commerce, National Bureau of Standards has issued *NBS Handbook 133* titled, "Checking the Net Contents of Packaged Goods" (January 2002). It replaces *NBS Handbook 67* "Checking Prepackaged Commodities," which served as the basis for most state weights and measures regulations. Gradually, the states have adopted *Handbook 133* as the official procedure for evaluating net contents. Most of the states have also adopted the portion of the National Conference on Weights and Measures *Uniform Laws and Regulations in the Areas of Legal Metrology and Engine Fuel Quality* (*NIST Handbook 130*, 2002 Edition) which is concerned with uniform weights and measures law. It requires that the average quantity of contents in the package of a particular lot shall at least equal the declared quantity, and no unreasonable shortage in any package shall be permitted. It also requires that variations from the declared quantity shall not be unreasonably large. Maximum allowable weight variations are defined in *Handbook 133*.

The National Institute of Standards and Technology has also published *Handbook 44* (2002) titled *Specifications Tolerances and Other Technical Requirements for Weighing and Measuring Devices*. All 52 states have adopted *Handbook 44*.

*NBS Handbook 133* separates sampling plans in two categories. The plans in Category A (See Table 11-1.) are used by weights and measures officials when the severity of the consequences for the packager or retailer of a lot not passing the test is relatively great. This means that if the official anticipates a possible regulatory action (court appearance), then Category A should be used.

Category B plans are used for meat and poultry products, and as an audit for screening purposes. In the event the audit reveals shortages, a federal or state agency might elect to follow up immediately with a Category A sampling. The plans resemble Military Standard 105D sampling plans for sizes over 50, but the decision criteria are different.

Maximum allowable variations (MAV) are listed in *Handbook 133* for packages labeled by weight, by volume, by count, and by area. The number of sample measurements exceeding the MAV may not exceed the MAV columns listed in the sampling plans for the appropriate category. The tables are fairly detailed, and a few of the entries are shown in Table 11-2 as examples of MAV for individual packages labeled by weight.

**Table 11-1. Sampling Plans for Category A and Category B**

| Inspection lot size | Sample size | Sample correction pkg factor | No. of minus errors allowed to exceed MAV | Initial tare sample size | |
|---|---|---|---|---|---|
| | | | | Glass | Other |
| Sampling plans for category A | | | | | |
| 2 | 2 | 8.984 | | | |
| 3 | 3 | 2.484 | | | |
| 4 | 4 | 1.591 | 0* | 2 | 2 |
| — | — | — | | | |
| — | — | — | | | |
| 12 to 250 | 12 | 0.635 | | | |
| 251 to 3200 | 24 | 0.422 | — | | 3 |
| Over 3200 | 48 | 0.2911 | 1 | | |

| | | Initial tare sample size | No. packages allowed to exceed MAVs |
|---|---|---|---|
| Sampling plans for category B (For use in USDA-inspected meat and poultry plants only) | | | |
| 250 or less | 10 | 2 | 0 |
| 251 and greater | 30 | 5 | 0 |

*Maximum allowable variation for individual packages
Applies only to underfills (shortages).

**Table 11-2. Maximum Allowable Variations (Meat and Poultry)**

| Less than 85 g or 3 oz | | 10% of labeled quantity |
|---|---|---|
| Homogeneous fluid when filled | All other products | Lower limits for individual weights |
| 85 g or more to 453 g 3 oz or more to 16 oz | | 7.1 g 0.016 lb (0.25 oz) |
| More than 453 g More than 16 oz | 85 g or more to 198 g 3 oz to 7 oz | 14.2 g 0.031 lb (0.5 oz) |
| | More than 198 g to 1.36 kg 7 oz to 48 oz | 28.3 g 0.062 lb (1 oz) |
| | More than 1.36 kg to 4.53 kg More than 48 oz to 160 oz | 42.5 g 0.094 lb (1.5 oz) |
| | More than 4.53 kg More than 160 oz | 1% of labeled quantity |

For Category B, the sample average must be equal to or greater than the label weight. For Category A, the sample average must be equal to or greater than the label weight minus the adjustment for sampling error. Sampling error for *Handbook 133* is defined as two standard deviations from the mean. An additional allowance for moisture is calculated if a regulatory agency has assigned a value, and this allowance is subtracted from the "nominal gross weight."

Later additions of *Handbook 133* have slightly revised this statement to read: "Allowance must be made for moisture loss during the course of good distribution when State or Federal regulations provide for them."

To prepare an effective program, governmental inspectors are encouraged to sample for net weights at three locations:

1. *Retail.* Testing at this point does not necessarily represent the net weights of large lots. Follow up inspections of a single lot number or label is usually required.
2. *Warehouse.* The lot consists of packages with identical labels and with the same manufacturer's lot symbol or code. It is a reliable method of evaluating a single lot or code.
3. *Point of pack.* Since a large number of packages of a single product are available for testing, an inspector can verify that the packer is following good packaging practices.

Samples from the lot size selected are removed from the lot by use of random numbers selected from over 30 pages of tables supplied with the handbook. Samples for tare measurements are also selected by use of random numbers. The number of tare samples for Category A varies as shown in Table 11-1; for Category B, 2 tares are required for all lot sizes. Special commodities require special methods of handling, measuring, or calculating. In the case of coffee, for example, the variation of the tare weight of a one-pound can may be so excessive relative to the product weight that special techniques are required. Additionally, since coffee is vacuum packed after filling, the weight of the air removed must be added to the test weight. Other special commodities are frozen foods, glazed seafood, milk, mayonnaise, and salad dressing.

To summarize the steps recommended by *Handbook 133*:

1. Identify and define the inspection lot.
2. Select the sampling plan.
3. Select the random sample.
4. Measure net contents in the sample.
5. Evaluate compliance with MAV requirements.
6. Evaluate compliance with the average requirement.

The two requirements for compliance:

1. The number of underweights may not exceed the MAV.
2. The sample average (corrected) must be greater than the label claim.

## EVALUATION OF NET CONTENT PERFORMANCE

So far, we have looked at how the government can audit for conformance to net content. Can the same method be used by industry as a quality control line procedure? Not too effectively. The sampling plans for Category B are designed with a 50/50 risk of acceptance/failure for lots whose average is at the labeled weight. This 50% risk may be excessive for the packer who is producing lots complying with the regulation. Category A sampling plans are more discriminating (99% rather than 50%) but provide only after-the-fact control. By the time the required 125 samples from an hour's lot of 10,000 units have been checked, for example, the entire lot may have to be scrapped; whereas the use of statistical quality control charting procedures would highlight a problem for immediate correction.

On the other hand, it is possible, and in fact, recommended that available quality control data be statistically analyzed periodically to provide an objective basis for evaluating the costs of overfilling and the risks of noncompliance under current production conditions. Most companies have much of this data readily available. Performance data may be in the form of quality control charts or records either as daily sheets or summary reports, audit reports, plant performance studies, production volume records, and standard ingredient costs. Specific data such as the following have to be located or assembled:

- average net weight (or volume)
- net content variation
- number of units produced
- average cost per ounce of ingredients

Inspectors are allowed to utilize audit tests in order to speed the process of detecting possible net content violations; spot tests using non-random samples, smaller sample sizes, or use of the manufacturer's tare data. Although this might indicate the need for further investigation, audit tests and other shortcuts will not result in enforcement action.

## INTERPRETING NET CONTENT CONTROL

Regardless of the size of the company, or the number of products, the procedure for evaluation is essentially the same. If the company is large and there are many products, it would be very difficult to conduct a complete analysis at the outset. It would be far more advisable to start with one product at one plant, or perhaps a couple of products at two plants, so that comparison between plants could be made at the same time. The example which follows is based on unpublished notes from lectures by C. Kloos at the University of California, April 1988. In this example, we have four plants producing packages of two varieties of nuts in various sizes. The total number of variables to be compared is 7, a fairly manageable number. These are recorded in columns 1, 2, and 3 in Table 11-3.

**Table 11-3. Net Weight Evaluation**

| Col 1 | Col 2 | Col 3 | Col 4 | Col 5 | Col 6 | Col 7 | Col 8 | Col 9 | Col 10 |
|---|---|---|---|---|---|---|---|---|---|
| Product | Label weight (oz) | Plant state | Units produced (millions) | Cost ($/oz) | Avg. Wt. (oz) | Sigma $(R̄/d_2)$ | Overfill (Col 6 – Col 2) | Sigmas (Z) overfilled (Col 8/Col 7) | Cost ($M) (Col 4 × Col 5 × Col 8) |
| Walnut | 16 | CA | 420.1 | 0.03 | 16.05 | 0.30 | 0.05 | 0.17 | 630.2 |
|  | 32 | GA | 302.0 | 0.03 | 32.20 | 0.35 | 0.20 | 0.57 | 1812.0 |
|  | 64 | NY | 56.9 | 0.03 | 64.31 | 0.40 | 0.31 | 0.78 | 529.2 |
| Subtotal |  |  | 779.0 |  |  |  |  | 0.37 | 2971.4 |
| Pecan | 16 | CA | 650.0 | 0.04 | 16.03 | 0.25 | 0.03 | 0.12 | 780.0 |
|  | 32 | GA | 315.5 | 0.04 | 32.16 | 0.29 | 0.16 | 0.55 | 2019.2 |
|  | 64 | NY | 106.0 | 0.04 | 64.50 | 0.36 | 0.50 | 1.39 | 2120.0 |
|  | 128 | CA | 22.5 | 0.04 | 128.90 | 1.20 | 0.90 | 0.75 | 810.0 |
| Subtotal |  |  | 1094.0 |  |  |  |  | 0.38 | 5729.2 |
| Total |  |  | 1873.0 |  |  |  |  | 0.38 | 8700.6 |

We shall start by listing the total number of units produced over the past sea-son (or other period) in Column 4. These figures should be readily available from production records if they are not accumulated on quality control records. It may be more difficult to obtain the cost figures to be entered in Column 5. If the total costs, or labor costs, or overhead costs are difficult to obtain from the production or accounting departments, use the more readily obtainable cost of raw ingredi-ents (and label the column accordingly). If the results of this analysis should show that a saving of 0.5 ounces per can is possible, and if the raw ingredients cost $0.01 per can, then the conclusion would be that a $0.005 saving per can is attain-able. If labor, packaging materials, overhead, advertising, and other costs have also been included, there might be considerable discussion as to whether the savings also apply to these factors. Certainly, few would argue the fact that a savings of raw materials as a result of net weight savings is a real benefit. If the costs of manufacturing can be used, combine them with raw materials and enter this figure in Column 5.

Next, prepare Column 6 by extracting the net weights for the season (or period of the study) from quality control records. If running averages are maintained, this should be easy; if not, select data from production records for an arbitrary inter-val, such as every 10th day, and calculate the average. As a guide, take about 30 days worth of data. It may include "good" days and "bad" days, but they should not be culled out of the data because, almost by definition, the average net weight is made up of good, bad, and indifferent days. Column 7 data (standard devia-tions) may be most readily obtained from the Range Charts ($R$ bar/$d_2$). Although small differences in sigma may not have any appreciable effect on the outcome of this analysis, average sigmas should be used if they have been revised for chart use during the period of this study. By including minor quality control modifica-tions, this might eliminate distracting arguments during presentation of the study at its conclusion.

Calculating the overfill for Column 8 is a simple subtraction of Column 2 from Column 6. As an example, using the data for 32-ounce pecans packed in New York, the average weight is 32.16, which is 0.16 ounces over label claim.

Column 9 is the calculated standard deviation from the mean being overfilled, or $Z$. (You should recall that the value for $Z$ at the average of the distribution = 0.) This statistic is calculated by dividing the amount of overfill (Column 8) by the net-weight standard deviation (Column 7). It will be used later in determining the risks of being out of compliance. As an example, take the 16-ounce containers of walnuts produced in California. The formula for $Z$ is:

$$Z = (X \operatorname{bar} - \mu)/\sigma = (16.05 - 16)/0.30 = 0.17$$

This means that 16-ounce walnuts in the California plant were overfilled by 0.17 standard deviations. What we have accomplished here is the conversion of overfill in ounces to overfill in standard deviations. This conversion will permit us to calculate the average overfill for the entire company. That, in fact, is what the average figure 0.38 at the bottom of Column 9 signifies: the weighted average of overfill for all products is 0.38 standard deviations ($Z = 0.38$). As a matter of interest, most companies overfill by nearly 1.0 standard deviations.

At this point, it might be wise to note that these calculations assume that the net weights are all normally distributed. Although it may not be absolutely accurate to make this assumption, most distributions for filling containers are very nearly normal, with slight skewing on the low side for companies using automatic checkweighers, and slightly on the high side for those who do not. The slight aberrations normally observed for these distributions have only a modest effect on the outcome of the calculations.

The final step in the evaluation of the performance data is the calculation of the overfill costs listed in Column 10. The calculation consists of multiplication of the volume produced (Column 4) by the average amount of overfill (Column 8), times the cost-per-ounce (Column 5). For the period selected, the overfill costs for walnuts was $2971.40; for pecans, $5729.20—with a total overfill cost of $8700.60.

Having established the costs related to overfilling, we now proceed to the analysis of risk associated with the weight control. The first step is to calculate the percent of individual containers whose net weights are below the stated label weight and enter in Column 11. This figure may be found by defining $Z$ as the number of sigmas overfilled (see Column 9) and determining the area under the normal curve below $(-Z)$. As an example, the 16-ounce walnut containers produced in California have a $Z$ value of 0.17. By using a table for the areas under the normal curve, it is observed that 43% of the area is below $(-Z)$; that is, 43% of the containers for that particular product are below label weight. This might be more easily seen from Figure 11-1.

Continuing with the calculations for Column 11, the weighted average percent below label weight for all of the items listed amounts to 36%. (Note that the weighted average is obtained by multiplying each value in Column 11 by the corresponding number of units produced, as shown in Column 4. The total of these calculations is divided by the total number of units produced to obtain the weighted average.)

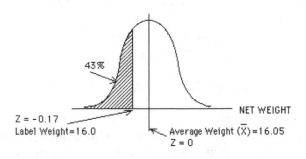

43%

$Z = -0.17$
Label Weight=16.0

NET WEIGHT

Average Weight ($\bar{X}$)=16.05
$Z = 0$

**Figure 11-1.**   Percent below label weight (16 ounces of walnuts, California).

**Table 11-4.   MAV for Two Varieties of nuts**

| Labeled weight (pounds) | MAV (decimal ounces) |
|---|---|
| 0.94–1.08 | 0.68 |
| 1.88–2.14 | 1.12 |
| 3.90–4.70 | 2.00 |
| 7.90–9.40 | 3.00 |

*Handbook 133* has defined unreasonable underfills with the establishment of values for maximum allowable variation (MAV) for various classes of product weights. Values for the four weights in our example have been selected from Handbook #133 and are listed in Table 11-4.

The minimum allowable weight is determined by subtracting the appropriate MAV from the label weight. For example, the 16-ounce package is between 0.94 and 1.08, and has a MAV of 0.68. By subtracting 0.68 from 16, the minimum value of 15.32 results. This is entered in Column 12 of Table 11-5.

Column 13 lists the percent of individual containers which is below the minimum allowable net weight. The calculations are similar to those used to determine Column 11, in which $Z$ determines the area under the normal curve.

$$Z = \frac{\text{Minimum allowable net weight (Col 12)} - \text{Average net weight (Col 6)}}{\sigma(\text{Col 7})}$$

As an example, the first entry in Column 13 is calculated for 16-ounce walnuts produced at the California plant. The individual containers could be as low as 15.32 ounces before being declared below minimum. The percentage of such containers may be found by first solving for $Z$ in the equation above.

$$Z = \frac{15.32 - 16.05}{0.30} = -2.43$$

The area under the normal curve is found in the $Z$ tables and is shown to be 0.7%. By continuing these calculations for all of the products, and taking their weighted averages, it is shown that the company has produced, on average, 0.23% of its containers below the allowable minimum.

The next calculations will determine the probability of failing to meet the compliance criteria under Plan A and Plan B of *Handbook 133*. Under Plan A, of the 30 samples selected, no more than one container may fall below the MAV, and the sample average must be no less than two standard deviations of the mean below the label weight. Under Plan B, of the 10 samples selected, no samples may be below the allowable minimum, and the sample average must be equal to or greater than the label declaration. A sample calculation for Plan A and for Plan B follows.

A defect is defined as a container whose net weight is below the allowable minimum. The probability of finding two or more defects in a sample of 30 is

**Table 11-5. Risk Analysis**

| Col 1<br>Product | Col 2<br>Label weight (oz) | Col 3<br>Plant state (Avg %) | Col 11<br>% below label (oz) | Col 12<br>Minimum (%) | Col 13<br>Below MAV (%) | Col 14<br>Plan A Pr(C>1) (%) | Col 15<br>Plan B Pr(C>0) | Col 16<br>Plan A Limit | Col 17<br>Plan A n=30 | Col 18<br>Plan B n=10 |
|---|---|---|---|---|---|---|---|---|---|---|
| Walnut | 16 | CA | 43 | 15.32 | 0.7 | 1.9 | 6.8 | 15.89 | 0.2 | 29.9 |
| | 32 | GA | 28 | 30.88 | 0.008 | 0.0 | 0.08 | 31.87 | 0.0 | 3.5 |
| | 64 | NY | 22 | 62.00 | 0.0 | 0.0 | 0.0 | 63.85 | 0.0 | 0.7 |
| Subtotal | | | 36 | | 0.38 | 1.0 | 3.7 | | 0.1 | 17.5 |
| Pecan | 16 | CA | 45 | 15.32 | 0.2 | 0.2 | 2.0 | 15.91 | 0.4 | 35.2 |
| | 32 | GA | 29 | 30.88 | 0.0 | 0.0 | 0.0 | 31.89 | 0.0 | 4.1 |
| | 64 | NY | 8 | 62.00 | 0.0 | 0.0 | 0.0 | 63.87 | 0.0 | 0.0 |
| | 128 | CA | 23 | 125.00 | 0.12 | 0.01 | 0.6 | 127.56 | 0.0 | 0.9 |
| Subtotal | | | 36 | | 0.12 | 0.1 | 1.2 | | 0.2 | 22.1 |
| Total | | | 36 | | 0.23 | 0.5 | 2.2 | | 0.2 | 20.2 |

expressed as $\Pr(C > 1, n = 30)$ and will be entered in Column 14. Similarly, the probability of finding one or more defects in a sample of 10 is expressed as $\Pr(C > 0, n = 10)$ and will be entered in Column 15.

These two calculations represent the risk of being out of compliance on the basis of individual containers in the test sample. They are based on the binomial probability distribution where $p$ (the percent defective) is that which is shown in Column 13 for each product. As an example, refer to the first item in Column 13, where $p = 0.7\%$, or 0.007. These values are then inserted into the binomial formula:

$$\Pr(C > 1, n = 30) = 1 - [\Pr(C = 0) + \Pr(C = 1)]$$
$$= 1 - [(0.993)^{30} + 30(0.007)^1 (0.993)^{29}]$$
$$= 1 - 0.981$$
$$= 0.019 \text{ or } 1.9\%. \text{ This is the first entry in}$$
Column 14 for 16-ounce walnuts produced in California.

$$\Pr(C > 0, n = 10) = 1 - \Pr(C = 0)$$
$$= 1 - (0.993)^{10}$$
$$= 1 - 0.932$$
$$= 0.068 \text{ or } 6.8\%. \text{ This is the first entry in}$$
Column 15 for 16-ounce walnuts produced in California.

Continuing for each of the seven products, we find that the company has a 0.5% chance of finding two or more defects in a sample of 30 (Plan A), and a 2.2% chance of finding one or more defects in a sample of 10 (Plan B).

Now that the risks of exceeding the maximum allowable variation have been determined, the next step is to find the risks associated with failing to reach an acceptable average weight. Under Plan A, the sample average can be no lower than two standard deviations of the mean $S_{\bar{X}}$ below the label weight. To determine the allowable limit under Plan A, first determine the value of $S_{\bar{X}}$. This is found from the equation:

$$S_{\bar{X}} = \frac{\sigma}{\sqrt{n}}$$

where $S$ is sigma for the product (see Column 7), and $n$ is the sample size. For 16-ounce walnuts from California, we find

$$S_{\bar{X}} = 0.30/\sqrt{30} = 0.055$$

The allowable limit (Column 16) is found by subtracting $2S_{\bar{X}}$ from the label weight. For 16-ounce walnuts produced in California:

$$\text{Allowable limit} = 16.0 - 2(0.055) = 15.89$$

Similarly, the remaining product allowable limits are calculated and entered in Column 16. This data may then be used to complete the calculation for Column 17, the risk of finding a mean below the limit for Plan A; and Column 18, the risk of finding a mean below the limit for Plan B.

For Plan A, calculate the number of standard deviations of the mean $S_{\bar{X}}$ being overfilled. This is done by using the equation:

$$Z_{\bar{X}} = \frac{[\text{Limit (Col 16)} - \text{Average net weight (Col 6)}]}{\sigma(\text{Col 7})/\sqrt{n}}$$

The risk of finding a mean below the limit (Column 17) is then defined as the area below $Z_{\bar{X}}$ under the normal curve. For 16-ounce packages of walnuts from California,

$$Z_{\bar{X}} = \frac{15.89 - 16.05}{0.30/\sqrt{n}} = -2.92$$

The probability that a sample mean of 30 containers of this product is therefore the area below $Z = -2.92$. By using the table for areas under the normal curve, we find the area corresponding to $Z = -2.92$ is 0.002, or 0.2%. The remainder of the products are similarly evaluated, and the weighted averages calculated to give an overall risk of company noncompliance of 0.2%.

The final calculations for this analysis are for Plan B compliance for the averages, Column 18. Under this plan, the sample average must be at or above the label weight. Therefore, the limit for the sample average is the same as the label weight. Calculate the number of standard deviations of the mean $S_{\bar{X}}$ being overfilled above the declared label weight. Define $S_{\bar{X}}$ by using the equation:

$$S_{\bar{X}} = \frac{\sigma}{\sqrt{n}}$$

$$= \frac{\sigma(\text{from Col 7})}{\sqrt{10}}$$

The number of standard deviations of the mean being overfilled is found using the equation

$$Z_{\bar{X}} = \frac{\text{Label weight} - \text{Average net weight (Col 6)}}{\sigma(\text{col 7}/\sqrt{10})}$$

$$= \frac{16.0 - 16.05}{0.30/\sqrt{10}} = -0.53$$

The probability that a sample mean of 10 packages of this product is therefore the area below $Z = -0.53$ under the normal curve, which is found in the tables to be 0.299, or 29.9%. When the remainder of the product data is calculated and the weighted averages determined, it is found that the overall company risk of noncompliance with Plan B requirements is 20.2%.

## PROCEDURES FOR SETTING FILL TARGETS

The two sets of calculations shown in Tables 11-3 and 11-5 have been based on hypothetical quality control data to reveal:

- The cost of overfilling (Column 10)
- The chance of being out of compliance with the existing quality control systems for net weight
- Under Plan A:
  failure to meet MAV requirements (Column 14)
  failure to meet sample average weight (Column 17)
- Under Plan B:
  failure to meet MAV requirements (Column 15)
  failure to meet sample average weight (Column 16)

As vital as this information is to top management, it still leaves three questions:

1. Are these levels of cost and risk satisfactory?
2. At what levels should the company be operating?
3. Can the company meet these new levels?

It is unlikely that a quality manager would have the gall to attempt to answer the first question. In the first place, top management would like to see both cost and risk levels at zero, and probably realizes that both goals are unattainable. Only management is in a position to answer this question, and it is quite probable that they won't. It is equally probable that management will ask instead "can't we do better?"

Question 2 can be calculated, and will be demonstrated below. This brings us to Question Number 3. Since the answer depends on the value of the standard deviation for each of the packaging lines, the answer can be determined by conducting a series of production floor tests to establish the effect of production variables (line speed, product uniformity, density, humidity, equipment condition, modified equipment, etc.) on the standard deviation of the weights attained. Since these tests are somewhat costly to perform, it would be well to calculate first the theoretical goals sought, and then check to see if the lines are capable of operating under those conditions. If not, then tests should be designed to explore methods of meeting the calculated standards.

An illustrative procedure for calculating the target net weight using the data for 16-ounce walnuts produced in California (the first line of Table 11-3) is shown below. The data required for this example are as follows:

| | |
|---|---|
| Label weight. From Column 2. | 16 av. ounce |
| Sigma. From Column 7. | 0.30 av. ounce |
| Target 5 below label weight. From management. | 25% |

Assuming that the weights are distributed normally, the target value ($\mu$) is set such that 25% of the area under the normal curve is below the label weight

(16.0 ounces). By referring to the table of areas under the normal curve, it is shown that $Z$ (the number of standard deviations) is equal to $-0.675$. Solving for $\mu$ in the equation:

$$Z = \frac{\text{Label weight} - \mu}{\sigma}$$

$$-0.675 = \frac{16.0 - \mu}{0.30}$$

$$\mu = 16.20$$

Therefore, a target net weight as low as 16.20 oz will result in no more than 25% below the label weight as illustrated in Figure 11-2.

Selection of a 25% maximum is one way to set target weights, but it is purely arbitrary. Another way is to select a risk of failing to comply with *Handbook 133* requirements. This, too, is arbitrary in that the risk is a management decision, but it is more meaningful.

As stated previously, *Handbook 133* has two requirements for compliance: (1) the maximum allowable variation (MAV); and (2) the average net weight. The first calculation will be to determine the target net weight needed to assure a minimum risk of producing an unacceptable number of short-filled containers below the MAV. Defining "minimum risk" is a management decision, but should management be unwilling to provide a figure, some reasonable low risk must be assumed in order to proceed. This is not as difficult as it may appear to be. The figure 98% compliance is occasionally heard in relation to risk. This implies that of the samples selected for net content evaluation by the authorities, no more than 2% will be found not in compliance. From a practical standpoint, if an individual food processor's products were sampled ten times a year, it would take 10 years to accumulate 100 samplings, and no more than two of those examinations might be out of compliance. It is not unreasonable to assume that with a 2% risk, a processor might have no violations for 10 times that period: 100 years!

It is possible that a company might not be selected for any sampling for several years. On the other hand, it is only fair to surmise that once a violation has been observed by the authorities, that company's products will come under close scrutiny for some time.

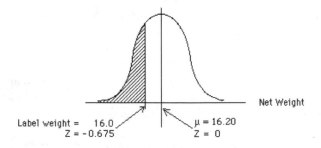

**Figure 11-2.** Selection of target weight.

For this example, management is willing to take a 2% risk of failing the compliance requirement for individual containers. Under Plan A, compliance requires, at most, one defect (container below the MAV) in a sample of 30. This is shown as

$$\text{Risk} = \Pr(C > 1, n = 30) = 1 - [\Pr(C = 0) + \Pr(C = 1)]$$
$$0.02 = 1 - [(1 - p)^{30} + 30(1 - p)^{29} \cdot (p)]$$

Solving for $p$ (percent defective), it is found that $p = 0.007$ or 7%. To solve for $\mu$ such that only 0.7% fall below the MAV (15.32 ounces), the appropriate $Z$ value must first be found, using the table for areas under the normal curve. Having found $Z = -2.455$, solve for $\mu$ using the equation

$$Z = \frac{\text{MAV} - \mu}{\sigma}$$

$$-2.455 = (15.32 - \mu)/0.30$$

$$\mu = 16.06$$

This means that if the target is set at 16.06, there is a 2% risk of non-compliance for individual underfills under Plan A of *Handbook 133*.

Similar calculations are performed for Plan B.

$$\text{Risk} = \Pr(C > 0, n = 10) = 1 - P(C = 0)$$
$$0.02 = 1 - (1 - p)^{10}$$

It might be more suitable to use logarithms to solve this equation:

$$(1 - p)^{10} = 0.98$$
$$10 \log(1 - p) = \log(0.98)$$
$$\log (1 - p) = 0.00202$$

solving for antilogarithms,

$$(1 - p) = 0.9980$$
$$p = 0.002, \text{ or } 0.2\%$$

The appropriate $Z$ value for 0.2% below the MAV is $-2.88$. Now the calculation for $\mu$ is completed by using the equation

$$-2.88 = (15.32 - \mu)/0.30$$
$$\mu = 16.18 \text{ ounces.}$$

This is the target to assure 2% compliance risk for underfills under the Plan B of *Handbook 133*.

Now that the calculations for MAV compliance have been calculated, the next step is to calculate compliance for average weight. The averages are normally distributed, with standard deviations defined as:

$$S_{\bar{X}} = \frac{\sigma}{\sqrt{n}}$$

The target $\mu$ is set so that the area under the normal curve below the minimum for the average is equal to the desired level of risk. As above, a 2% risk will be used. Under Plan A, $n = 30$, and the minimum sample average must be less than 2 standard deviations:

$$\text{Minimum} = \text{Label weight} - 2S_{\bar{X}}$$

$$= 16.0 - 2\left[\frac{0.30}{\sqrt{30}}\right]$$

$$= 16.0 - 0.11$$

$$= 15.89$$

The appropriate Z value ($Z_{\bar{X}}$) for a 2% risk is $-2.054$. Therefore, the target net weight ($\mu$) can be found by solving the equation

$$-2.054 = \frac{15.89 - \mu}{0.30/\sqrt{30}}$$

Hence, $\mu = 16.00$ ounces (see Figure 11-3).
Under Plan B of *Handbook 133*, the sample size is usually 10, and therefore,

$$S_{\bar{X}} = \sigma/\sqrt{10}$$

Note that there are no allowances for the sample mean to be below the label weight under Plan B. The desired target weight for a 2% risk can be found by solving the equation:

$$-2.054 = \frac{16.00 - \mu}{0.30/\sqrt{30}}$$

Hence, $\mu = 16.11$ ounces (see Figure 11-4).

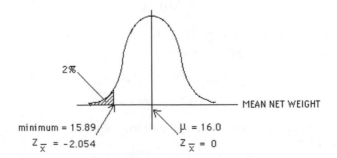

Figure 11-3. Compliance under Plan A.

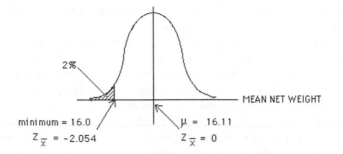

**Figure 11-4.** Compliance under Plan B.

**Table 11-6.** Target Net Weights for 16-Ounce Walnuts, California (to comply with *Handbook 133*)

| Criteria | Plan A | Plan B |
|---|---|---|
| 25% below the label weight 2% Risk of noncompliance | 16.20 oz | 16.20 oz |
| (a) Individual underfills | 16.06 oz | 16.18 oz |
| (b) Sample average | 16.00 oz | 16.11 oz |

At this point, it might be wise to tabulate the results of all of the above target weight calculations. A comparison of the two compliance programs is shown in Table 11-6. From this table, management should be in a position to select one program. The final decision may be based on an analysis of the cost/risk trade-off. To derive the relationship between the costs of overfilling and the risks of noncompliance, a company should determine the overall costs for one sigma overfill. This is useful since the overfill policy is usually expressed in terms of a certain number of sigmas overfill, and provides for the evaluation of alternative overfill policies without considering each product separately. Again, we shall use the 16-ounce package of walnuts processed in the California plant. The data on which the calculation are based are as follows:

| | |
|---|---|
| Label weight | 16.0 ounces |
| Average net weight | 16.05 ounces |
| Sigma ($\sigma$) | 0.30 ounces |
| Volume | 420,100,000 units produced |
| Cost | $0.03 per ounce |

The cost of one sigma is found by determining the total cost of overfills, and dividing that number by the level of sigma overfill:

$$420,100,000 \text{ units} \times 0.05 \text{ ounces} \times \$0.03 \text{ per ounce} = \$630,150$$
$$\text{Amount overfilled/sigma} = 0.05/0.30 = 0.17$$
$$\text{The cost of overfilling} = \$630,150/0.17 = \$3.71 \text{ million}$$

Similarly, calculations for the costs for overfilling all products will provide a rational means of selecting a cost/risk policy.

All of the above calculations are based on a system for complying with the requirements of *Handbook 133*, and selecting a risk and a cost level for each product. There are other techniques which are used successfully, but to obtain the assurance that the company is in compliance, it would be advisable to run periodically through these calculations. Some managements may believe that mere compliance with the law is not sufficient for their product lines, and it is perfectly conceivable that filling techniques can be improved to the point that weights are well within the compliance limits. In general, this may be accomplished by progressively reducing the sigma value for each production line through experimentation, training, maintenance, more uniform product, and improved equipment. Simply comparing the control limits of $X$-bar and $R$ charts as these improvements are introduced will point to the direction for tighter weight control. It has been demonstrated that this technique can cut the value of sigma to the point where the target value of a 16-ounce container can be a scant 0.016 ounces above target weight with no loss in productivity. The reduction of associated costs related to product giveaway on a high speed line are often spectacular.

# 12    Design of Experiments

## INTRODUCTION

Design of experiments consists of a series of techniques used to explore various operating conditions of a process with the goal of finding lower costs of raw materials, process, and product; improved product performance; less variability in the product; or higher production capacity to meet increasing demand. In large companies with specialized departments, experimental design is the province of engineers, food scientists, or the research and development department. Quality control personnel may become involved with the performance, and possibly some of the evaluation of industrial experimentation. Consequently, an understanding of the techniques of this discipline is valuable. In smaller companies where the task of industrial experimentation often falls to the quality control manager, a knowledge of statistical design of experiments is imperative.

It might be well to consider again the meaning of "quality." Amongst others, the clothing industry commonly uses the word to signify superior appearance or performance. For our purposes, throughout the book, we have referred to "quality" as "a level," not necessarily a superior level or the highest level. There are many classes of products which need not be of high quality to perform their function; in fact high quality may be a deterrent to the marketability. Disposable tools supplied for assembly of light-duty toys or furniture may be made of stamped steel sheet, rather than heavy-duty, forged, chrome steel alloy. Pie-grade fruit to be further processed into pies, jellies, or fillings need not consist of unbroken uniform pieces. Salt used to melt sidewalk ice would gain nothing but cost if it were manufactured to the same purity and quality level as food-grade salt. Plastic film packaging certainly does not exhibit the superior protective quality of hermetically sealed metal containers, but many foods with short shelf life, and many with desirable appearance which can be advantageously displayed in clear plastic need not, and perhaps should not, be sealed in more expensive containers. Does statistical design of experiments have any place in industries such as these? Of course it does. It may be possible to improve the quality level at no additional cost, thus providing a marketing advantage even to throwaway products. It may also be possible to design experiments which lead the way to maintaining the quality level at higher rates of production, with less expensive or alternative raw materials, or with less labor.

At the other end of the quality scale, a series of experiments might be designed to eliminate defects in costly food products, such as discoloration in blanched almonds used for stuffing prunes. Or new process methods may be explored to improve the color retention of freeze-dried vegetable products. Or modification of cooking methods and packaging might improve the sensory characteristics and extend the life of shelf-stable meals. The list is endless, and the opportunities to increase profits while improving quality level are unlimited.

Occasionally the need for an experimental effort is thrust upon a processor. A frequent driving force to industrial experimentation is the threat of a changing market: consumers' tastes change, a competitor modifies his product, or introduces a new one. Occasionally a raw material source dries up or becomes more costly due to weather, an increased demand from other markets, or other conditions. Disparaging press reviews of suspected environmental damage caused by a product or its packaging may force a company into a crash program of experimentation.

The main goal of a statistically designed experiment is to obtain unambiguous answers to specific questions. Modifying a production facility or running a non-standard experimental process is expensive. There is lost production, wasted wages and overhead, and the likelihood of creating unusable scrap material. So it is of great importance that the experiment be planned sufficiently to provide specific answers, and avoid the necessity of repeating the procedure to correct mistakes made. Consider a test casually run with a modified formula, after which the line foreman discovers that it was produced on the wrong machine at a different temperature. Even if the results of the test showed an improved product, the foreman is left with doubts about the reasons for the success: was it the formula? the machine? the temperature? There is no room for careless planning or execution in a statistically designed experiment.

There is another reason for careful planning: it is rarely possible to introduce a single variable in an experiment. For example, supposing a test is suggested which will result in lower moisture product from a dryer by the increase of dryer air flow. It would not be surprising to find that an increase in air flow caused shorter product retention time in the dryer, and to compensate for the resulting decrease in BTU transfer to the product, the temperature would also have to be raised. It is also rarely possible to introduce variables in a process which will deliver a single change in the final product. In the above example, it is likely that not only will the moisture be affected, but the product color will probably be darker from the added temperature, and the product density might well be altered appre-ciably. These new conditions may or may not be undesirable. The point to be made here is that careful planning prior to running the experiment should eliminate sur-prises in the outcome, and should reduce the necessity of conducting additional experiments because of unanswered questions. The results of an experiment to show the efficiency obtained from using large pineapples as a raw material would indicate that a major saving in the labor force could result from eliminating many of the trimmers and replacing them with a large-fruit mechanical trimmer. But this ignores the real world fact that *all* pineapples are not large, and that trimmers will still be required for the normal-crop-size raw materials received during the canning season.

The two goals of statistical design of experiments (unambiguous answers and low-cost testing) are reached through the elimination of extraneous variables, and the ability to handle many factors simultaneously. Extraneous variables commonly encountered are the day-to-day differences found in any production facility: different employees, varying motivation of the workers, environmental changes (temperature, humidity, light, ventilation, vibration, odor), day of the week, proximity to holidays or payday, and even local news events. For the most part, these variables are related to people. There are also variables due to machines which appear to be identical but which have unique operating characteristics. When the output of two filling machines is compared, rarely are they found to produce identical weight dispersions. The causes may be the result of machine speed differences, wear, product flow, or machine location. Raw materials used in the process may vary from supplier to supplier, and from hour to hour. In some industries this variable may be controllable: a hopper car of plastic pellets used for molding may not vary in composition from top to bottom; but a truckload of tomatoes may have drastic quality differences due to causes as simple as location in the pile: crushed on the bottom, sunburned on the top, cold in the front. Then there are field-to-field chemical variations and maturity differences. Finally there are frequently uncontrolled differences between the technicians conducting the tests and the evaluators who interpret the results.

Often there are conditions under which it is not possible either to eliminate these variables or to hold them constant. There are also circumstances under which it would be undesirable to attempt to hold them constant. This might be the case where a test is under consideration to establish the ruggedness of a process when external and uncontrollable variables are commonly present. Such a test might be designed to determine which raw ingredients may be used that are resistant to uncontrollable process fluctuations, and can be depended upon to result in a uniform quality product. It is thus important to avoid confusing *essential external variables* with *experimental variables*. Usually uncontrolled variables may be handled by randomly selecting them for inclusion in the test design.

There are three objectives in designing experiments. (1) All of the design factors and all of the response variables associated with the problem being investigated should be included. (2) The design factors should be appropriately balanced so that no factors are either overemphasized or overlooked. (3) The reproducibility of results should be both precise and accurate. *Precision* refers to the variability or uniformity of results, and is generally characterized by the standard deviation, $\sigma$. *Accuracy* is a measure of the difference between the measured and the theoretical results, and is expressed as $(\overline{X} - \mu)$. As an example, a filler scale might produce highly precise weights—all containers within 0.001 ounce—but the accuracy might be unacceptable at 0.75 ounce overweight.

The number of variables in a design depends on the complexity of the investigation, the production time available, and the cooperation of production and other managerial personnel. Tests designed to explore a single variable are usually the easiest to sell to management, since they are readily understood and appear to be the least costly to undertake. Some of the limitations have already been discussed. Experiments may be conducted off-line. There is no universal definition of off-line, but it generally refers to experiments conducted in the pilot plant, in the laboratory, on commercial equipment at an equipment manufacturer's facility, or on the plant equipment when the plant is not in normal production. It is intended to eliminate interference with production schedules, but may not necessarily be most cost-effective.

In addition to conducting experiments with only a single variable, testing with slight changes in just a few small operating conditions at a time is frequently a successful way to design experiments. Small changes probably will not affect production output quantity or quality appreciably, but will show the direction for further improvement. Procedures of this type have been formalized as EVOP, or evolutionary operations. And a third general experimental technique examines several different operating conditions, usually over a relatively short period of time.

## ELIMINATION OF EXTRANEOUS VARIABLES

### Example
Which of four recipes makes the best cherry pie (A, B, C, or D)? In order to evaluate the products, the sensory testing panel requires four pies of each recipe. The laboratory is unable to bake more than four pies per day.

The most obvious solution would be to bake one pie of each type (A, B, C, and D) on one day, and submit them to the testing panel each day for four days. The problem with this setup is that it does not take into account the day-to-day variations which may occur amongst the judges, or the changes in the laboratory personnel making up the pies. These are *extraneous variables*. The conclusions from such a test might improperly disclose apparent differences between pies which are truly differences between days.

A method which minimizes these day-to-day differences has been devised by statisticians and is known as a *randomized block design*. Under this plan, different recipes are made up each day, but not necessarily one of each type. Over the four-day period, however, note in the plan below that each pie has been baked four times.

|   | Day |   |   |
|---|---|---|---|
| 1 | 2 | 3 | 4 |
| B | A | C | C |
| D | C | A | D |
| A | B | D | A |
| C | D | B | B |

In a randomized block design, all treatments are assigned to a block. A block is defined as a homogeneous set of conditions for comparison of the treatments. In the above example, Days are Blocks. Again, the main purpose of this plan is to avoid confounding of extraneous variables with primary variables. A secondary purpose is to improve the precision by removal of effect from experimental error.

Another example of a randomized block design is one in which four bread formulas are compared for loaf volume when baked in a microwave oven. The primary variable here is the bread formula; the extraneous variable is the type of oven used. Both the formula data and the oven data are referred to as design variables, or design factors (variables over which the designer has control); the loaf volumes are referred to as the *response variables (outcome variables)*.

The first step in designing this test is to develop a protocept. This is a laboratory benchtop formulation which satisfies the general requirements for the product to be tested. Some of the considerations might be size, weight, texture, flavor or color. Having established the attributes of the product, the protocept is then ready to be subjected to a series of tests. It is suggested that the first tests be designed as simply as possible.

Note the outcomes in the table below:

| Formula | Ovens | | | |
|---------|------|------|------|------|
|         | GE   | TA   | SA   | AM   |
| 1       | 156  | 143  | 160  | 137  |
| 2       | 150  | 151  | 157  | 129  |
| 3       | 112  | 137  | 126  | 109  |
| 4       | 162  | 151  | 175  | 149  |

It appears that the smallest-volume loaves are produced from Formula #3. Regardless of which oven is used, this formula produces the lowest numbers for volume. Formula #4 produces higher volumes than any other formula for a given oven; but also note that the 151 volume in the TA oven is smaller then the 156 volume in the GE oven. This is extraneous information. The goal was to compare formulas, not microwave ovens. This is a one-factor design: bread formula.

An alternative plan for the four-recipe pie-baking problem above is called the *Latin Square*. It is designed to control for two blocking factors: Day and Sequence. You will note in the Latin Square below that each pie is baked only once on each day, and that the first pie baked is different each day. This eliminates the possibility of drawing incorrect conclusions in the event that the first pie baked is always different than subsequent pies due to lack of oven prewarming. The two extraneous variables here are (1) the day the pies were baked, and (2) the sequence of baking.

|          |   | Day |   |   |
|----------|---|-----|---|---|
| Sequence | 1 | 2   | 3 | 4 |
| 1        | A | D   | C | B |
| 2        | B | A   | D | C |
| 3        | C | B   | A | D |
| 4        | D | C   | B | A |

When a large number of variables must be compared at the same time, a modification of the randomized block design may be used. Assume that we wish to compare seven pie recipes. The sensory panel judges can evaluate no more than three pies before losing their ability to detect flavor differences. The design which satisfies these requirements is called an *Incomplete Block Design*. The seven recipes for pies are shown as A through G below. Note that each judge evaluates only three pies, and that each pie is sampled by three different judges. The design is carefully balanced so that each group appears the same number of times, and each formula appears the same number of times. Note also that the pair AE occurs only once, as does the pair AC, EC, etc.; that is, each pair of recipes appears only once. In this manner, each pie recipe can be compared independently of judge-to-judge variation. Prepared tables of incomplete block designs are available in the literature.

|   |   |   | Judge |   |   |   |
|---|---|---|-------|---|---|---|
| 1 | 2 | 3 | 4     | 5 | 6 | 7 |
| A | G | A | C     | B | C | D |
| E | F | B | G     | E | D | G |
| C | A | D | B     | F | F | E |

The following are several commonly used examples of incomplete block designs (Table 12-1).

Statisticians have developed many types of block designs to handle experiments other than the simple ones shown above. A list of the most commonly used block designs follows:

- Randomized block
- Balanced incomplete block
- Latin squares
- Graeco-Latin squares (for 3-block variables)
- Hyper Latin squares (for more than 3 variables)
- Incomplete Latin squares.

## Table 12-1.  Incomplete Block Designs

| Block | Treatments | Block | Treatments | Block | Treatments |
|-------|-----------|-------|-----------|-------|-----------|
| **(1) 4 Treatments, 2 per block** | | | | | |
| 1 | a  b | | | | |
| 2 | c  d | | | | |
| 3 | a  c | | | | |
| 4 | b  d | | | | |
| 5 | a  d | | | | |
| 6 | b  c | | | | |
| **(2) 5 Treatments, 2 per block** | | | | | |
| 1 | a  b | 6 | a  d | | |
| 2 | c  d | 7 | b  c | | |
| 3 | b  e | 8 | c  e | | |
| 4 | a  c | 9 | a  e | | |
| 5 | d  e | 10 | b  d | | |
| **(3) 5 Treatments, 3 per block** | | | | | |
| 1 | a  b  c | 6 | a  b  d | | |
| 2 | a  b  e | 7 | a  c  d | | |
| 3 | a  d  e | 8 | a  c  e | | |
| 4 | b  c  d | 9 | b  c  e | | |
| 5 | c  d  e | 10 | b  d  e | | |
| **(4) 6 Treatments, 2 per block** | | | | | |
| 1 | a  b | 6 | d  f | 11 | b  d |
| 2 | c  d | 7 | a  d | 12 | c  f |
| 3 | e  f | 8 | b  f | 13 | a  f |
| 4 | a  c | 9 | c  e | 14 | b  c |
| 5 | b  e | 10 | a  e | 15 | d  e |
| **(5) 6 Treatments, 3 per block** | | | | | |
| 1 | a  b  e | 6 | b  c  d | | |
| 2 | a  b  f | 7 | b  c  e | | |
| 3 | a  c  d | 8 | b  d  f | | |
| 4 | a  c  f | 9 | c  e  f | | |
| 5 | a  d  e | 10 | d  e  f | | |
| **(6) 7 Treatments, 3 per block** | | | | | |
| 1 | a  b  d | | | | |
| 2 | b  c  e | | | | |
| 3 | c  d  f | | | | |
| 4 | d  e  g | | | | |
| 5 | a  e  f | | | | |
| 6 | b  f  g | | | | |
| 7 | a  c  g | | | | |

## HANDLING MANY FACTORS SIMULTANEOUSLY

Looking at the cherry pie formulation from another viewpoint, suppose we are interested in evaluating the effect of three factors:

- Butter (B) versus margarine (M)
- Cherry supplier A or B
- Amount of sugar—high or low.

The number of pies which can be baked is restricted to four. A typical non-statistical method which might be used would be to change one variable at a time. The first pie would contain butter, cherries from supplier A, and low sugar. Next, a pie would be baked using margarine, but otherwise the same as the first one. The third pie would be the same as the first, but with cherries from supplier B; and the fourth pie would be the same as the first, but with a high level of sugar. This arrangement is shown below:

Nonstatistical plan

| Shortening | Cherries | Sugar |
| --- | --- | --- |
| B | A | LO |
| M | A | LO |
| B | B | LO |
| B | A | HI |

This plan adequately compares some of the pie variables with the first pie, but does not cover all of the interactions. For example, there is no formula to compare the first pie with one made with margarine and high sugar; nor is there a pie with margarine and cherries from supplier B. To include all of the combinations would require many more pies than could be baked at one time, but with the four combinations above, it is obvious that the three variables are not treated uniformly. Only one test uses margarine, and three use butter; only one test has high sugar, and three have low.

By using a statistical plan, some of these deficiencies may be overcome. In the statistically designed plan below, each variable is examined the same number of times: two butters and two margarines; two high sugars and two low sugars; two supplier A and two supplier B. Note also that each variable is tested one time against each of the other variables: for example, butter (B) is tested with supplier A and supplier B, and with high and low sugar. By examining the plan below, it is found that all of the variables are similarly compared once with each other.

One example of the advantage of the statistical plan over the nonstatistical plan is the fact that the statistical plan will compare the effect of butter over margarine for each supplier, and at each level of sugar. The nonstatistical plan does not

Statistical plan

| Shortening | Cherries | Sugar |
| --- | --- | --- |
| B | A | LO |
| B | B | HI |
| M | A | HI |
| M | B | LO |

compare the effect of butter over margarine either for supplier B, or for high sugar. Another advantage of the statistical plan is that all of the formulas are used in estimating the effects. The shortening effect is the difference between the average of the first two formulas and the average of the last two formulas. By being able to use averages, the standard error is reduced by a factor of $\backslash R(2)$. This small standard error enables the statistical plan to detect small differences.

The types of multifactor experimental plans which handle many factors simultaneously to be discussed are:

- Full factorial designs
- Fractional factorial designs
- Response surface design
- Mixture designs.

## FULL FACTORIAL DESIGNS

A full factorial design will include all possible combinations of all of the conditions. In the case of the cherry pie tests above, if there had been no restrictions on the number of pies baked at one time, a full factorial design might have been used, and would appear as shown below:

| Shortening | Cherries | Sugar |
| --- | --- | --- |
| B | A | HI |
| B | A | LO |
| B | B | HI |
| B | B | LO |
| M | A | HI |
| M | A | LO |
| M | B | HI |
| M | B | LO |

The simplest factorial design is called a 2 by 2 design. It considers the responses of a test in which two factors (variables) are examined at two levels (values). In the following example, two variables (temperature and concentration) will be studied at two levels (low and high). The response to be evaluated shall be the yield of the process. A diagram representing the experiment is shown in Figure 12-1.

A total of four conditions will include all of the possibilities for conducting this test design. (See Table 12-2.)

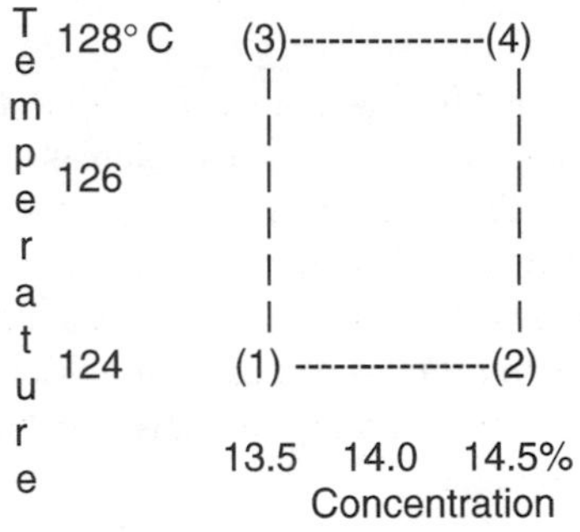

```
T
e  128° C     (3)---------------(4)
m               |               |
p                |               |
e  126          |               |
r                |               |
a                |               |
t                |               |
u  124         (1) ---------------(2)
r
e              13.5  14.0  14.5%
                   Concentration
```

**Figure 12-1.** A 2 × 2 design.

**Table 12-2.  2 × 2 Factorial Experimental Design Conditions**

| Condition number | Concentration (%) | Temperature (°C) |
|---|---|---|
| 1 | 13.5 | 124 |
| 2 | 14.5 | 124 |
| 3 | 13.5 | 128 |
| 4 | 14.5 | 128 |

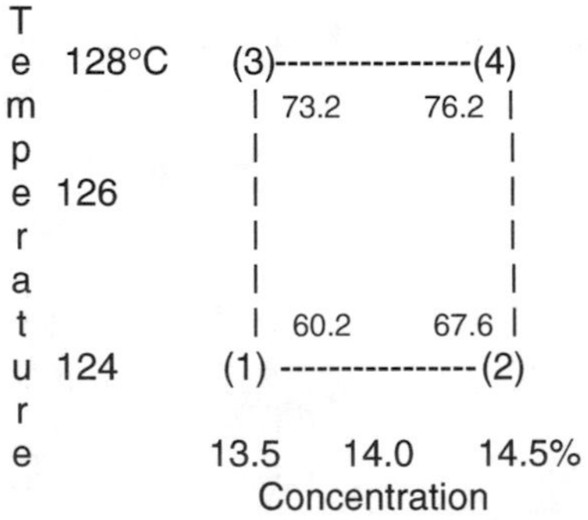

```
T
e  128°C     (3)----------------(4)
m             | 73.2      76.2 |
p             |                |
e  126        |                |
r             |                |
a             |                |
t             | 60.2      67.6 |
u  124       (1) ----------------(2)
r
e            13.5    14.0    14.5%
                  Concentration
```

**Figure 12-2.** Data from a 2 × 2 experiment.

**Table 12-3.   2 × 2 Factorial Experimental Design Data**

| Condition number | Concentration (%) | Temperature (°C) | Yield |
|---|---|---|---|
| 1 | 13.5 | 124 | 60.2 |
| 2 | 14.5 | 124 | 67.6 |
| 3 | 13.5 | 128 | 73.2 |
| 4 | 14.5 | 128 | 76.2 |

The experiment is subsequently conducted under these four sets of conditions, yielding the results shown in Figure 12-2 and Table 12-3.

Examining the effects of increasing the concentration from 13.5 to 14.5%:

Concentration effect

| | | |
|---|---|---|
| 67.6 − 60.2 = 7.4 | @ | temperature 124 °C |
| 76.2 − 73.2 = 3.0 | @ | temperature 128 °C |
| 71.9 − 66.7 = 5.2 | | Average yield |

The yield increases at either temperature, but to a greater extent at 124 degrees. If the process were to be operated at the higher concentration, we might expect that the lower the temperature (between 124 and 128), the better the yield.

Temperature effect

| | | |
|---|---|---|
| 73.2 − 60.2 = 13.0 | @ | concentration 13.5 |
| 76.2 − 67.6 = 8.6 | @ | concentration 14.5 |
| 74.7 − 63.9 = 10.8 | | Average yield |

Similarly, increasing the temperature while holding the concentration at 13.5%, the yield increases by 13.0%. At the high concentration, increasing the temperature increases the yield by 8.6% From this simple design, a surprisingly large amount of useful information emerges.

The interaction effect can be quantified as well. The difference between the concentration effect from high temperature to low temperature is 7.4 − 3.0 = 4.4. The difference between the temperature effect from high concentration to low concentration is 13.0 − 8.6 = 4.4. It might be expected that one would measure the interaction as 4.4, but the definition used by statisticians is based on 1/2 of this difference, and would be expressed as 2.2 ± 2.2. To explain the interaction effect, consider the following:

If there were no interaction effect, the main effects of both variables would be the same at both levels of the other variable. That is:

Concentration effect = 5.2 @ temperature = 124 °C and 128 °C
Temperature effect = 10.8 @ concentration = 13.5% and 14.5%

If we now consider the actual effects at the high level of the other factor:

Concentration effect = 3.0 @ temperature = 128 °C
Temperature effect = 8.6 @ concentration = 14.5%

And the interaction results in the following:

Interaction effect = 3.0 − 5.2 = −2.2 and 8.6 − 10.8 = −2.2

The graph in Figure 12-3 answers the question "does it matter what the temperature level is when the concentration changes are examined?" It is obvious that there is a large difference in yield when moving from 13.5% to 14.5% concentration on the 124-degree temperature graph; and there is a smaller difference in yield when moving from 13.5% to 14.5% concentration on the 128-degree temperature graph. By changing the graphed axes to yield and temperature, the effect is again observed: there is a larger effect at 124 degrees than at 128 degrees (Figure 12-4).

As a matter of interest, the yield values 60.2, 67.6, 73.2 and 76.2 in the above example were taken from an actual test, and although only a single value

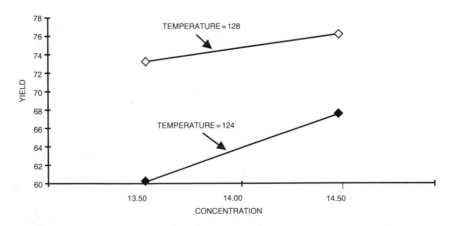

**Figure 12-3.** Interaction plot of concentration by temperature.

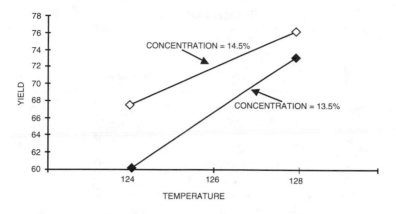

**Figure 12-4.** Interaction plot of temperature by concentration.

was shown in order to simplify the principle, actually two sets of values were generated from duplicate tests. A summary of the above discussion of data, calculations, interactions and main effects is shown in the following:

Main effect A = 5.2      Main effect B = 10.8      Interaction effect AB = −2.2

| Condition number | Replication | | Average | Standard deviation | Variance |
|---|---|---|---|---|---|
| | 1 | 2 | | | |
| 1 | 59.1 | 61.3 | 60.2 | 1.56 | 2.42 |
| 2 | 68.2 | 67.0 | 67.6 | 0.85 | 0.72 |
| 3 | 72.4 | 74.0 | 73.2 | 1.13 | 1.28 |
| 4 | 74.9 | 77.5 | 76.2 | 1.84 | 3.38 |
| | | | | Total | 7.80 |

Average variance = 7.80/4 = 1.95
Average standard deviation = 1.396 ~ 1.40
Standard error of the effect = 1.4 $\sqrt{2}$ = 1.0
Main effect A is between 5.2 ± 2 (3.2 to 7.2)
          B is between 10.8 ± 2 (8.8 to 12.8)
Interaction effect AB is between −2.2 ± 2 (−0.2 to −4.2)

An example of a full factorial design with statistical analysis is the examination of the effect of a fungicide and two pesticides on the egg production of pheasants. Four observations are available for each combination.

**Pheasant Egg Counts**

|  | Pesticide levels | | | | | |
|---|---|---|---|---|---|---|
|  | None | | Level A | | Level B | |
| No fungicide | | | | | | |
|  | 15 | 29 | 7 | 15 | 17 | 19 |
|  | 18 | 30 | 9 | 18 | 16 | 21 |
| With fungicide | | | | | | |
|  | 12 | 15 | 13 | 10 | 9 | 11 |
|  | 10 | 9 | 6 | 8 | 14 | 13 |

An analysis of variance shows that there is an effect from both the fungicide and from the pesticide.

| | | Analysis of variance | | |
|---|---|---|---|---|
| Source | df | M.S. | $F$ | |
| Fungicide | 1 | 294.00 | 16.04 | $(P<0.01)$ |
| Pesticide | 2 | 87.17 | 4.75 | $(P<0.05)$ |
| $F \times P$ | 2 | 36.50 | 1.99 | |
| Error | 18 | 18.33 | | |

There is an effect both from the fungicide and from the pesticide.

Full factorial designs are used if there are only two or three factors to explore. They are also useful in the event of higher order interactions. That is, some factors have little effect on outcomes of an experiment until a threshold is reached, after which the effect may be significant. It is interesting to note that the statistician may be unaware of these interactions, and must depend on the experience of quality control or line personnel to provide information on their importance so that they may be incorporated into the experimental design. Another application requiring full factorial design is for analyzing qualitative independent variables, such as the number of different suppliers of raw materials. If there are four suppliers available, all four must be included in the test plans.

## FRACTIONAL FACTORIAL DESIGNS

Fractional factorial designs are used when there is a large number of possible combinations of variables, and are generally applied when the factors may occur at two levels. The fraction selected depends on assumed interactions, and must come from a statistical design. An example might be a process with six variables occurring at two levels. The total number of interactions is 64, and it is highly unlikely that management would agree to shutting down a line in order to run 64 separate experiments! Information for all of the two-factor interactions would require only 32 tests. Only 16 tests would be required to investigate some of the two-factor interactions; and if no two-factor interaction information were required, only eight tests would be needed. In common practice, an experimenter needs information on only a few of the interactions, and a complete series is rarely run. Fractional factorial plans are useful when the higher order interactions referred to above are of no particular importance. This type of plan may be effective when only the main effects of the variables are of interest; that is, when "screening" tests are required.

The eight-test screening plan generates sufficient information to identify which of the variables (e.g., ingredients) have the greatest significance on the outcome of the problem being evaluated. This should further reduce the number of detailed tests required to isolate exact differences. If these eight tests had disclosed that of the six ingredients in the formula, two were of importance and four had virtually no significance, then subsequent tests need be concerned only with those two ingredients.

The following example of a fractional factorial was designed to measure the response (Y) of variations in eight ingredients at two levels of concentration in a cat food mix (Table 12-4). Note the balanced nature of this test plan: the low levels of A (17), which are matched with the low levels of B (7), appear four times; the high levels of A (26), which are matched with the high levels of B (13), also appear four times. Likewise, there are four tests each for the high–low and the low–high combinations of ingredients A and B. Similar two-level pairs are found throughout the test plan for the eight ingredients and the 16 tests.

From the analysis of variance (Table 12-5), it becomes apparent that ingredients A, D, and E are the ones that truly have a large effect on the response. This screening test has therefore revealed the three ingredients which will require further experimentation.

**Table 12-4.   Eight-ingredient Cat Food Test Levels and Responses**

| Ingredients | | | | | | | | Response |
|---|---|---|---|---|---|---|---|---|
| A | B | C | D | E | F | G | H | Y |
| 17 | 7 | 5 | 2 | 0 | 0 | 6 | 3 | 0.92 |
| 26 | 13 | 11 | 5 | 0 | 0 | 6 | 3 | 1.44 |
| 17 | 7 | 5 | 2 | 0.5 | 25 | 12 | 4 | 1.43 |
| 26 | 13 | 11 | 5 | 0.5 | 25 | 12 | 4 | 1.91 |
| 26 | 13 | 5 | 2 | 0 | 25 | 6 | 3 | 1.74 |
| 26 | 13 | 5 | 2 | 0.5 | 0 | 12 | 4 | 1.28 |
| 17 | 7 | 11 | 5 | 0 | 25 | 6 | 3 | 1.77 |
| 17 | 7 | 11 | 5 | 0.5 | 0 | 12 | 4 | 1.32 |
| 26 | 7 | 5 | 5 | 0 | 0 | 12 | 3 | 2.03 |
| 26 | 7 | 5 | 5 | 0.5 | 25 | 6 | 4 | 1.46 |
| 17 | 13 | 11 | 2 | 0 | 0 | 12 | 3 | 1.45 |
| 17 | 13 | 11 | 2 | 0.5 | 25 | 6 | 4 | 1.00 |
| 26 | 7 | 11 | 2 | 0 | 0 | 6 | 4 | 1.65 |
| 26 | 7 | 11 | 2 | 0.5 | 25 | 12 | 3 | 1.17 |
| 17 | 13 | 5 | 5 | 0 | 0 | 6 | 4 | 1.67 |
| 17 | 13 | 5 | 5 | 0.5 | 25 | 12 | 3 | 1.29 |

Table 12-6 consists of several commonly used examples of Fractional Factorial designs.

### Table 12-5.   Analysis of Variance

| Source | df | M.S. | F | |
|--------|----|------|---|---|
| A | 1 | 0.14251 | 9.55 | $P<.05$ |
| B | 1 | 0.00766 | 0.51 | |
| C | 1 | 0.00276 | 0.18 | |
| D | 1 | 0.23281 | 15.60 | $P<.01$ |
| E | 1 | 0.74391 | 49.86 | $P<.01$ |
| F | 1 | 0.00681 | 0.46 | |
| G | 1 | 0.00051 | 0.02 | |
| H | 1 | 0.00331 | 0.22 | |
| Error | 7 | 0.01492 | | |

### Table 12-6.   Fractional Factorial Designs

| a b c d e f g h | a b c d e f g h | Effects | df |
|-----------------|-----------------|---------|----|
| *1/2 rep of a $2^4$ factorial* | | | |
| 0 0 0 0 | | Main | 4 |
| 1 1 0 0 | | 2-factor | 3 |
| 1 0 1 0 | | Total | 7 |
| 1 0 0 1 | | | |
| 0 1 1 0 | | | |
| 0 1 0 1 | | | |
| 0 0 1 1 | | | |
| 1 1 1 1 | | | |
| AB = CD, AC = BD, AD = BC | | | |
| *1/4 rep of a $2^5$ factorial* | | | |
| 0 0 0 0 0 | | Main | 5 |
| 1 1 0 0 0 | | 2-factor | 2 |
| 0 0 1 1 0 | | Total | 7 |
| 1 0 1 0 1 | | | |
| 0 1 1 0 1 | | | |
| 1 0 0 1 1 | | | |
| 0 1 0 1 1 | | | |
| 1 1 1 1 0 | Main effects have 2-factor aliases | | |
| AC = BD, AD = BC | | | |
| *1/2 rep of a $2^5$ factorial* | | | |
| 0 0 0 0 0 | 1 0 0 0 1 | Main | 5 |
| 1 1 0 0 0 | 0 1 0 0 1 | 2-factor | 10 |
| 1 0 1 1 1 | 0 0 1 1 0 | Total | 15 |
| 0 1 1 1 1 | 1 1 1 1 0 | | |
| 0 0 1 0 0 | 1 0 0 1 0 | | |

**Table 12-6.** (*continued*)

| a | b | c | d | e | f | g | h | a | b | c | d | e | f | g | h | Effects | df |
|---|---|---|---|---|---|---|---|---|---|---|---|---|---|---|---|---|---|
| 0 | 1 | 1 | 0 | 0 |   |   |   | 0 | 1 | 0 | 1 | 0 |   |   |   |   |   |
| 0 | 0 | 0 | 1 | 1 |   |   |   | 0 | 0 | 1 | 0 | 1 |   |   |   |   |   |
| 1 | 1 | 0 | 1 | 0 |   |   |   | 1 | 1 | 1 | 0 | 1 |   |   |   |   |   |

*1/8 rep of a $2^6$ factorial*

| a | b | c | d | e | f | g | h | Effects | df |
|---|---|---|---|---|---|---|---|---|---|
| 0 | 0 | 0 | 0 | 0 | 0 |   |   | Main | 6 |
| 1 | 0 | 1 | 0 | 0 | 1 |   |   | 2-factor | 1 |
| 1 | 0 | 0 | 1 | 1 | 0 |   |   | Total | 7 |
| 0 | 1 | 1 | 0 | 1 | 0 |   |   |   |   |
| 0 | 1 | 0 | 1 | 0 | 1 |   |   |   |   |
| 1 | 1 | 1 | 1 | 0 | 0 |   |   |   |   |
| 1 | 1 | 0 | 0 | 1 | 1 |   |   |   |   |
| 0 | 0 | 1 | 1 | 1 | 1 |   |   | Main effects have 2-factor aliases | |

AB = CD = EF

*1/4 rep of a $2$ factorial*

| a | b | c | d | e | f | g | h | a | b | c | d | e | f | g | h | Effects | df |
|---|---|---|---|---|---|---|---|---|---|---|---|---|---|---|---|---|---|
| 0 | 0 | 0 | 0 | 0 | 0 | 1 |   | 1 | 0 | 0 | 0 | 0 |   |   |   | Main | 6 |
| 1 | 1 | 1 | 0 | 1 | 0 | 0 |   | 0 | 1 | 0 | 1 | 0 |   |   |   | 2-factor | 7 |
| 1 | 1 | 0 | 1 | 0 | 1 | 0 |   | 0 | 0 | 1 | 0 | 1 |   |   |   | Error | 2 |
| 0 | 0 | 1 | 1 | 1 | 1 | 1 |   | 1 | 1 | 1 | 1 | 1 |   |   |   | Total | 15 |
| 1 | 0 | 1 | 1 | 0 | 0 | 1 |   | 0 | 1 | 0 | 0 | 1 |   |   |   |   |   |
| 1 | 0 | 0 | 0 | 1 | 1 | 1 |   | 0 | 0 | 1 | 1 | 0 |   |   |   |   |   |
| 0 | 1 | 1 | 0 | 0 | 1 | 0 |   | 1 | 1 | 1 | 0 | 0 |   |   |   |   |   |
| 0 | 1 | 0 | 1 | 1 | 0 | 0 |   | 1 | 0 | 0 | 1 | 1 |   |   |   |   |   |

AC = BE, AD = BF, AE = AC, AF = AD, CD = EF, CF = DE, AB = CE = DF

**1/16; rep of a $2^7$ factorial**

| a | b | c | d | e | f | g | h | Effects | df |
|---|---|---|---|---|---|---|---|---|---|
| 0 | 0 | 0 | 0 | 0 | 0 | 0 |   | Main | 7 |
| 1 | 1 | 1 | 1 | 0 | 0 | 0 |   | Total | 7 |
| 1 | 1 | 0 | 0 | 1 | 1 | 0 |   |   |   |
| 1 | 0 | 1 | 0 | 0 | 1 | 1 |   |   |   |
| 1 | 0 | 0 | 1 | 1 | 0 | 1 |   |   |   |
| 0 | 1 | 1 | 0 | 1 | 0 | 1 |   |   |   |
| 0 | 1 | 0 | 1 | 0 | 1 | 1 |   |   |   |
| 0 | 0 | 1 | 1 | 1 | 1 | 0 |   |   |   |

*1/8 rep of a $2^7$ factorial*

| a | b | c | d | e | f | g | h | a | b | c | d | e | f | g | h | Effects | df |
|---|---|---|---|---|---|---|---|---|---|---|---|---|---|---|---|---|---|
| 0 | 0 | 0 | 0 | 0 | 0 | 0 |   | 1 | 0 | 1 | 0 | 0 | 1 | 0 |   | Main | 7 |
| 0 | 0 | 0 | 0 | 1 | 1 | 1 |   | 0 | 1 | 0 | 1 | 0 | 1 | 0 |   | 2-factor | 7 |
| 1 | 1 | 1 | 1 | 0 | 0 | 0 |   | 1 | 0 | 1 | 0 | 1 | 0 | 1 |   | Error | 1 |
| 1 | 1 | 1 | 1 | 1 | 1 | 1 |   | 0 | 1 | 0 | 1 | 1 | 0 | 1 |   | Total | 15 |
| 1 | 1 | 0 | 0 | 0 | 0 | 1 |   | 1 | 0 | 1 | 0 | 1 | 0 | 0 |   |   |   |
| 0 | 0 | 1 | 1 | 0 | 0 | 1 |   | 0 | 1 | 0 | 1 | 1 | 0 | 0 |   |   |   |
| 1 | 1 | 0 | 0 | 1 | 1 | 0 |   | 1 | 0 | 1 | 0 | 0 | 1 | 1 |   |   |   |
| 0 | 0 | 1 | 1 | 1 | 1 | 0 |   | 0 | 1 | 0 | 1 | 0 | 1 | 1 |   |   |   |

AE = BF = CG, AF = BE = DG, AG = CE = DF, BG = DE = CF, AB = CD = EF,
AC = BD = EG, AD = BC = FD

**Table 12-6.**   (*continued*)

| a | b | c | d | e | f | g | h | a | b | c | d | e | f | g | h | Effects | df |
|---|---|---|---|---|---|---|---|---|---|---|---|---|---|---|---|---------|-----|

*1/6 rep of a $2^8$ factorial*

| a | b | c | d | e | f | g | h | a | b | c | d | e | f | g | h | Effects | df |
|---|---|---|---|---|---|---|---|---|---|---|---|---|---|---|---|---------|-----|
| 0 | 0 | 0 | 0 | 0 | 0 | 0 | 0 | 1 | 0 | 0 | 1 | 1 | 0 | 1 | 1 | Main | 8 |
| 1 | 1 | 1 | 1 | 0 | 0 | 0 | 0 | 1 | 0 | 0 | 1 | 0 | 1 | 0 | 1 | 2-factor | 7 |
| 0 | 0 | 0 | 0 | 1 | 1 | 1 | 1 | 0 | 1 | 1 | 0 | 1 | 0 | 1 | 0 | Total | 15 |
| 1 | 1 | 1 | 1 | 1 | 1 | 1 | 1 | 0 | 1 | 1 | 0 | 0 | 1 | 0 | 1 | | |
| 1 | 1 | 0 | 0 | 1 | 1 | 0 | 0 | 1 | 0 | 1 | 0 | 1 | 0 | 0 | 1 | | |
| 1 | 1 | 0 | 0 | 0 | 0 | 1 | 1 | 1 | 0 | 1 | 0 | 0 | 1 | 1 | 0 | | |
| 0 | 0 | 1 | 1 | 1 | 1 | 0 | 0 | 0 | 1 | 0 | 1 | 1 | 0 | 0 | 1 | | |
| 0 | 0 | 1 | 1 | 0 | 0 | 1 | 1 | 0 | 1 | 0 | 1 | 0 | 1 | 1 | 0 | | |

AE = BF = CH = DG, AG = BH = CF = DE, AC = BD = EH = FG, AB = CD = EF = GH.

AF = BE = CG = DH, AH = BG = CE = DF, AD = BC = EG = FH

# RESPONSE SURFACE DESIGNS

After conducting fractional factorial experiments to determine which factors are of importance, the next step is to explore the interrelationships of these factors by use of response surface designs. The goal here is to find the best combination of the variables.

As an example, consider the pressure, time, and temperature needed to apply a plastic cover onto a container to maximize burst strength. In the chapter covering regression analysis (Chapter 12), only two variables were involved: the relation between $Y$ and $X$. In the case of three variables, the model equation becomes more complicated.

Variables:

$X_1$   Temperature

$X_2$   Time

$X_3$   Pressure

Outcome:

$Y$   Burst strength

Model:

$$Y_{ijk} = B_0 + B_1X_1 + B_2X_2 + B_3X_3 + B_{11}X_1^2 + B_{22}X_2^2 + B_{33}X_3^2 + B_{12}X_1X_2 + B_{13}X_1X_3 + B_{23}X_2X_3 + E$$

$E$ = Experimental error.

The *central composite design* (see Figure 12-5) is the classic design for response surface designs. Each side of the cube represents one of the three variables.

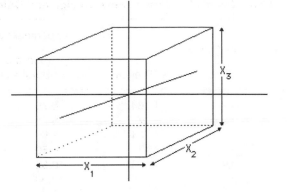

**Figure 12-5.**   Central composite design.

**Table 12-7.   Plastic Lid Response Curve Conditions**

|  | Plastic lid application | | | | |
|---|---|---|---|---|---|
|  | Star | Low | Medium | High | Star |
| Temperature | 300 | 325 | 362.5 | 400 | 425 |
| Time | 0.5 | 0.9 | 1.5 | 2.1 | 2.5 |
| Pressure | 60 | 64 | 70 | 76 | 80 |

The conventional procedure places the lowest level of $X_1$ at the left; the lowest level of $X_2$ at the front, and the lowest level of $X_3$ at the bottom. For example, if $X_1$ were reduced, the front face would be narrowed vertically.

In previous designs, we have considered only low and high levels of the variables. In the case of response curves, an additional data point is added half way between the high and low levels, and another set of extra points is added a short distance above the high level, and also a short distance below the low level. The extra levels protrude from the faces of the cube, and are referred to as *star points*. As an illustration, the low, medium, and high levels are shown with the star points as selected from the data columns in Table 12-7.

If the data columns are examined further, it may be noted that the star points are tested against the midpoints of each of the other variables (Table 12-8). To illustrate: the TIME star points of 0.5 and 2.5 are specified in tests where the TEMP value of 362.5 and the PRESSURE value of 70 are the midpoints of each of these variables. There are instances where the variables cannot be run at the star points because of physical or practical conditions; under these circumstances, the design is modified as needed. A statistician should be consulted when this occurs.

Table 12-8.   Plastic Lid Experimental Design and Outcome

| Experiment conditions | | | Experiment outcome | |
| --- | --- | --- | --- | --- |
| | | | Y (duplicate tests): | |
| $X_1$ Temp | $X_2$ Time | $X_3$ Pressure | Burst strength | |
| 325 | 0.9 | 64 | 8.75 | 7.5 |
| 325 | 0.9 | 76 | 8.25 | 9.0 |
| 325 | 2.1 | 64 | 7.50 | 7.0 |
| 325 | 2.1 | 76 | 8.50 | 9.5 |
| 400 | 0.9 | 64 | 10.50 | 9.75 |
| 400 | 0.9 | 76 | 10.00 | 10.0 |
| 400 | 2.1 | 64 | 12.00 | 13.75 |
| 400 | 2.1 | 76 | 13.00 | 12.5 |
| 300 | 1.5 | 70 | 4.00 | 7.0 |
| 425 | 1.5 | 70 | 12.75 | 12.0 |
| 362.5 | 0.5 | 70 | 8.75 | 8.5 |
| 362.5 | 2.5 | 70 | 11.00 | 10.25 |
| 362.5 | 1.5 | 60 | 9.25 | 9.0 |
| 362.5 | 1.5 | 80 | 9.75 | 9.5 |
| 362.5 | 1.5 | 70 | 9.25 | 9.5 |

After running the tests and obtaining the outcome information, a linear regression analysis is utilized to determine the equation relating the variables. From the above data, a stepwise linear regression analysis produced the following model:

$$Y = 8.925 - 0.00238*Temp - 11.059*Time + 0.0333*Temp*Time$$

The remaining Bs were nonsignificant. The pressure variable had no effect on the burst strength. Using the equation, it is possible to determine the burst strength for any combination of time and temperature selected. Note that although only five different levels of variables were examined in the test procedure, the resulting equation permits calculating the burst strength for *any* combination of factors. By constructing a graph of this equation, the values of all variables can readily be selected (Figure 12-6).

Using the chart, select a temperature of about 300° and a time of 2 s, the burst strength will be about 6; a temperature of 350° and a time of 1.5 s will produce a burst strength of 9; or 375° for 0.7 seconds will produce a 9 burst strength. (The curves are read in the same manner as contour maps.)

Whenever a series of experiments is performed to produce a response surface, it is wise to select the best combination of points and retest them to prove that the data are reliable, and that the calculations were correct. In some instances, the response surface analysis will show that more than one set of conditions appear

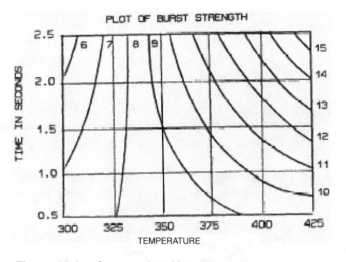

**Figure 12-6.**  Contour plot of burst strength.
*Credit:* Wendell Kerr, Calreco, unpublished manuscript.

to be optimum, and it is again important to verify this information in "the real world."

Response surface plans are used where functional relationships are desired, and where the independent variables are quantitative. Variables such as "Supplier A" and "Supplier B," or machine #37 versus machine #12, cannot be considered suitable for this design. If there are more than five variables in the system, it is advisable first to run a screening design to eliminate some of the variables; otherwise, the testing becomes too large to handle.

Some applications for response surface designs may appear relatively trivial, but may result in substantial reduction in costs or improvement in quality. In the potato processing industry, peeled potatoes pass over an inspection belt where inspectors remove grossly defective potatoes, followed by trimmers who remove defects after the potatoes have been cut into french fries. If a fixed number of workers is available for the process of removing defects, the question arises: how should the workers be allocated between inspection and trimming? Of 12 workers, should 3 be used as inspectors and 9 as trimmers, or should it be 6 and 6? By use of response surface analysis, the optimum allocation may be determined. The variables in such a test would be the numbers of inspectors and trimmers; the outcome would be the number of defects per pound after both operations.

## MIXTURE DESIGNS

Mixture design is a special response surface type since the total of the variables must equal 100%. In the temperature/time/pressure example above, there is no such requirement. Where the variables must equal 100% ($X_1 + X_2 + X_3 = 100$), the general equation

$$Y_{ijk} = B_0 + B_1X_1 + B_2X_2 + B_3X_3 + B_{11}X_1{}^2 + B_{22}X_2{}^2 + B_{33}X_3{}^2 + B_{12}X_1X_2$$
$$+ B_{13}X_1X_3 + B_{23}X_2X_3 + E$$

becomes greatly simplified to:

$$Y = B_1X_1 + B_2X_2 + B_3X_3 + B_{12}X_1X_2 + B_{13}X_1X_3 + B_{23}X_2X_3 + E$$

In the following example, three types of fish are available for use in preparing fish patties: mullet, sheepshead, and croaker (Table 12-9). The experimental design is:

Variables:

$X_1$ Mullet

$X_2$ Sheepshead

$X_3$ Croaker

Outcome:

$Y$ Mixture flavor

The fitted model is:

$$Y = 4.2X_1 + 5.1X_2 + 3.9X_3 - 3.6X_1X_3 + 1.8X_1X_3 + 0.6\ X_2X_3$$

As previously, after running the tests and obtaining the responses, the data are analyzed to arrive at a model equation relating the variables. A graph of the model

Table 12-9.  Fish Flavor Experiment Design and Outcome

| Experiment conditions | | Experiment outcome | |
|---|---|---|---|
| $X_1$ % Mullet | $X_2$ % Sheepshead | $X_3$ % Croaker | $Y$ Flavor |
| 100 | 0 | 0 | 4.2 |
| 0 | 100 | 0 | 5.1 |
| 0 | 0 | 100 | 3.9 |
| 50 | 50 | 0 | 3.8, 3.6 |
| 50 | 0 | 50 | 4.4, 4.5 |
| 0 | 50 | 50 | 4.5, 4.7 |
| 33 | 33 | 33 | 4.4, 4.2, 4.3 |

is read as follows: the flat side of the triangle opposite the labeled apex is the zero level of that variable (see Figure 12-7), and the apex is the 100% level for that variable. The center point of the triangle is a mixture containing equal amounts of each of the three fish in the blend. Examining either the model or the chart, it becomes apparent that the better flavored blends contain higher sheepshead proportions.

One of the limitations of this type of analysis is that rarely can each ingredient be used at levels from zero to 100%. It is not possible to go from 0% to 100% sugar in hot cocoa mix, for example. The final cocoa mix must contain cocoa, milk solids, fats, salt, and many other components.

In some instances it is possible to apply limits; in other instances the total formula might be reduced to some of the key ingredients for purposes of experimentation. In such cases, the final trial run of some of the key findings becomes imperative, in order to test their validity. The need for the final test may be demonstrated in designing tests of dry soup mix, where the salt content may be permitted to vary from zero to 8%. In the event that the "best" mixture for soup has a salt content of 6%, it might be found by actual testing that this level is too high because of interfering flavor characteristics of the other ingredients. It is unlikely, of course, that such a situation would exist near the set points of the actual experiment, since each of these is evaluated for flavor as the "outcome." At intermediate points, however, there is always a possibility of unusual flavor combinations which should be subsequently confirmed.

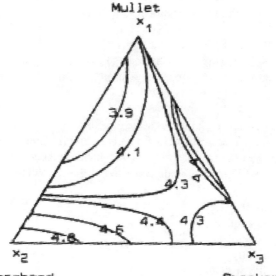

**Figure 12-7.**   Contour plot of flavor.
*Credit:* John A. Cornell, *Experiments with Mixture Designs, Models, and the Analysis of Mixture Data* (New York: John Wiley & Sons).

The graphical representations for response surfaces become increasingly complex as the number of variables increases, and this, among other reasons, accounts for the limitation of five variables for this type of analysis. For example, if a fourth fish were added to the fish patty experiment, the resulting chart would be a three-dimensional tetrahedron. This is dramatically more difficult to interpret.

The following example represents a practical problem which illustrates a central composite design experiment leading to a response surface analysis. The problem is to set the color specifications on a processed tomato product, using the trichromatic color system. The system currently in use measures reflectance of the product and expresses the results on a scale of white-to-gray which does not always represent satisfactorily the red color perceived by the eye. The trichromatic technique of measuring color provides numerical values for Chroma (brightness), Hue (redness), and Value (lightness to darkness), and is thought to be more reproducible and more characteristic of visual perception. In addition to these objectives, management requested that the specifications be set as simply as possible, preferably by use of a single range of numbers, rather than the three number sets normally used with the newer system. To assist in developing realistic specifications, management has directed that consumer testing be utilized to establish likes and dislikes of product color.

In the examples above, the equation terminology has been based on $B$ and $X$ relationships. There is no complete agreement on nomenclature. The researcher for this color problem used a different group of identifiers, but note that the form of the mathematical model is identical.

$$CR = f(\text{Hue, Chroma, Value})$$
$$CR = A_0 + A_1 \text{ Hue} + A_2 \text{ Chroma} + A_3 \text{ Value} + A_{12} \text{ Hue} \times \text{Chroma}$$
$$+ A_{13} \text{ Hue} \times \text{Value} + A_{23} \text{ Chroma} \times \text{Value} + A_{11} \text{ Hue}^2$$
$$+ A_{22} \text{ Chroma}^2 + A_{33} \text{Value}^2$$

In order to develop a continuous scale of consumer response, it was considered necessary to provide each subject with a nine-point hedonic rating scale for evaluation, rather than the two or three points which have been used in some of the previous samples. The rating scale consisted of the following:

- Dislike extremely
- Dislike very much
- Dislike moderately
- Dislike slightly
- Neither like or dislike
- Like slightly
- Like moderately
- Like very much
- Like extremely

The central composite design selected for this experiment requires 15 samples to fit the model:

| | |
|---|---|
| $2^3$ factorial points | 8 |
| Star points above each face | 6 |
| Center point | 1 |
| Total | 15 |

A pictorial model (Figure 12-8) shows the location of the 15 points.

The experimental design is constructed in the same manner as the plastic lid application example above. Compare the selection of matched variables in the two examples (Table 12-10).

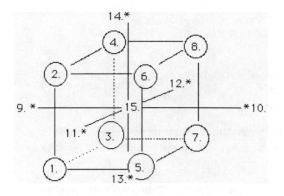

**Figure 12-8.** Central composite design.

**Table 12-10.   Tomato Product Color Experimental Design**

| Point | Hue | Chroma | Value |
|---|---|---|---|
| 1 | 34 | 44 | 20 |
| 2 | 34 | 44 | 30 |
| 3 | 34 | 56 | 20 |
| 4 | 34 | 56 | 30 |
| 5 | 46 | 44 | 20 |
| 6 | 46 | 44 | 30 |
| 7 | 46 | 56 | 20 |
| 8 | 46 | 56 | 30 |
| 9 | 31 | 50 | 25 |
| 10 | 49 | 50 | 25 |
| 11 | 40 | 41 | 25 |
| 12 | 40 | 59 | 25 |
| 13 | 40 | 50 | 17.5 |
| 14 | 40 | 50 | 32.5 |
| 15 | 40 | 50 | 25 |

If it were possible, each consumer would be asked to rate all fifteen samples and the control; however, fatigue would reduce the sensitivity of the test, and a simpler arrangement is required. By using a balanced incomplete block design, each consumer evaluates 8 samples, as shown in Table 12-11. The number of consumers needed to balance the design is governed by the relationship: $B \times K = R \times T$, where

B = number of consumers
K = number of samples evaluated by each consumer
R = number of times each sample is evaluated
T = number of samples being tested

For this test,
$K = 8$, $T = 15$, $R$ was specified at 50

$$B \times 8 = 50 \times 15$$

$$B = 93.75$$

We chose $R = 48$, which yields:

$$B \times 8 = 48 \times 15$$

$$B = 90$$

In order to meet the required number of consumers ($B = 93.750$), it will be necessary to repeat the above design six times ($15 \times 6 = 90$), randomizing the order

### Table 12-11.  Consumer Responses

| Consumer No. | Sample presentation order | | | | | | | |
|:---:|:---:|:---:|:---:|:---:|:---:|:---:|:---:|:---:|
|  | I | II | III | IV | V | VI | VII | VIII |
| 1 | 11 | 4 | 2 | 5 | 10 | 3 | 14 | 15 |
| 2 | 4 | 1 | 3 | 15 | 13 | 11 | 6 | 9 |
| 3 | 9 | 2 | 14 | 4 | 7 | 1 | 10 | 13 |
| 4 | 15 | 3 | 1 | 10 | 8 | 13 | 7 | 5 |
| 5 | 7 | 13 | 10 | 12 | 11 | 2 | 3 | 6 |
| 6 | 6 | 10 | 12 | 1 | 4 | 14 | 8 | 3 |
| 7 | 12 | 9 | 13 | 14 | 15 | 10 | 11 | 8 |
| 8 | 10 | 11 | 9 | 7 | 6 | 8 | 5 | 4 |
| 9 | 5 | 14 | 7 | 3 | 1 | 9 | 12 | 11 |
| 10 | 8 | 15 | 6 | 2 | 3 | 7 | 9 | 14 |
| 11 | 1 | 7 | 11 | 8 | 2 | 4 | 15 | 12 |
| 12 | 2 | 6 | 15 | 9 | 5 | 12 | 1 | 10 |
| 13 | 13 | 8 | 5 | 11 | 14 | 6 | 2 | 1 |
| 14 | 14 | 5 | 4 | 6 | 12 | 15 | 13 | 7 |
| 15 | 3 | 12 | 8 | 13 | 9 | 5 | 4 | 2 |

Note: $T = 15$ samples; $k = 8$ samples/consumer; $B = 15$ consumers.

of sample presentation. The test was completed and the average response for each sample was calculated and tabulated. The consumer hedonic rating for each sample is tabulated in Table 12-12.

To simplify the statistical analysis, it is necessary to eliminate one column of variables. The variables removed from the analysis for the present are the data for values. This, in effect, reduces the cube model to a square, with nine responses to consider instead of the 15 available.

These nine responses (plus the five center-point replicates) are selected and shown in Table 12-13 as a subset. Following the columns of data, the consumer response mathematical model is shown with the value variables omitted.

### Table 12-12.   Average Consumer Responses

| Sample no. | Hue | Chroma | Value | Average response |
|---|---|---|---|---|
| 1 | 34 | 44 | 20 | 4.18 |
| 2 | 34 | 44 | 30 | 3.19 |
| 3 | 34 | 56 | 20 | 5.49 |
| 4 | 34 | 56 | 30 | 5.08 |
| 5 | 46 | 44 | 20 | 4.38 |
| 6 | 46 | 44 | 30 | 4.69 |
| 7 | 46 | 56 | 20 | 5.43 |
| etc. | etc. | etc. | etc. | etc. |

### Table 12-13.   Simplified Model with Value Responses Omitted

| Sample no. | Hue | Chroma | Response |
|---|---|---|---|
| 1 | 34 | 34 | 4.18 |
| 2 | 34 | 56 | 5.49 |
| 3 | 46 | 44 | 4.38 |
| 4 | 46 | 56 | 5.43 |
| 5 | 31 | 50 | 3.23 |
| 6 | 49 | 50 | 6.21 |
| 7 | 40 | 41 | 3.74 |
| 8 | 40 | 59 | 6.10 |
| 9 | 40 | 50 | 5.62 |
| 10 | 40 | 50 | 6.10 |
| 11 | 40 | 50 | 5.66 |
| 12 | 40 | 50 | 5.61 |
| 13 | 40 | 50 | 5.76 |
| 14 | 40 | 50 | 5.91 |
| 15 | 40 | 50 | 5.91 |

The simplified model is then solved using the data:

$$\text{Consumer response (CR)} = A_0 + A_1 \times \text{Hue} + A_2 \times \text{Chroma} + A_{11} \times \text{Hue}^2$$
$$+ A_{12} \times \text{Hue} \times \text{Chroma} + A_{22} \times \text{Chroma}^2$$

Using a regression analysis computer program, the following equation is obtained:

$$\text{Consumer response (CR)} = -52.23 + 1.16 \times \text{Hue} + 1.20 \times \text{Chroma}$$
$$-0.01336 \times \text{Hue}^2 - 0.01090 \times \text{Chroma}^2$$

or

$$0.01336 \, (\text{Hue} - 43.4)^2 + 0.0109 \, (\text{Chroma} - 55)^2 = (5.9 - \text{CR})$$

Since $\text{CR} = 5 = $ neither like nor dislike, the right-hand term above becomes approximately $6 - 5$ which $= 1$. Therefore,

$$\frac{(\text{Hue} - 43.4)^2}{67.37} + \frac{(\text{Chroma} - 55)^2}{82.57} = 1$$

The best reponses are Hue at 43.4 and Chroma at 55.0.

As we have seen before, one of the advantages of this type of analysis is that the intermediate points of consumer preference may be extrapolated without conducting another series of experiments. The easiest way to do this is graphically. The graph of the response surface is approximated in Figure 12-9. The intersection of the two axes has a value of 43.4 Hue and 55.0 Chroma, with a hedonic response of 6.5 (the maximum preference rating of all of the consumers). The inner ring would have a value of 6.0; the intermediate ring, 5.5; and the outer ring 5.0. It is therefore possible to interpret consumer preferences at any combination of values.

Figure 12-9 shows a small portion of the total response surface, and may be thought of as the upper surface, with the observer standing above it, and looking down. A computer printout of the entire surface is shown in Figure 12-10, and represents a three-dimensional picture.

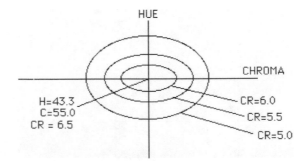

**Figure 12-9.**  Surface response–elliptical contours.

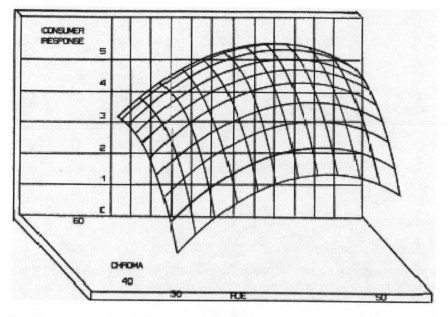

**Figure 12-10.**   Complete response surface.
*Credit:* Ransall Hamlin. *The Place of Design in Process Improvement* (Hunt Wesson Foods, Inc. 1987), K-116.

Returning to the initial statement of the experimental design goal, a single range of numbers may be selected as a color specification. An initial candidate for such a specification might be: if a consumer rating is above 5.5, accept the sample; if the score is below 5.5, resample. If resample is above 5.5, accept—otherwise reject.

The question of what to do if the value variable were also significant should be mentioned. If value were included along with hue, chroma and response, a mathematical solution would require a four-dimensional model—extremely difficult to depict! It would be wise to consult with a statistician, who might propose a compromise. The compromise method is to use only two of the variables and the response, as we have done here, then extract a slice from the model which includes the required specification levels, and then proceed to repeat the experiment (in whole, or in part) with value, response, and one of the other variables. For a complete test, a second replicate test would be required with the remaining variable. If the color problem is a critical one, the entire series of tests will be necessary; otherwise, a few spot values may be tested to "rough out" the location of specification levels for all the variables.

Elliptical contours are particularly useful since they define maximum responses. There are many other contours which are less desirable. One of these is the so-called saddle point response (Figure 12-11). The intersection of the axes is not the maximum response, but an intermediate value from which moving in

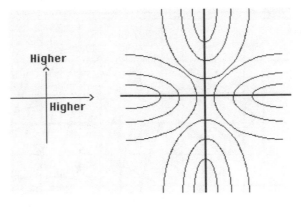

**Figure 12-11.**   Saddle point contours.

a vertical or right horizontal direction shows increasing responses, and conversely, moving down or to the left shows decreasing responses.

## EXPERIMENTAL DESIGN ANALYSIS BY CONTROL CHART

The classical techniques of analyzing the results of the tests are based on analysis of variance. It is also possible to use a different approach in which quality control techniques, such as control chart analysis, may be used to analyze the outcomes. To illustrate, let us look at a microwave bread test similar to the example shown previously. Four types of formulations will be tested in four different microwave ovens. All of the ovens selected for the test are 600+ W rated. As before, the outcome of interest is the volume of the loaves.

The test procedure called for baking five loaves of each formulation in each oven, for a total of 20 loaves per oven. (As a matter of interest, the number of samples was selected as five since the distribution of the means of five samples is very close to normal, regardless of how skewed the distribution of the individual samples might be. This is less true for smaller-size samples; and larger sample sizes do not appreciably improve the normality of the averages.) After baking, the average volume of the individual loaves is to be measured, and the averages of each group of five samples is to be calculated. In addition, the standard deviation of each group of five will be calculated. To summarize, the means and standard deviations for each group of five are to be calculated for each microwave, and all 16 sets of data analyzed. This now represents the outcomes of the baking tests for four different formulations in four different microwaves (Table 12-14). A final series of calculations is the overall average for each formulation and for each oven.

Merely looking at the data, it appears obvious that formula #4 has the highest volumes in each oven. What is less obvious is the reliability of this conclusion.

**Table 12-14. Quality Control Chart Analysis of One-factor Design Test. Effect of Microwave Oven on Bread Formula Volume. Average Loaf Volume (cubic inches), $n = 5$ loaves**

| Formulation | GE | TA | SA | AM | Average |
|---|---|---|---|---|---|
| 1 | 156/19.6 | 143/14.7 | 160/21.3 | 137/25.4 | 149.0 |
| 2 | 150/11.9 | 151/12.3 | 157/15.9 | 129/9.1 | 146.8 |
| 3 | 112/30.1 | 137/24.6 | 126/25.9 | 109/19.0 | 121.0 |
| 4 | 162/19.1 | 151/18.7 | 175/15.4 | 141/20.1 | 157.2 |
| Oven average | 145.0 | 145.5 | 154.5 | 129.0 | 143.5 |

Would one be sure that if the tests were repeated several times that the volume of formula #4 would always be greater than the others? Or, to express the question differently: can one state with any degree of confidence that these results for formula #4 are significantly different, reproducible, and higher than the others? Some kind of an analysis must be conducted at this point. $t$-Tests are useful for comparing differences between two means, or comparing a mean to a standard. When there are three or more to compare, the analysis of variance technique may be used. This utilizes the $F$ ratio (analysis of means) which is defined as:

$$F \text{ ratio} = \frac{\text{Variance between means}}{\text{Variance between each sample}}$$

If the variance between means is greater than the variance within the means (between samples), then there are significant differences. This requires many statistical calculations. Note that for oven TA, the volume means for Formulas #2 and #4 are equal at 151, and the variation for Formulation #2 (12.3) is much less than the variation for Formula #4 (18.7). In other words, if only one oven (TA) had been used for this test, the #4 formulation would not have been obviously best, as had been stated in the above paragraph. This again emphasizes the need for a balanced test.

Elimination of the differences between microwaves might be accomplished by the use of block designs. This has already been examined. But there is another method, perhaps simpler, which uses a control chart. Obviously, we are not dealing with a continuous process which lends itself to a control-chart technique, but we are dealing with a number of values, the relationship of which can be demonstrated graphically. The uniformity of the data can be represented for visual, rather than mathematical evaluation. This technique was suggested years ago before the common availability of computers, and was presented by Ellis Ott in his book, *Interpretation of Data*. published in 1979. This technique is still an excellent way to examine data.

A control chart is constructed (Figure 12-12), using the data obtained in Table 12-14. The center line is the overall average loaf volume: 143.5 cu. in.

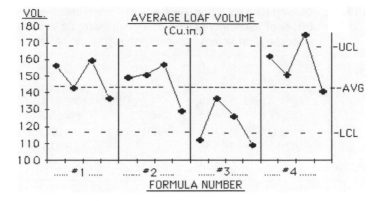

**Figure 12-12.** Control chart analysis of four ovens.

Limits are calculated which are the equivalent of three standard deviations, such as are used on the conventional control chart. The calculation of the limits will be explained shortly. Let us interpret the chart data.

First, examine the data for uniformity (consistency) of loaf volumes for each oven. Formula #1 is fairly uniform, with all loaf volumes within the control limits, and equally distributed on both sides of the average. Formula #2 appears to be satisfactory. However, two of the four loaves baked with Formula #3 are below the lower control limit, indicating that this may be an unsatisfactory formula. The fourth formula shows three readings within the control limits, and one loaf volume higher than expected from chance alone. Based on these observations, formula #4 appears to be the candidate to be taken to the next stage of development.

However, could one expect to find similar results if this test were repeated? A statistical test is required to determine whether the selected formula is significantly different than the others. This is accomplished by performing the analysis of means. The calculations are as follows:

$$\text{Average } \overline{X} = \left(\sum X\right)/N = 143.5 \text{ cu. in.}$$

$$\text{Average variance } S = \sqrt{\frac{s_1^2 + s_2^2 + s_3^2 + \cdots + s_{16}^2}{16}} = 19.7$$

Each cell represents ($n = 5$) loaves of bread; therefore, there are $n - 1$ or 4 degrees of freedom in each cell. There are 16 sets of 4, or 64 degrees of freedom.

Control charts use 3 standard deviations as the control limits. Determining 3 standard deviations for process control would normally require the use of the $Z$-table, which assumes an infinite number of observations. Here we shall use a slight modification of that principle because, in this example, several means (16) are involved, but the data is limited—it does not represent an ongoing process.

The control limits to be used are expressed as "decision lines" and are found using the following formulae

$$\text{Limits} = X \pm H_\infty \hat{\sigma}_{\bar{x}} \qquad \text{where } \hat{\sigma}_{\bar{x}} = \frac{\sigma}{\sqrt{n}}$$

$$= 143.5 \pm 2.94(19.7/\sqrt{5})$$

$$= 143.5 \pm 25.9 \text{ or } 117.6 \text{ to } 169.4$$

The value for $H_\alpha$ is found in the $H_\alpha$ table in the appendix (Table A-12) by extrapolating for $k = 16$ and df $= 64$. The table includes values for $H_\alpha$ at a producer's risk of rejecting a good lot $(\alpha)$ of 5%; that is, there is a 5% risk that there is a significant difference between means when there is none.

As has been mentioned previously, the usual method of analyzing data when a computer is available is to have a technician enter it into an analysis of variance program and let the computer do the work. By using the graphical method shown above, rather than a computer analysis, the "feel for the data" is not lost. The pictured data shown on a graph should be much more meaningful to the experimenter than the numerical computer printout.

There is more information which may be extracted from the data. It is possible to determine if the differences observed between formulae and the differences observed between ovens are statistically significant. The process of calculating the significant difference level is called the Multiple Range Test for the Just Significant Difference Value (JSV), and uses the formula:

$$\text{JSV} = q\sqrt{(s^2/n)}$$

where $q$ is found in the table: The Upper 5% of the Points of the Studentized Range $q = (x_n - x_1)/s$ which is in the Appendix (Table A-11).

$$\text{JSV} = 3.74\sqrt{19.7^2/5} = 32.9$$

The differences among ovens must be at least 32.9 cu. in. to be significant. The largest differences between ovens as shown in Table 12-14 are:

$$\text{For Formula \#1: SA} - \text{AM} = 160 - 137 = 23$$
$$\text{\#2: SA} - \text{AM} = 157 - 129 = 28$$
$$\text{\#3: TA} - \text{AM} = 137 - 109 = 28$$
$$\text{\#4: SA} - \text{AM} = 141 - 175 = 34$$

It may be concluded, therefore, that there is a significant difference between oven SA and oven AM, but no others. Or, in other words, formulation #4 is more adversely affected by choice of ovens than are the other formulations.

Another comparison of importance is that of the significance of differences between formulations.

$$\text{For oven GE: } 4 - 3 = 162 - 112 = 50$$
$$\text{TA: } 2 - 3 = 151 - 137 = 14$$

$$SA: 4 - 3 = 175 - 126 = 49$$
$$AM: 4 - 3 = 141 - 109 = 32$$

The conclusion here is that Formulation #3 has a significantly lower volume than Formula #4 for three of the four ovens. It is not clear whether Formula #4 is better than the other two formulations. The selection of which protocept to select for subsequent testing would probably be #4, but it is indicated that Formulations #1, 2, and 4 should be tested once more in a single oven (to rule out the effect of ovens on the outcome). From a marketing point of view, oven SA would be selected, since it appears to contribute to the highest volume for Formulation #4.

The discussions of the several paragraphs above show the advantage of using this Multiple Range Test instead of the Analysis of Variance (ANOVA) to evaluate the formulations and the ovens. Both types of analysis permit the use of a number of variables in a single test, thus showing the interactions of variables, saving considerable time and money when compared to the more traditional single-variable testing procedures. ANOVA tells the experimenter whether the variation between means is significant or not; but it fails to show which of the means are the significantly different ones. The Multiple Range Test accomplishes both of these objectives. An additional advantage of the Multiple Range Test approach is that the final report to management is both pictorial and tabular, and is easier to present and to understand.

To summarize the information required to design an experiment:

1. Definition of the purpose and scope.
2. Identification and specification of the experimental variables.
   (a) Primary variable
   (b) Background variables (control by blocking)
   (c) Uncontrolled variables (control by randomizing)
   (d) Variables held constant
3. Estimates of repeatability (estimate of error or standard deviation).
4. Desirability of conducting experiment in stages.
5. Prior knowledge about results.
6. Constraints on the variables.

We have explored a few of the more commonly used experimental designs, but have merely skimmed the surface of the accompanying mathematical manipulations of data. Although computer programs are available to perform this portion of the designs, it is suggested that the manual procedures be mastered in order to understand which programs are applicable to the problem at hand.

# 13    Vendor Quality Assurance

It is difficult to generalize about costs of raw materials in the food industry since the products vary from flavored water to caviar. The cost of the inexpensive drink's flavoring, sweetener, acidifier, colorant and packaging might be as little as 15% of the factory door cost. Restaurant operations may estimate that 40% of the total cost of operation is for purchased foods. High-ticket food items could exceed 50%. A high-cost raw material focuses attention on the need for uniform acceptable quality of that material to a greater extent than does a modestly priced unprocessed food ingredient. From a quality control standpoint, this is an unfortunate concept, since the final quality of the finished product depends not on the price of the raw ingredients, but on their quality level and uniformity. With this in mind, it appears that a large part of the responsibility for the successful production of a uniform quality food item, manufactured with a minimum of scrap and rework, rests with the purchasing department. In fact, the responsibility rests with all of the processor's departments, each of which should be in close contact with the suppliers.

There is an attempt by some to refer to the quality at all steps in the construction of a product as the responsibility of each worker who performs some function which affects the finished quality, and that each worker, therefore, may be considered to be a vendor or supplier to the worker at the next step. This same philosophy suggests that all workers along the chain, from raw material to the final product, are also customers of the vendors in the preceding operation. This concept of "quality is everyone's business" may be perfectly valid, but for the sake of clarity, we shall consider the vendors (or suppliers) as the organizations who supply the raw materials to the processors (or vendees, or customers, or manufacturers). The vendors may be farmers, or they may be manufacturers of cans, corrugated board, process machinery, kraft bags, food additives, or any other material or service which is involved with the flow of a product through the processor's plant.

A typical cycle showing the flow of raw materials to the processor (Figure 13-1) may assist in identifying some of the costly flaws of commonly used systems.

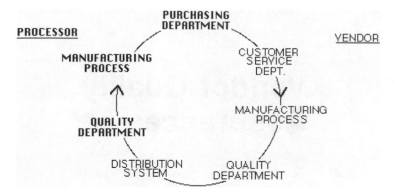

**Figure 13-1.** Movement of purchased goods from vendor to processor.


Ordinarily, the manufacturing department will issue a purchase requisition for raw materials or supplies, usually with a description or specification. This is where quality control problems might originate. Unless the material quality requirements are clearly defined when forwarded to the purchasing department, the possibility exists that unusable material will be ordered. The purchase order which is then prepared flows to the vendor's customer service department which reviews (or at the very least, glances at) the requirements, and prepares either a ship-from-stock order or a production schedule for the vendor manufacturing facility. In the course of production, the vendor's quality department identifies product which does not conform to the requirements, and directs it to rework or scrap. All else is shipped to the processor with a bill of lading, and possibly a certificate of quality. When received by the processor, the material is usually inspected for compliance with the specification before releasing it for production. The inspection may range from merely counting the number of cases received and comparing this number with the quantity ordered, all the way to sampling inspection and even 100% inspection and analysis before approval. Again, unsatisfactory material is either reworked or returned to the vendor for credit.

This simple path has been used for years by food processors of all sizes, and many of them are well aware of the inadequacies of the procedures. Accepting less than satisfactory quality raw materials practically guarantees that the finished product will be of less than satisfactory quality as well, or that it will have satisfactory quality at higher production cost because of sorting, scrap, or rework. Frustrations and costs can arise from incomplete specifications, even with duplicate testing at the supplier's and processor's plants. Consider a purchase order which calls for another shipment of 50,000 cans with the usual red company logo. Perhaps the vendor's quality control lab notices that the color appears to be a little lighter than they remember it from the last run, but assumes it is close enough. When the cans are received by the processor, their quality control lab notices that the color is a little less red than the last run, but since there are no more cans available to run, rather than shut down the line, the production department decides to use them. Two weeks later, when the off-color cans are shipped from the plant to the field warehouses to the distribution centers and finally to the stores, the processor's salesmen see the cans for the first time, and angrily demand that these substandard orange cans be replaced with the correct red-colored product. When stacked on the grocers' shelves next to normal colored cans, the contrasting appearance strongly indicates that the company is unconcerned with controlling their quality, and sales are suffering.

This hypothetical problem might have been stopped at the outset by specifying the color requirement in numerical terms (with acceptable limits) when the purchase order was written. And it might have been stopped at either the vendor's quality control laboratory, or at the processor's quality control inspection at the receiving dock, or as the cans were first fed to the production lines. Several people had an opportunity to take effective action after observing the defect. It also might have been stopped if an open line of communications existed between the purchasing department and the customer service department; or between the vendor and the processor quality control departments. Perhaps the production department should not have the authority unilaterally to decide whether it is more economical to run substandard materials or to shut down a production facility.

There is another lesson to be learned here as well. If the vendor's quality control laboratory is required to examine the cans, why is it necessary to duplicate this effort at the processor's quality control laboratory? The answer for this example is simple: because the two laboratories do not talk to each other. A spot check and an occasional audit by the processor's laboratory might be all that is required, provided that the vendor knows exactly what is required, and has a free line of communication to the processor's laboratory in the event of nonstandard product. If this plan could be implemented, it might be possible to reduce the size of the quality control inspection crew presently assigned to incoming inspection. Simply by improving communications, we have already improved productivity!

## VENDOR–VENDEE RELATIONS

Obviously, testing and retesting does not necessarily prevent quality problems, nor does it cement relations between vendor and customer. There are opportunities to improve both of these situations. When properly designed, a program known as "vendor certification" can be an effective tool. This relationship cannot be proclaimed and used overnight. It must be developed slowly. Philip Crosby has invented the "Quality Management Maturity Grid" (*Quality is Free*, McGraw-Hill, 1979) which lends itself to this growth of trust and understanding between suppliers and customers.

*Stage 1*. Uncertainty
    Orders are placed with the vendor and accepted without question by the purchaser. Little but the most obvious receiving inspection is performed.

*Stage 2.* Awakening

The purchaser becomes aware that much material is rejected on the line due to drifting quality of the raw material. He suspects that there must be a better way of getting uniform and acceptable quality.

*Stage 3.* Enlightenment

Sampling inspections are increased, and quality engineering studies are conducted to explore improvement of raw material quality uniformity. Exploratory discussions start between vendor and supplier.

*Stage 4.* Wisdom

The purchaser formalizes incoming inspection plans. Some progress is made toward vendor certification of analyses of shipped lots. The purchaser is aware of the possibility of selection and certification by audit of candidate supplier's plants to insure that the supplier's plant is in statistical control, and is capable of producing acceptable product.

*Stage 5.* Certainty

Acceptance of shipments is conducted solely by the terms of the purchase order.

According to Kaoru Ishikawa, there are ten quality control principles for good vendor–vendee relationships.

1. Both vendor and vendee are fully responsible for the quality control applications with mutual understanding and cooperation between their quality control systems.
2. Both vendor and vendee should be independent of each other and respect the independence of the other party.
3. The vendee is responsible to bring clear and adequate information and requirements to the vendor so that the vendor can know precisely what he should manufacture.
4. Both vendor and vendee, before entering into business transactions, should conclude a rational contract between them with respect to quality, quantity, price, delivery, terms and methods of payment.
5. The vendor is responsible for the assurance of quality that will give satisfaction to the vendee, and he is also responsible for submitting necessary and actual data upon the vendee's request.
6. Both vendor and vendee should decide the evaluation method of various items beforehand which will be admitted as satisfactory to both parties.
7. Both vendee and vendor should establish, in their contract, the systems and procedures through which they can reach amicable settlement of disputes whenever problems occur.
8. Both vendee and vendor, taking into consideration the other parties' standing, should exchange information necessary to carry out better process/quality control.
9. Business control activities (production, inventory planning, ordering, reporting systems) should be performed sufficiently—to maintain their relationship on an amicable and satisfactory basis.

10. Both vendor and vendee, when dealing with business transactions, should always take into full account the consumers' interests.

"Always have a mutual confidence, cooperation, and a degree of live-and-let-live in vendor–vendee relations."

—Kaoru Ishikawa

## SPECIFICATIONS FOR RAW MATERIALS, INGREDIENTS, SUPPLIES

By now it should be clear that vendor quality starts with specifications. Many food-related specifications exist, occasionally accompanied by detailed test methods, in governmental and commercial areas. Some of these standardized specification sources are listed below.

Federal Sources
    Food and Drug Administration
    Bureau of Standards
    U.S. Dept. of Agriculture
    Military Specifications
    General Services Administrations
International Organizations
    Food and Agriculture Organization (FAO)
    World Health Organization (WHO)
    Codex Alimentarius
International Organization for Standardization (ISO)
Industry Specifications (Standards agreed upon by producers in an industry)
    American Dairy Products Institute
    American Egg Board
    American Frozen Food Institute
    American Institute of Baking
    American Meat Institute
    American Spice Trade Assn.
    Associations for Dressings and Sauces
    Can Manufacturers Institute
    Composite Can and Tube Institute
    Flexible Packaging Association
    Glass Packaging Institute
    Grocery Manufacturers of America
    Institute of Food Technologists
    International Coffee Organization
    National Coffee Association
    National Dairy Council
    National Soft Drink Assn.
    Processed Apples Institute

Technical Assn. of the Pulp and Paper Industry
The Food Processors Institute
The Vinegar Institute
Special Purchasing Specifications
Standards agreed upon between vendor and vendee.

Some care should be used when selecting standardized specifications for the first time. They may be more stringent than needed, or they may not be sufficiently definitive to satisfy the requirements of a particular process. A moisture specification worded "not to exceed…%" may be required to insure satisfactory keeping quality, but perhaps a lower limit might also be required to permit rapid rehydration, or to prevent clumping when the material is used in a mix. With the exception of governmental standards, where the entire specification must be adhered to, a specification for a raw material should be as simple as possible, and should include only those attributes which are considered critical to the user's process. As a general rule, the fewer critical specifications demanded by the purchaser, the lower the price from the vendor. There might be little point to specifying the whiteness of corn sugar used in a mix destined for black licorice candy ropes.

A number of factors affect the longevity of raw material specifications. Some become completely obsolete as substitutes or new varieties become available. Public attitude and demands change, customers change their preferences for foods they wish to buy, new technology makes it possible to produce new foods or modify existing types, and modified raw materials may become available as established ones become scarce.

The "perfect" specification can be created without any familiarity with the product; but to be of any value, a specification must be practical and workable. It is relatively difficult to write a specification which can be used on the factory floor to produce goods, or which can be used to purchase raw materials on a widely available basis. Where possible, specifications should be developed mutually between the vendor and the vendee, so that the test methods, shelf life, sampling methods, and storage conditions are compatible to both parties' operations.

Many of the requirements for preparing specifications were covered in Chapter 7. For the sake of completeness, a checklist of frequently used specification parameters for raw materials follows:

- Bacteriological Standards
   SPC, Yeast, Mold, Coliforms, *E. coli, Salmonella, Lactobacillus*, Thermophilic Sporeformers, etc. Usually expressed in maximum numbers per gram, or presence or absence.
- Physical Standards
   Particle Size, Flowability, Visual Color, Solubility, Bulk Density, Flavor, Odor, Sediment, etc.
- Analytical/Chemical Standards
   Fat, Moisture, Water Activity, Volatile Oil, Starch Gelatinization, pH, Sugar Content, Residual Pesticides, Sodium, Protein, etc.

- Sanitation Standards
  Insect fragments, Rodent Hairs, Foreign Material, Special Processing
  Conditions, Personnel Hygiene, Procedures for Plant and Equipment
  Sanitation, Hygiene, Pest Control, etc.
- Other Standards
  Shelf Life, Storage Conditions, Product Coding, (Lot Numbers, Item
  Numbers, Dates), Handling Procedures, Formulas, etc.
- Sampling Procedures and Test Methods
  Type of sampling (random, stratified, periodic, etc.) Number of samples
  per lot, per period of time, per shipment, Specific test methods (AOAC,
  FDA, non-standard, etc.), Lot acceptance sampling and evaluation.

Specifications for packaging materials require dimensional data, as well as packaging composition (board, laminates, film, adhesives, cans, jars), printing detail (graphics, logo), and sometimes chemical and physical standards (oxygen permeability, moisture barrier, product transfer resistance, other chemical barriers).

## QUALITY ASSURANCE OF PURCHASED GOODS

When one considers the astronomical number of materials, suppliers, processors, products and materials in the food processing industry, it is no wonder that no single set of quality control procedures has been evolved to satisfy the vendor, the processor, and the final customer. Plans have been made available on a regular basis for many years, starting as early as 1948 with H.F. Dodge's discussions on avoiding lot rejections triggered by sampling plans. (H.F. Dodge, Administration of a Sampling Inspection Plan, *Industrial Quality Control* vol. 5, No. 3, pp. 12–19, Nov. 1948.) Even the basis for inspection—should it be by variables or by attributes—has been the subject of heated debate since the early 1940s. Some have maintained that attribute inspection is easier, requiring relatively unskilled personnel, speedy examinations, and ironclad "accept or reject" criteria. Others argue that there are at least three classes of defects for each attribute, and that considerable judgment is required to distinguish between them. Unsolved is the problem of when it is more economical to return a substandard shipment, to subject a substandard shipment to 100% screening inspection, or to use it and charge back to the vendor the added costs of using the substandard material on the production line. On the other hand, even though variables analysis of a shipment may require skilled personnel, much time, and many more examinations than would an attribute-based system, the results of variables testing are concrete, indisputable numbers requiring no judgment on the part of the technician, and rarely cause friction between the vendor and the vendee.

One would expect that a logical analysis balancing the costs and the savings of each alternative would produce a series of plans at various end-product quality levels which would satisfy both the vendor, the vendee, and the final consumer. In the real world, this is merely the starting point! The vendor wishes to maximize profits; the processor wishes to minimize costs; and the customer looks for perfection in the product. It is not possible to negotiate, and usually risky, to compromise the level of product quality offered to the customer. But negotiations between vendor and vendee are the only realistic tools available. There has to be a starting point from which continual improvements can be made.

There are hundreds of acceptance plans available, and the choice is based upon mutual agreement between the seller and the buyer. One attribute plan in use for years is the one based on MIL STD 105-E of the U.S. Department of Defense. It offers three levels of inspection for acceptance: normal, tightened, and reduced, and provides detailed switching procedures between the levels based on lot rejection experience. It requires that lot inspection initially be set at normal, changed to tightened when 2 out of 5 consecutive lots or batches have been rejected; returned to normal when 5 consecutive lots are again found acceptable; lowered to reduced when 10 consecutive lots are found acceptable, etc. As cumbersome as this appears, it has worked successfully for years in the quality control acceptance function of military and civilian industries alike. It has found its way into process control as well as material control, and has been used successfully as a reliable (although not particularly rapid) indicator of quality level changes.

One of the goals of acceptance sampling is to avoid raw material testing at the processor's plant entirely. The thought behind this goal is that the vendor is far more experienced in testing the line of products he sells than is the processor, and may be in a better position to correct defects as they are uncovered. If this generates added costs to the vendor, then they may be passed along to the processor, frequently at less cost than adding a second level of quality inspection. The processor would then be in a position to receive certified shipments, along with quality control documents showing test data, control charts, and final reports as pertain to each shipment. He would reduce his raw material quality inspection system to an occasional audit, preferably at the vendor's operation. A word of caution: where critical defects might be found in a raw material, certification is desirable, but it is not enough evidence of compliance. Audit samples must be run on each shipment, regardless of efforts on the part of the supplier. The responsibility for a safe food product rests with the processor, not the raw material vendor.

An interesting plan showing the interaction between vendor and vendee is suggested by Ishakawa (Table 13-1), and shows the developing and maturing relationship.

At stage one, quality control is least developed. The vendor ships the raw material as soon as it is manufactured or assembled, without any inspection or testing. The vendee accepts the material without inspection and sends it to the manufacturing operation. Manufacturing in turn has to 100% inspect the material to select that portion which is suitable for production. In some cases, manufacturing may not even inspect the material, but may use it blindly. Since this generates unsatisfactory finished product, low productivity, and excessive scrap and rework, the vendee proceeds to stage number 2: 100% inspection at the receiving dock. If this inspection procedure is effective, only acceptable materials now reach the manufacturing operation, but the added inspection costs may equal or even exceed the costs of lower productivity in stage 1.

**Table 13-1.   Development of Vendor–Vendee Relationship**

| | Vendor | | Vendee | |
| --- | --- | --- | --- | --- |
| Phase | Manufacturing dept. | Quality dept. | Quality dept. | Manufacturing dept. |
| 1 | — | — | — | 100% inspection |
| 2 | — | — | 100% inspection | — |
| 3 | — | 100% inspection | 100% inspection | — |
| 4 | — | 100% inspection | Sampling or checking inspection | — |
| 5 | 100% inspection | Sampling inspection | Sampling or checking inspection | — |
| 6 | Process control | Sampling inspection | Checking or no inspection | — |
| 7 | Process control | Checking inspection | Checking or no inspection | — |
| 8 | Process control | No inspection | No inspection | — |

Prompted by the volume of returned goods from the vendee's receiving inspection, the vendor now engages in 100% inspection of the finished goods, and the vendee's quality control notes an improvement in the quality of the materials received. As a result, the vendee reduces receiving inspection from 100% to a statistical sampling plan, such as 105-D (stage 4). If he is more confident, and if the vendor is willing, the vendee will now be satisfied to merely review the vendor's quality control records. Otherwise, the vendee might try reduced inspection or skip-lot sampling.

The vendor is now encouraged by the reduced costs brought about with relatively minimal effort at quality control, and expands the coverage (stage 5) to include 100% inspection in his own processing department, while reducing the finished material sampling to an acceptance sampling plan prior to shipment. The vendee, meanwhile, continues minimal incoming inspection.

Although he may have reached a rewarding level of shipping all satisfactory material, the vendor should explore the possibilities of reducing his own scrap and rework by using some of the tools of experimental design and quality engineering. By studying and improving his process capability, the vendor should be able to replace 100% process inspection with process control techniques, while still retaining the lot acceptance method of finished goods sampling method for lot acceptance. The vendor has reached stage 6. The vendee continues to check the incoming material quality, but may have reached the point where shipments, accompanied by a certificate of analysis, can be accepted.

In stage 7, the effectiveness of process control at the vendor's operation allows reduction of end-of-the-line sampling inspection and lot acceptance to checking a few samples. The vendee may now accept materials which are accompanied by the vendor's process control charts, letters of analysis or similar certificates.

Stage 7 is probably attainable; stage 8 is more a goal than an actual condition. Stage 8 assumes an ideal situation where process capabilities have been improved, and reliable day-to-day process control has been implemented. Under these conditions, shipping inspection by the vendor is no longer required. It is as if the vendor's manufacturing line had become attached to the start of the vendee's processing line. It is unlikely that this condition can be maintained, if reached at all, since conditions change, employees or management may change, people become comfortable at their work and make mistakes, and the requirements for new and improved raw materials are always demanded by processors. By maintaining constant contact with the vendor, these conditions can be predicted before they develop, and corrective steps taken.

It should become apparent, however, that as both vendor and vendee strive to approach stage 8, both parties may drastically reduce the number of personnel required for the task of inspection. This reduction, as well as the reduction in cost from the elimination of defects, is accompanied by a rise in productivity, a lowering cost, and the establishment of a reliable vendor quality assurance.

Ishikawa's philosophies as discussed above are somewhat detached from people. The statements appear as absolutes and as goals to be attained by "vendors," "vendees," "companies," "departments," etc., although this may be a misinterpretation by Ishikawa. In many companies, there exists a "them and us" attitude which is extremely counterproductive. The quality control and production departments frequently have conflicting goals ("produce according to the specifications or shut down" versus "produce at least cost and get it out the door on time"). We have already looked at the conflicting goals of the vendor and his customer. One might expect that the background, training, interests, and goals of purchasing personnel are of a far less technical nature than those of quality control professionals. Somehow, all of the suspicions, antagonisms, conflicts and jealousies of these parties have to be put aside in order for a quality control system to work. One can imagine the devastating impact a statement such as "you, Mr. Vendor Quality Manager, really made a big blunder on that last shipment to us, but we Vendee Quality people discovered it in time." One can easily visualize Mr. Vendor Quality Manager thinking about some form of retribution which would darken the reputation of his counterpart at the vendee's plant.

In order to improve productivity by purchasing defect-free raw materials, one key relationship is that between quality and purchasing. If the purchasing department has not been exposed to the details and the power of $X$-bar and $R$ charts, they are unlikely to look upon them favorably when offered by quality control. If the quality department is unaware of the difficulty of negotiating lowest prices with independent vendors while still bowing to the company management's intuitions, prejudices and favorite suppliers, they may not be anxious to work with the quality department of a less-than-desirable vendor selected by the purchasing department. Situations such as these are common, and though not completely avoidable, they may be smoothed out if both parties earnestly work toward the common goal of productivity.

## SELECTING AND NURTURING A SUPPLIER

The responsibility of vendor selection has historically rested with the purchasing department. Somewhat archaic criteria are still being used in the process: service, reputation, and lowest price. Over the years, selections have been affected to varying degrees by promotional efforts on the part of potential suppliers. Although it looks ridiculous in print, decisions have been swayed by friendships, golf, gratuities, lunch and dinners. Practices based on these suspect principles continue because of the philosophy "if it ain't broke, don't fix it." If quality and productivity is affected adversely by selections based on such decisions, then the supplier system is "broke" and needs to be carefully reassessed. Fortunately for both the food industry and the consumer, these bleak practices are disappearing, and are being replaced by far more logical decision-making criteria, with the selection process including input from other concerned department managers.

If the price from a potential vendor appears completely out of line with his competition, it is suggested that before outright rejection, the reasons for the higher price be investigated. There is always the possibility that the price includes special services, higher or more uniform quality, warehousing, delivery concessions, etc. Price discussions used to be the first topic of discussion, but are now moved further down the line. The first questions should uncover the vendor's capability to supply precisely those raw materials, ingredients, or packaging supplies required by the vendee's specifications on a rigorous delivery schedule. Close cooperation between the two parties may be required in the event that a vendor is unable to supply material to the specifications of the processor. In such a case, it is desirable to have selected a vendor who would willingly develop a satisfactory material with the assistance of the vendee. An easily accessible line of communication should be available. The vendor's use of process control and the ability to respond to the need for enhancing process capabilities should be explored. Additionally, he should be able to control the volume of production, and have the financial ability to insure expansion if required to meet increasing vendee needs.

It is desirable for a vendor to have exhibited a stable management system that is respected by others in his industry. There should be an understanding of the vendee's management philosophy as well, and close cooperation between managements should be established and maintained. The vendor should maintain high technical standards and have the capability of dealing with future technological innovations. The vendor must not be in a position where corporate vendee secrets would be breached.

An assessment of the management philosophy of the vendor is closely linked to the company's quality control policy, and may be difficult to obtain. Person-to-person contacts will eventually disclose knowledge, capabilities, personalities, and quality understanding.

A vendor audit of quality control may be requested by the purchasing department to reinforce its initial selection of a vendor. Depending on the nature of the materials under consideration, an initial audit may be extremely detailed, or somewhat informal. In either case, the following are considered to be necessary areas for inclusion:

1. The quality philosophy of the vendor.
2. Quality records of other (non-competitor) vendees.
3. Corporate history and current trends in vendors industry.
4. Complete product list.
5. Details on equipment, processes, production capabilities.
6. The vendor's quality assurance system, including quality education and implementation programs.
7. Vendor's controls on their own procurement of raw materials and subcontracting.
8. Sanitation practices.

Maintaining a good relationship with vendors once they are on line is a necessary and frequently rewarding effort. The purchasing department maintains a relationship which may have different objectives than those of the quality department. The processor's quality records of materials acceptance from each major supplier should be routinely analyzed for trends. Nonroutine audits conducted at the vendor's plant will occasionally disclose practices or conditions which need to be brought to the attention of both the vendor and the processor's purchasing department. If needed, advice and assistance should be freely given to the supplier to eliminate problem areas and to further cement relationships. If indicated, inter-laboratory calibration tests might be run to show an interest in assisting the vendor. A system of rewarding suppliers for implementing and maintaining process controls and quality systems will promote effective defect prevention and continued satisfactory quality. Because of the mobility of our society, it would not be unlikely to lose a good vendor–vendee relationship temporarily through loss of key people from either company. The only choice is to start over again, building up the confidence and interchange of ideas necessary to keep the system operating.

Relationships may be improved in many unexpected ways. Suppose that a vendor's quality manager has been unsuccessful in convincing his CEO that there is a need to update laboratory equipment. No more than a word or two from the vendee quality manager to the CEO that his quality laboratory would benefit from certain modernized equipment, and chances are the vendor and vendee quality managers become long-term allies.

The terms "vendor" and "vendee" in the above discussion have been used generically. The impression is that they refer only to suppliers and purchasers of materials and products. However, the same principles apply equally to the relationship between processors and co-packers or processors and warehouses. The co-packers provide an extension to the production lines as a service, and are subject to the same scrutiny and assistance that would normally be provided within the processor's plant. By the same token, outside warehouses provide a service which can adversely affect quality unless monitored. Specifically, rotation of stock or inadequate storage temperature controls may both adversely affect shelf life. Sanitation, pest control and control of damaged goods are equally critical quality considerations.

The effect of a JIT (or just-in-time) contract with a vendor raises special quality considerations. Briefly, a JIT system of obtaining raw materials is based on a belief that no inventory of raw materials need be held at the processor's plant if the vendor can supply the correct material "just in time" to meet the production schedule for the day. The system is far from new—it has been in use for years in the canning industry, for example. Field crops such as corn, peas, tomatoes, peaches, etc. are delivered on a scheduled basis to the canning, dehydrating or freezing plant within minutes of their actual processing. In spite of the complications which arise in tight scheduling, the system works well, and permits processing crops at the moment of peak flavor. Similarly, for many years, the can, plastics, glass and paper packaging products manufacturers have supplied the day's packaging requirements to food processors starting on the morning needed, and continuing throughout the production period as scheduled. The savings in inventory storage costs and warehouse space needs at the processing plant are enormous, although in some instances, these needs might simply be moved back to the vendor's manufacturing plants where *they* generate the costs to be either absorbed or passed along.

Quality control in a JIT system becomes critical. There is no time to reject and replace a lot of defective incoming packaging materials. They must be produced right the first time. A successfully working quality control system must be in place at the vendor's manufacturing facilities before a JIT program is started. It is not uncommon to have a contractual arrangement between vendor and vendee on the disposition of costs associated with substandard shipments, and there are many options available. A shipment of substandard cans might be 100% inspected by the processor, with the added costs absorbed by the vendor. Or the cans might be run through the production line and the filled product sorted at the end of the line, with costs of inspection, scrap and rework passed back to the vendor. A post-manufacturing sampling inspection at the vendor's plant prior to shipment might be made by the vendor, the processor, or both before shipment is approved, hopefully eliminating the expensive difficulty of shipping defective materials. In the "real world," either the vendor or the processor or both generally maintain a small emergency supply of packaging materials to tide them over in the event of a major quality problem. Whenever a substandard shipment does manage to escape the quality control system, it is imperative that the cause of failure be immediately determined and corrected so that it can never happen again.

Before leaving the subject of vendors, a word or two concerning single source of supply would be in order. There has been a growing interest in this concept over the years. The justifications are that better control of quality and costs can be effected by the close relationship developed between single source and purchaser; that increased volume sales from the single source should provide quantity discounts; that fewer liaison personnel are required from purchasing, engineering, quality control and others; and that a sense of "family loyalty" develops between the two organizations, with better understanding of mutual needs.

On the other hand, the familiarity and friendships which develop between the two closely allied companies might lead to occasional difficulty in enforcing unpleasant decisions, such as refusing a shipment on the basis of quality or late delivery. Other inherent dangers of a single source are those involved in the event of labor disputes, inclement weather, loss of key personnel, power failures, transportation problems, all of which might interfere with continuous on-time delivery of raw material to the processor. Then there is the ever-present threat of company takeovers which could spell inconvenience or worse to either of the two companies. In the event that the vendor is bought out, the processor might find itself without a ready source of raw material; in the event of a processor buyout, the vendor might find itself without a customer. The vendor is placed in the uncomfortable position of depending solely on one major customer, and probably a greatly reduced sales force, since none may be needed under the arrangement. If the vendor is dropped for any reason by the processor, there may not be sufficient time available to rebuild a sales force to enable it to remain in business. Perhaps these possibilities can be solved by contractual agreements, but they are potential problems to single-source purchasing.

## PACKAGING SUPPLIER QUALITY ASSURANCE

Although the following discussion pertains to purchasing of packaging materials, the same principles can be applied to any raw material supplied by an outside vendor. The buyer-seller relationship of the past has occasionally been antagonistic. Worse still, the "friendly arrangements" between purchasing managers and some of their vendors were occasionally responsible for the questionable supplies arriving at the receiving dock. Although these battles and special dealings may never be completely eliminated, there has been a growing trend toward a relationship best defined as "partnering" between supplier and user. It is becoming increasingly clear that the seller, the purchaser, and the final product user have all benefited from this honest and open interaction.

An autonomous purchasing manager for packaging materials developed a quality assurance program with suppliers, which borders on a vendor certification program. Although the procedures put forth are not necessarily recommended, they have become fairly common in companies of all sizes, and in many cases are successful.

Contacts with the vendor's sales people have been handled generally by the purchasing department. This avoids the possibility of creating violations of contractual agreements which might not be known by, for example, quality control, engineering, or production. On the other hand, frequent meetings of the purchasing department with members of all other groups within his company are mandatory, so that improvements, suggestions, and problems can be aired. These may be identified as formal teams, with a rigorous schedule of carefully structured meetings, or they might be more informal.

Packaging materials are often as important to the quality of food products as are the other raw materials, and for some food manufacturers, the cost of packaging materials is second only to labor costs. Substandard packaging material may result in lost production time, loss of materials, claims, and customer dissatisfaction. In the past, the practice for preventing non-conforming materials from reaching the customer consisted of receiving inspection and tests, with the hope that any defective material would be detected before reaching the production line. Manufacturers have become increasingly aware of the weakness of this system. It is rarely economically possible to sample every lot of every packaging material received fast enough and reliably enough to prevent operational problems. Furthermore, as production facilities become increasingly automated, there are fewer operators on the line to detect substandard packaging materials which went undetected at receiving inspection.

The simple solution is to expect improved processes at the vendor facilities. The first step in this conversion of policy is the preparation of complete, accurate specifications so that the vendor has a clear understanding of the nature of "defective material." The vendor now knows what quality control testing is required for his customer, and should be able to provide satisfactory shipments with supporting data. In effect, this shifts the inspection and approval of packaging material from the processor's receiving dock back to the supplier. Certification papers (if required) include production data, size and frequency of samples, test results, and signed reports. Depending on experience with the vendor and the nature of the operation, some companies prefer advanced certification; others might accept certification with the shipment. Some of the responsibility continues to rest with the purchaser, and all shipments should still be subject to audit by them.

In order for specifications to be workable, they should clearly list minimum, and maximum measurements, as well as targets. They should be worked out through cooperation with vendors and purchaser jointly. In some instances, the vendor's process capability charts will provide realistic data which both parties find reasonable. Most packaging specifications can be constructed as fast as they can be typed into a computer. However, there are instances where critical performance or intricate design will call for patience and understanding between the two parties. The engineering drawing of a pour spout, for example, may appear to be precisely defined, but tiny deviations from the dimensions shown might produce catastrophic leakage. The vendor might provide modifications or compromises which can make the design realistic.

Some of the tests run routinely on paper and fiberboard packaging materials to avoid problems on the line are: material composition, inside and outside dimensions, tensile, tear, burst, weight, moisture, smoothness, caliper, moisture, brightness (TAPPI: Technical Association of the Pulp and Paper Industry). In addition there are hundreds of specific tests for composite cans (Composite Can and Tube Institute), flexible films (National Flexible Packaging Association), cans, glass, paper and fiber, molded plastics, etc.

Management expects that most directives to install a new company system will be complied with fairly quickly, and and that the new system will be essentially complete when introduced. For example, a system to prevent multiple disbursements of payables in the accounting department might require several months of planning, but will usually be effective the day it is implemented. Similarly, the introduction of new systems to eliminate incorrect time card punchins, or pilferage on the shipping dock, or overblown claims in advertising, or excessive absenteeism—these and other problem areas are normally corrected within a few months. Not so with quality control systems. Here, the enormous number of factors which must be right, and the even more enormous number of factors which cannot be permitted, may take three or four years to uncover and specify. For example, a packaging converter may mistakenly manufacture a double bulk bag (one inside another) as few times as once every 100,000 bags. This could represent as few as one defective bag in a month's production at the purchaser's plant, but could result in several hours of downtime to clean up spilled product and jammed equipment. With a defect this rare, it is unlikely that, at the outset, the packaging specification would initially include a tolerance for double bags, but after a year or two's experience with this problem, it would no doubt be added to the specification.

Nor would one expect to have the weight of glue on the side seam of a folding carton included on the initial packaging specification. This could be recognized as a critical measurement for the purchaser, but would be a difficult characteristic for him to measure. On the other hand, if made aware of the problem, the vendor would be in a position to readily set up a control on the amount of wet glue applied to the cartons during manufacture, and could devise a fairly simple test to ensure that no cartons are manufactured without adequate glue.

One of the best sources of information leading to improved specifications is the worker on the line. Given the opportunity, he / she might say one of the following:

1. Every time we run film no. 17 from the supplier in Pittsfield, the right side of the film tears at the package former. This doesn't happen with other suppliers.
2. The carton sleeves all work fine on a new pallet load, but as soon as we get down to the bottom third of the pallet (layer no.6), they are always deformed, and we have to reject a lot of product.
3. Near the end of each shift, blue ink from the paper printing builds up on the forming tube, and we have to shut down and clean it off. Can't you guys find a better blue ink?
4. Last week we had a skidload of K.D. cases which ran through the shift without a single mis-feed. It sure made my job easier. I jotted down the shipment number: ACB69-340.

Most vendors have quality control systems in place. If these programs are not completely satisfactory to the customer in their present form, many customers are now willing, if requested, to assist the supplier in further developing their programs. Some of the requirements which vendors might not normally have in place are: documented procedures of their production process, handling and segregation of nonconforming materials, correlation of production samples with specific times of manufacture and with lot numbers of each shipment, periodic audit samples supplied to customer which duplicate vendor's production samples. On the other hand, for less critical products, perhaps simple checklists of vendor operations could be satisfactory documentation for the purchaser. If, for example, the checklists indicated a number of culling problems throughout a production run, the purchaser would also expect to find a series of inspection reports accompanying any shipment involved, which indicated appropriate culling of defects at the vendor's plant. In the event that this becomes a continuing problem, the vendor would be expected to provide a more formal quality control system. Where informal checklist systems are used, some purchasers find it desirable to receive the checklists in advance of the shipment, since they might consider it necessary to have the vendor conduct further defect removal. Formalized quality control documents, on the other hand, usually accompany shipments.

In examining shipment documentation, it should be apparent that the immense volume of paper could not reasonably be digested by purchasing, quality control, receiving, production, and engineering. As the program evolves, the record flow should indicate a road map where certain documents are directed to specific departments for review before they are collected at some central office, usually in purchasing or quality control. As a guide, the paperwork should be reviewed by the department which needs the information to operate, and by an individual who understands how to read and interpret the data. A control chart which shows eight points all in a row with the same value might be interpreted by the untrained reviewer as an example of superb process control. In fact, it might indicate that somebody was falsifying test measurements.

Even under the best quality control systems, it is possible that over a long period of time, as many as three shipments per thousand might possibly be substandard. Here is where the huge data file is most valuable. By careful detective work, it is frequently possible to pinpoint the cause of the reject. In some cases, the problem might not be observed in every portion of the shipment, but might be isolated to the left side of machine #9 during the last half of the second shift on nearly every Tuesday (for example). With detailed supplier data such as this, process improvements may be indicated, and specific products can be culled from the offending shipment, rather then outright reject-and-destroy. Costs involved in this inspection are usually negotiated amicably.

The question arises: who should be responsible for holding a shipment which is accompanied by questionable data? Or, for that matter, who should be responsible for rejecting an incoming shipment when either the data or an audit shows it to be unsuitable? Also, who, if anyone, should have the authority to override a shipment rejection? There is no single answer to these three questions, and yet, they are critical to the control process. Each company finds its own workable solution, depending in part on the function of the departments involved and, to a larger extent, on the abilities of individuals within each company. Where quality control, purchasing and operations are all equally capable of handling the decision unilaterally, it might be preferable to appoint one department to handle simple, routine rejections or approvals, and to construct a small team consisting of the three departments to act as a material review board. Needless to say: the simpler the better! Obviously, this works best in relatively small organizations.

Carried to an extreme, there is a point at which too much record keeping can evolve from a relatively simple system of reporting. If the vendor is a paper converter, should he also supply documentation from all of *his* suppliers? The paper mill would normally provide documentation with each shipment; the other raw material companies might supply theirs: ink, staple wire, coatings, glues, metal stampings, etc. Perhaps a statement from the converter which indicates that he has inspected all of these materials and found them within their specifications should suffice. There may be instances where the customer might wish for copies of out-of-specification sample test results, along with the specific causes, and information on steps taken by the vendor to correct the process.

Occasionally the customer will find the need to modify an existing specification, or perhaps create an entirely new specification to accommodate a process, product or packaging change. These changes require close cooperation between vendor and customer well in advance of the contemplated production. Agreement should be reached on quality control basics: sampling location, size, frequency, methods of test and reporting. Advance samples from the vendor should be subject to both laboratory audit and production pretest well in advance of the scheduled full production date.

Close relationships with the vendors processes and quality systems can lead to an occasional ethical dilemma. If a vendor should experience a quality problem which requires considerable time and expense to correct, does the purchaser have the right to pass this information on to competitive vendors? Although this might appear to be unfair, there is a workable solution that some manufacturers have found to be acceptable. During the early discussions of closer working relationships between vendor and purchaser, this subject of sharing proprietary information with other vendors needs to be explored in detail. Most vendors will accept the idea, with the understanding that by sharing their own efforts with other vendors, they may have as much to gain in return. If a vendor wishes to exempt specific areas of trade secrets from this agreement, it should probably be expressed in writing.

The "partnering" concept of the vendor-purchaser relationship tends to lead toward the single source principle. As a supplier and the purchaser grow closer together over time, understanding needs and capabilities, the purchaser might come to treat the supplier as part of his company—a partner. With this close interchange, the purchaser might well decide that there is no need for a second or third source. On the other hand, there is no one best way to manufacture corrugated shipping cases. Each supplier of corrugated has developed his own production techniques with his own specialized equipment, and has, over the years, discovered solutions to countless production problems, many of which are unknown to his competitors. This is one of the compelling reasons why so many purchasing managers prefer two or three sources.

## Record Keeping

Quality documents pertaining to shipments should be accumulated by either purchase order number or shipment number. In addition, vendors should supply calibration information related to their test equipment and reagents. By using available computer tools, the paperwork can be kept to a minimum. For example, scanner outputs, fax-modems, floppy disks and CDs containing data can be transferred electronically to a central filing system with minimum effort, and keyed to shipment numbers.

The use of specialized universal product coding, in addition to the product UPC information which appears on shipping cases (for example), can be tailored to fit specific needs of both vendor and purchaser.

The major contribution to the partnership concept of quality control between the packaging material vendor and purchaser was expected to be a combination of reduction in defects on the processor's production line, as well as lower costs of purchased materials by elimination of manufacturing interruptions at the vendor's plants. A greater benefit has been the ability of the processor to adopt a "Just-In-Time" delivery schedule for purchased materials. As the defect level of incoming materials disappears, it is no longer necessary to maintain high levels of backup stock on the warehouse floor. As one pundit described this change: we have replaced "Just-In-Case" production with "Just-In-Time." Eliminating the need for this cushion of materials also frees up a significant amount of capital. Obviously, for those with uncertain delivery operations resulting from occasional severe weather, unpredictable harvest conditions, unscheduled labor conflicts, extremely long supply routes or other traffic problems, "Just-In-Time" cannot be taken too literally. Allowances must be made for these extraordinary circumstances.

A note of caution: this approach appears to depend on one purchasing manager and one supplier sales representative. Although this system can be quite effective, the danger of losing either of these individuals through promotion or otherwise might severely handicap the business relationship and effectiveness.

## SUPPLIER CERTIFICATION PROGRAMS

In much of the above discussion, the word "certified" has been used to describe a guarantee of product quality or a recognition of a supplier's quality qualifications. The USDA has offered a certifying service for many years. Since the 1970s, the United States Department of Agriculture has furnished an inspection service which provides certification of food processing plants and the quality of processed fruits and vegetables, honey, molasses, nuts, sugar, sirups, tea, coffee, spices and condiments. The regulation is found in USDA CFR TITLE 7, Chapter 38 Food Safety and Quality Service, Part 2852. Plant certification requires a plant survey by the USDA to determine whether the plant and methods used are suitable in accordance with their regulations, as well as those of the FDA Good Manufacturing Practice in manufacturing, processing, packing or holding. If the plant meets the requirements of the survey, product inspection services may be performed. Certificates of sampling and loading are issued. Official grade stamps are affixed to those lots packed under either continuous inspection of the USDA or packed under the Quality Assurance Program of the USDA. Additional details of the regulations, including sampling plans and acceptance levels have been published in a manual titled *Regulations Governing Inspection and Certification of Processed Fruits and Vegetables and Related Products*, USDA, Food Safety and Quality Service, Fruit and Vegetable Quality Division, Processed Products Branch, Publication 0-310-944/FSQS-378.

Many companies have customized certification procedures as part of their total quality management (TQM), in contrast to the packaging purchasing manager's example discussed above.. These programs appear under such other titles as Supplier Improvement, Vendor Partnership, Supplier Qualification, Preferred Supplier, Supplier Development, Vendor Excellence, Supplier Quality Management, Select Supplier. By spreading the responsibilities and goals over many company departments, the shortcomings of the packaging materials example above are reduced. Some food companies will certify suppliers without audit provided that they have been inspected by recognized external organizations such as the American Institute of Baking, or by qualified food industry consultants, or have become ISO 9000 certified. This is certainly a short cut to certification, since large companies have reported that it can require three to five years to achieve fully qualified approval. Some of the vendor's major quality control areas which are examined by various companies include:

Continuous improvement
Data management
Experimental design techniques
Fault tree analysis
Good Manufacturing Practice
Hazard Analysis and Critical Control Points
Kosher certification
Laboratory capabilities
On-time delivery system
Contamination control
Process capability studies
Process control
QC audit system
Recall procedures
Record retention system
Sampling procedures
Sample retention system
Sanitation—macro and micro
Specifications
Test methods—accuracy and precision
Traceability
Training programs

In the simplest form of a supplier certification program, the quality control manager prepares a questionnaire for the potential vendor, listing titles of quality control procedures considered essential to providing satisfactory materials to his company. The questionnaire is filled out by the vendor and returned to the company's purchasing department where a decision is made to "certify" whether the vendor is acceptable.

Another simplified system requires that each vendor's shipment be accompanied by a quality certificate, with or without supporting test data and control charts demonstrating compliance with specifications. These certified shipments may be audited by the purchaser's quality control department.

Although these simplified methods appear somewhat elementary, they may be satisfactory where the products involved are ordinary and uncomplicated. In some cases they may also be all that a potential supplier will tolerate from a small lot buyer.

At an intermediate level of program sophistication, some customers will conduct plant visits to rate potential suppliers on several specific criteria before certifying them: statistical quality control procedures, product quality, process capability, service, and price. Selecting the initial candidates for certification is usually based on the dollar value, critical nature of the product, volume of material supplied, or perhaps single source suppliers. Generic product suppliers (salt, oils, sugar, corn syrup, simple maintenance parts for machinery) can be included after the critical product suppliers are in the program. These plant visits can be most fruitful if the supplier receives a clear understanding of how his product is to be used. They may provide suggestions to improve the use and decrease the cost of their product.

Larger companies have developed more detailed programs. More than one company has formed a team to visit the supplier's plant to directly review the routine quality control procedures as well as the quality control tools in use for trouble shooting and product improvement. The team is composed of members from purchasing, production, and quality control, and starts its investigation with the vendor's top management. If indicated, the team may request that he attend a seminar on total quality management. Obviously, this last step must be accomplished with much tact and good judgment.

The common goal of the certified supplier programs is the assurance of receiving problem-free materials or services on time, all the time. Some companies will contribute to this goal by providing technical assistance to the supplier for installing quality control systems. A second common goal is to reduce cost. This can be accomplished by reducing losses from defective materials, from eliminating receiving inspection and returned shipments, and from supplier process improvements, possibly aided in part by the purchaser's technical personnel. A third goal is to improve the quality level of the product purchased through improved operations by the supplier. A fourth goal expressed by a few companies is to develop a virtual partnership between vendor and purchaser to their mutual advantage. With the exception of the USDA program mentioned above, purchasers will generally certify specific products, not the vendor's complete line of products or entire operation.

The most complete formal programs, generally in large companies, are sometimes considered proprietary, and nondisclosure agreements must be signed.

These programs generally include stages of certification which demonstrate that a supplier has furnished product or service which:

Consistently reaches an established quality level
Complies with all governmental regulations
Meets Just-in-Time requirements
Continually improves product quality
Improves price / value relationship

Approvals generally are accomplished in three to five stages (or more), and suppliers may be identified by a number of classifications, as shown in the following table:

| Stage | Classification | Suggested Descriptions |
|-------|----------------|------------------------|
| I. | Potential Restricted Provisional Limited | Might become a supplier after product or service has been sampled and reviewed |
| II. | Conditional Authorized | May furnish products on limited basis, subject to receiving inspection |
| III. | Acceptable Approved Accredited | Preceded by plant visit Certificate of analysis required Selected to participate in certification process |
| IV. V. | Qualified Select Certified | Delivers consistent, excellent quality Preferred Meets all of the TQM process, product and service requirements |
| VI. | Continuous | Periodic audits of plant and product |

As complex as these certification programs appear to be, there is increasing evidence that they are effective in both cost savings and quality improvements. It is expected that this business technique will continue to grow in popularity.

# 14 Implementing a Quality Control Program

## MANAGEMENT COMMITMENT

When top management directs that a quality program be created and adhered to, the quality manager's job is relatively straightforward. He arranges for staff and sets about organizing a system starting with specifications, manual, methods of sampling, testing, reporting, auditing, methods improvement, cost reduction, and training. Usually the first efforts are centered around production. Assuming initial success and also assuming that the directive is still in force, the next logical step is to install quality control programs in other departments within the company: shipping, purchasing, accounting, sales, marketing, personnel, engineering, etc.

The chances are that this is not exactly what top management had in mind. Prior to the 1940s, quality control meant "inspect the raw materials and the finished product, and don't let anything bad get by." There are still organizations with this attitude, and there may even be some where the product and process is so simple, and the employees so dedicated to high quality production, that it may work satisfactorily. For some older food companies, the main goal for years has been to get as many cases of product as possible out the door, and using quality control to assist production to attain this goal would be acceptable to company personnel; but quality control should not be allowed to "waste their time" in other departments.

The concept of total quality control, that is, quality control encompassing all of the company's functions, is still fairly new. It is often difficult to install because of the closely-guarded-turf syndrome: "I'm the accounts payable department manager, and nobody from quality control or anywhere else is going to tell me how to run my business." Yet, the quality of work in the accounts payable department may be very poor. Payments may be so slow that thousands of dollars of prompt-pay discounts are lost; X% of the checks may be sent out with the wrong amount, to the wrong account, or duplicated. Y% of the invoices may be misfiled or destroyed. Z% of the invoices may be for unsatisfactory services or perhaps those not performed at all. Certainly, the statistical quality control principles using sampling plans and statistical analysis can be used to pinpoint weaknesses in the system, and provide measuring devices for process improvements. It would not be unexpected for the accounts payable manager to tell the quality control manager, "Don't try to educate me with your quality control principles until you've solved all the problems of poor product quality control on the production line. That's what you were hired for." And in many cases, he could be right!

With this in mind, we will restrict our discussion of implementing a quality control system to those areas of a food processing company directly involved with operations.

## GETTING STARTED

To begin with a more optimistic note, a statistical quality control program can be created in a company with or without the direct support of top management. Support merely makes the job easier. If management has selected a philosophy such as statistical or motivational, the first step has already been taken; and from there on, every step should refer to that philosophy.

It is possible to buy books, video tapes, or consultant services which supply preformed quality control programs prepared for other companies. Some of these might be applicable, but one must recognize that although the principles might work for all, the details may not fit into the product safety, the cultural atmosphere, the product line, the governmental regulations or the physical processing equipment of every company. Trying to educate managers in a darkened room with video tapes aimed at another type of industry will likely put some of them to sleep. A program aimed at certain hardware industries may be built around such principles as "reliability," "mean time between failure," and "warranty control." Much of this does not apply to food products. The quality control program of a food product cannot be built around a concept that even suggests that the food's reliability makes it safe to eat 87% of the time; or that if it should produce sickness it may be returned for "repair" under a warranty certificate. Some of the programs have zero defects as a goal. In reality, the only way to reach zero defects is to widen the specifications. As soon as the line worker realizes this, the program loses its punch. This does not mean that there is no value in a zero defects goal. It is an admirable way to present progress charts. Where attribute acceptance programs are used, an improvement in quality control which permits dropping the acceptable quality level (AQL) to the next more stringent table—closer to zero—should be heralded with loud applause throughout a company bent on improving the quality level.

Avoiding blind acceptance of hardware or electronic quality control programs cannot be emphasized enough. In an article published in *Quality Magazine* (November 1988, p. 20), a production manager of Hewlett-Packard Co., the world's current leading test and measuring equipment manufacturer, admitted to a major problem with product quality. The article describes how, in 1981, there was a 50% chance of product failure resulting in the equipment being returned for warranty repairs "within certain areas of the company." By instituting a new stringent quality control program, the annualized failure rate was reduced stepwise so that by 1985 the failure rate had dropped to 14%. After a few swings in either direction, the product failure rate leveled off at 4% by 1989. The target failure rate for the 1990s was estimated to reach about 3% to 4%. Certainly, a quality control program such as this one, which has been claimed a huge success for a company in the electronics industry, cannot be used without some major modifications in the food industry. In the first place, it is inconceivable that a food product with a 50% failure rate would ever be considered ready to market. Furthermore, there is no "repair under warranty" concept for the food industry's consumers. Finally, the thought of a 3% failure rate as a goal for a food product is ridiculous. If 3% of carbonated soft drink cans were failures (leaking cans permitting the carbon dioxide to escape, pull tab openers which fell off when attempting to open the can, product contaminated with cleaning fluids, jagged edges on the pour opening, moldy product, wrong flavor, missing formula ingredients such as acid, color or flavoring), then one of every 17 six-pack cases sold would result in a dissatisfied customer. This is totally unacceptable for any food product.

Some programs are aimed at quality control personnel and tend to be highly mathematical. If presented in digestible doses to those personnel, they may be trained quickly and efficiently. On the other hand, a mathematics based program, loaded with statistical terms and dull tables, will not readily be embraced by most managers or supervisors whose backgrounds make this material difficult to absorb. The lesson to be learned here is that any potential program should be screened before presentation, and if not applicable to the company, another program must be found or created.

## AN IN-HOUSE PROGRAM

One major food processor tried such a formal mathematical program and found that it was poorly received. They then attempted to put together small teams of operations and line personnel at which cost savings projects were discussed and at which fragments of statistical training programs were taught. This approach worked. With specific company goals to keep their interest, the teams almost immediately introduced methods modifications for product improvement and cost savings. At the same time, they were made aware of the meaning of statistical control in a relatively painless fashion, so that when they encountered $X$-bar and $R$ charts on the plant floor, they were familiar with their meaning. As expected, an effort of this type must be closely monitored since it will, by its own nature, generate costs: lost time, lost labor, lost production.

These teams were not modeled after Japanese quality circles. The teams consisted of management, supervisory, and hourly personnel, and the discussions were guided along specific problems. This gave everybody an opportunity for input without wasting time with unsound engineering approaches to complicated systems. In order to start a team in the right direction and to show how team efforts are effective, a simple demonstration was conducted. In this example, a team leader went through the preparation of a peanut butter sandwich, but without plans, tools, or materials readily available. He crisscrossed the front of the room looking first for the bread, then a plate, then a knife, and then the peanut butter. He stopped to perform other actions which had no bearing on the project, such as getting a drink of water, stopping to open up a pack of chewing gum, and looking for a wastebasket for the wrapper. After ten minutes of aimless and uncoordinated motions, the peanut butter sandwich was finished. Following this, a single member of the team was asked to list all the wrong actions and suggest correct methods to perform this simple task efficiently. This was followed with a team discussion to demonstrate how many additional and better ideas could be generated by group contributions. After a few days of practice with actual production line situations, the team came up with a few minor productive suggestions relating to their job. The individuals gradually realized that their input was being used, and employees at all levels were gratified to be considered a part of the production team. As they became aware of the positive results, some of the more technical and statistical information from the Juran and Deming principles was slowly introduced. Under these conditions, the teams were more receptive to statistics, and the education program was under way.

The next stage was to introduce outside professional training. Some of the employees were sent to community college statistical quality control courses. In addition, a private consultant was brought into the plant to conduct a series of courses to groups of employees who were now partly pretrained and ready for the heavier material. Several quality control employees were also selected for specialized five-day training in process control courses by the consultant. These specially trained personnel then presented four-hour classes each day over a period of three days to the production personnel in the plant. Rather than presenting detailed procedures, these courses included sound statistical principles as an overview, covering much of the subject matter contained in this book, but on somewhat of a theoretical approach: sampling, distributions, control charts for variables and attributes, data collection, and chart analysis. This endeavor turned out to be only partially successful. It was too general and theoretical, and failed to give the plant people the specific tools they needed. Management looked at this part of the program as "planting more seeds" of statistical quality control, with the expectation that if enough seeds were planted, sooner or later something would sprout.

At the outset, the production managers and the plant managers were invited to these consultant lectures as well, but they did not choose to attend. This was unfortunate, for the hourly plant people read this not only as a lack of interest on the part of the managers, but as an indication that they probably didn't believe in the idea of quality control. Unless the workers are assured that the manager thinks the program is important, the workers will not pay much attention at the classes.

One of the earliest statistical tools presented by this same company to line workers was Pareto analysis. It was offered at first as a dollar-saving tool. Half-hour sessions one day per week were scheduled for the team to teach them this technique, and eventually apply it. Pareto analysis is simple to present and understand, and it was shown to be powerful. The company which tried this approach is a large food processor, with plants nationwide. When the program was started, six teams were assembled, and due to their successes, their number was expanded to over 100 within four years. The 100 teams were credited with generating over 7 million dollars of saving a year. Although the program was originally under the direction of the quality control department, it grew to a point where it had to be transferred as a separate ongoing operations department function.

This company feels that it still has a long way to go before it considers the training program a complete success, but the progress has been steady, and in four years the spirit of cooperation has spread out more and more through all departments. There are still some supervisors who greet suggestions from the quality control technician with, "I'm busy; stop bringing your problems to me." Pressure applied indirectly from the technician to the quality control manager to the plant manager, and down through the various layers to the uncooperative supervisor, usually results in grudging compliance. In time, this supervisor will realize that the suggestions are of value in improving his department's productivity, and eventually cooperation will be easier to obtain.

As a matter of interest, this company used to depend upon a suggestion program to generate new cost-saving ideas. Employees who contributed successful programs were compensated for their ideas. That program has been all but abandoned as unsuccessful. The most serious problem with the suggestion program was its bureaucratic nature. A suggestion from a worker could take months to work its way up through levels of management and, if approved, more months back down again before any action could be taken. To add to the failure of the system, some of the approved ideas, for which the employees were rewarded, never were implemented. The new team programs also have a compensation and awards plan built in. Successful teams receive a cash award at the end of the year, calculated as a percent of the savings actually generated. Occasional interim awards in the form of gifts, jackets, pen sets, etc. are presented at special dinners or sports events. These presentations are made by senior members of the firm, usually vice presidents, to impress on the teams their importance to the company.

## TEAM QUALITY SYSTEMS

The preceding examples of team development and operation are each based on relatively unstructured beginnings, evolving into formal team programs after their usefulness has been demonstrated. Over the years, universities and consulting firms have refined systems for team structure, training and operation which have had successful implementations. It is unlikely that agreement will be reached on the "one best method" of introducing teams because of the wide differences in managements, culture and diversity within industries.

Team activities are performed in a wide variety of ways in food companies. Although often thought of as modified quality circles, the concept is not new.

In November 1932 McCormick and Company established a team known as the "Junior Board." They met periodically to discuss improvements in operations, and reported their suggestions to the senior board for action at their next meeting. During the first five years of their operation, over 2000 suggestions were accepted and implemented, with resulting improvements in plant production estimated at 30%. The specific methods of composition, operation and compensation have changed over the years, but the overall concept remains.

Articles in food trade magazines and newspapers report a steady stream of successful team projects in a wide variety of areas: new production techniques, refining accounting procedures, moving facilities to less expensive locations, absenteeism, accident reduction, processing customers' orders, reduction in line changeover time, defect reduction, material ordering practices, improve sanitation, reduce energy consumption, reducing breakage, customer-driven product specifications, improved stock rotation, community-plant relations, on-time delivery.

The team approach to improvements and problem solving may not work for everyone. Some of history's product and system inventive geniuses probably worked best without a team of "equals." Thomas Edison, Albert Einstein and Luther Burbank are examples. Where a company is fortunate enough to employ some near-geniuses, perhaps it would be best to allow them to work alone, if that is there expressed need. And there are many formalized team organizations directed by an autocratic coach, who cannot (and perhaps should not) encourage or accept empowerment of their individual members: sports teams such as football, baseball, basketball, soccer and hockey come to mind. The same is true of symphony orchestras, where the will of the conductor may not be challenged without destroying the unified perfection of the ensemble.

A quality improvement team cannot be expected to replace a quality control system. On the other hand, it is unrealistic to expect any single employee to have the ability to solve all quality and productivity problems, and provide a flow of improved quality and productivity ideas. Thoughtfully selected and trained employee teams can contribute to improvement in every company's operation and profitability. Note the wide variations in quality team operations:

1. *Team sizes.* These vary widely according to complexity of the project and the needs perceived by management—there are no rules for minimums, but it is wise to limit the size to avoid interminable discussions without conclusions. (A major candy company found an exception to this limitation: at one point in the development of a new candy product, they claimed to have had 150 people working on the project.)

2. *Leaders.* It is possible to have successful teams without leaders, but chances for direct action are improved with the appointment of a leader, either by management appointment or by team election. Some teams seem to work well with revolving leadership.

3. *Team assignments.* For greatest opportunities, these are often presented in general terms to allow for open-minded exploration of the entire subject by the team. More frequently than not, the final problem statement turns

out to be quite different than originally thought by management, and may result in more far-reaching quality improvements and cost-saving results.

4. *Rewards.* There is little agreement on the methods for rewarding teams for their successes, in part because of the possibility of fomenting team jealousies, and partly because of perceived uneven-handed recognition of groups over individual efforts. Agreement by labor unions to a reward plan has been a problem for some. The Human Resources Department is faced with a decision on whether to reward individuals as determined by classical evaluation methods, or to review the individual's performance in group activities.

5. *Schedules.* Different teams within the same company might have varying meeting schedules depending on complexity of the project, technical ability of the members, and availability of both company and outside information. In addition, the urgency of the project will influence the schedule.

It would appear from this discussion that quality improvement teams are all unique, but successful teams do have six things in common: management support, training in quality team techniques, a stated mission, interdepartmental membership, budgeted funds, and empowerment.

Just as quality improvement teams require input from various departments, the Research and Development Department's efforts eventually involve other departments before introduction of a new product. Many companies have formalized the procedure by creating cross-functional R&D teams. Initial ideas for product research may originate in the marketing department, or from a customer, or from any employee's interesting idea. An R&D team will receive input from marketing regarding pricing, packaging, portion size, and other product specifics. Production, engineering and quality control provide information regarding specifications, test requirements, equipment and procedure needs. Purchasing locates and prices ingredients and packaging materials. Cost accounting calculates prototype standard costs, and determines pricing. Outside sources of information may come from equipment manufacturers, customers, raw material suppliers, and the advertising agency.

There are many sources of detailed information on various types of team operations. Figure 14-1 illustrates four concepts of team formation. Type IV is an informal and ineffective choice, selected by management which is not particularly serious about the use of teams. Type III is a step in the right direction, where a manager realizes his shortcomings and solicits assistance from other departments to act as an informal team. Type II is a fairly effective method of team formation, and has been used to solve specific production or quality problems. Type I teams have the greatest potential for continued success.

Some authorities have segregated teams into various classifications, depending upon their longevity and scope. Examples are: (1) the project team, which is assembled with a single specific goal in mind, and usually consists of a small group of employees who are all familiar with the problem or system to be explored; (2) the permanent team, generally appointed by management to meet on

---

**I**

Management forms a team which represents every department. The goals are problem solving, process and quality improvement. Training is furnished to all.

**II**

A team is formed from several experienced workers in the affected departments who are directly involved with a specific problem. They are likely to cooperate in making it work, since they have a personal commitment to suggested solutions.

**III**

A department head has no training or experience in the financial or engineering skills (for example) required to solve a quality improvement problem. He assembles members of the financial and engineering departments to form a team which should fill in the voids of his qualifications.

**IV**

Nobody in management has a solution to a new quality problem. A team of management friends meets to look for useful ideas.

---

**Figure 14-1.** Team formation concepts.

a prescribed schedule, and to consider opportunities within a specific area, or group of areas; (3) the self-directed work team, which is empowered to consider any and all aspects of an organizations processes.

Once a plan has been formalized for team selection, the next step is training. For some quality improvement teams, it may be advantageous to teach the basics of control charts and elementary statistics; for others, this need can best be performed by experts in the company, or contracted from outside consulting services. In addition to providing guidance on principles of cooperation and ground rules for conducting work sessions, a number of techniques need to be taught to the members:

- Brainstorming
- Process flow diagramming
- Fishbone chart construction and analysis
- Pareto analysis
- Scatter diagram preparation and interpretation.

Finally, the Deming cycle: PLAN/DO/CHECK/ACT (also referred to as the TQC improvement cycle, or the PDCA cycle) must be understood by all members so that the proposals of the team can be finalized. Team recommendations in the food industry may require an extra step to assure that the end product complies with governmental regulations and safety standards.

## STEPWISE PROCEDURES FOR
## TEAM PROBLEM SOLVING

After the initial training programs for the various teams are well underway, each team is presented with a practical problem in its production area. Rather than attempt to solve a major problem at the outset, it is broken down into small projects which can be more easily handled. To make the project easier to understand, it is then shown to the team as a fishbone diagram. The problem causes are found by team brainstorming, and placed on the diagram. The next logical step is to analyze the importance of each of the potential causes by use of a Pareto analysis. It is worthwhile to spend the time starting with a numerical Pareto analysis, and then follow it with a dollar analysis, since the most frequently appearing problem may not be the most costly. This now gives the team a series of project priorities based on the value to the company.

Once the major problem has been selected, the process is studied in detail. The tools generally used for this are the process flow chart and the process flow diagram. At this point, the team is ready to brainstorm for possible solutions to the problem. The rules for brainstorming are simple: a facilitator or leader is chosen to maintain order. The facilitator asks for suggestions at each point on the fishbone diagram, and the team offers "brainstorm" thoughts in rotation, one team member at a time. All thoughts are written on a blackboard. From this point, the team proceeds to data collection.

The three main reasons to collect valid data are discussed by the team, with respect to the problem at hand:

1. Analyze the process to determine its capabilities and limitations, with the goals of improving the capabilities and eliminating defects.
2. Provide the basis for action to achieve a state of statistical control over the process.
3. Inspect parts or products for inspection or rejection.

It should be emphasized that collecting data when the solution to a problem is not fully understood should be aimed at "exploring," not "proving." Thorough study of the data should be undertaken with all of the tools available: frequency distributions, histograms, and line charts. As a training aid, a Quincunx might be introduced at this point to demonstrate the generation of normal distributions through random motions of beads fed first at one target, and subsequently at another. This demonstration should lead into an explanation of three-sigma process control. The theory is followed by fitting the data collected on the plant floor to a normal curve, and calculating the control limits. Finally, the data is plotted on $X$-bar, $R$ charts to further simplify discussion of the process. By the time the team instruction has reached this point, the members should have a fairly clear concept of problem solving and the tools of quality control.

The procedures for corrective action are introduced. Finally, the quality control reporting formats are presented and explained. For one particular company, the quality control manager prepares daily summary reports which are discussed at staff meetings. The major report is one page of HACCP summaries, pie charts showing defect causes, Pareto charts which highlight major problems, and control charts for plant operations. These reports can be prepared by hand, but are more commonly computer generated. Teams then review quality control recap reports which are much more detailed, and include summaries of all team projects. The recap reports form the basis for management decisions on the scheduling of future team projects, as well as decisions to install new equipment, new procedures, or new processes to implement team suggestions.

This completes the first cycle of team training, problem discovery, problem analysis, data collection and analysis, proposals for problem solutions, review and implementation—and then on to another problem discovery and solution.

An example of a cost-improvement project is found in maraschino cherry processing.

**Operation**:      Raw → Bleach → Color → Flavor → Glass
**Cost per pound**:  $0.40   $0.70      $1.00    $2.00

Historically, a plant had stationed four inspectors at the raw cherry-receiving tables to remove blemished, crushed, and spotted cherries. An additional eight inspectors were stationed along the bleach discharge belt to pick out defectives which had passed the first table. The reasoning behind the double-sized crew was that the bleached fruit showed up blemishes more clearly than the red-colored fruit at receiving. Finally, three inspectors were distributed between the color and glass operations. One of the improvement teams asked the logical question: "Would it not be more profitable to place most of the inspectors at the beginning of the line where the defects that would be discarded were worth only $0.40 per pound, as contrasted with throwing away defects after coloring which were worth $1.00 a pound?" The obvious "yes" answer leaves some doubt as to the effect this would have on final quality. No amount of discussion would resolve the question without hard data to support it. A series of tests was conducted with various numbers of inspectors at various locations, using statistically designed experiments. The final results showed that two inspectors were still needed after the color operation, and that only 10 inspectors stationed at the receiving tables could produce satisfactory product quality, thus saving two inspectors and thousands of dollars needlessly spent on bleaching and coloring defective material which would subsequently be scrapped. (From a practical standpoint, the savings, though substantial, were less than expected because markets existed for flawed product, although at substantially lower prices.)

## PROGRAMS WITHOUT MANAGEMENT SUPPORT

The program described above was blessed with top management support, and was spearheaded by a dynamic quality control manager who made it work successfully. But what happens to the team effort if this dynamic manager moves on to another company? If he is replaced by an equally competent leader, the program will continue as long as it receives continued support. If, on the other hand, no leader appears, it is possible that each team will become autonomous, and effectiveness is bound to suffer. When everybody is in charge, nobody is responsible. The teams will crumble.

Another ugly possibility is the loss of support due to a change of heart on the part of the chief executive, or worse still, a change of the chief executive to one who has other priorities. Changes are not merely possible- they are likely. Top managements do change from time to time, and the possibility of a merger bringing in a new quality philosophy is not unheard of. An equally unpleasant scenario, and one which may be fairly common, is the directive by the new chief executive to the quality control manager: "Here's $100,000. Go build a quality control department."

Under any of these conditions where top support is not available, the tendency of a new quality control manager might be to attempt repair of the first quality problem in sight. A poorly planned effort such as this could considerably shorten his career as a manager. The trap of moving first and planning later is sometimes referred to as the "Ready! Fire! Aim! syndrome." Without management support, any false move can scuttle the possibilities of a successful program.

There is another method for installing a quality control program in a company, with or without the support of management. This series of steps was first tried in the mid-1940s before there were any guidelines for food industry process control. It has since been introduced into several companies with major success. In three companies of considerable size, this quality control system was constructed step by step, starting with a single technician. As these systems generated profits, additional personnel were added, and the pace of development picked up.

Starting with Step 1, note the tentative approach of finding an area of quality control in need of assistance using either Pareto or *political* procedures. Why political? When management of one company was shown that eight packaging lines were losing possibly several thousand dollars per week each from overfills, they expressed righteous indignation and disbelief that they could be such poor and ignorant managers as to allow this to happen. In this kind of an egotistic atmosphere, it was evident that another approach was needed. A Pareto analysis placed excessive intermittent product dust content in the finished package far down the critical quality problem list from the overfill problem. However, it was a gnawing condition which regularly aggravated management because of weekly consumer complaints. Why was it far down the Pareto list? Because the total value of a year's worth of complaints amounted to less than $1000 worth of product. On the other hand, here was an opportunity to relieve production management of a constant abrasive conflict with the sales department by eliminating the excessive dust. After a few weeks of sampling and statistically analyzing the dusty product line, it was observed that the excessive dust was caused by build-up in the ends of elongated filler hoppers. This showed up in the capability analyses as a process with two sets of dust quality levels. The design was changed to provide pyramidal discharge hopper designs which completely eliminated the intermittent dust deliveries, and made the fledgling quality control technician a minor hero to the production management.

With this minor success, it was now possible to find cooperation with solving the major quality problem of overfill. Step 2 was the use of statistical techniques to calculate the process capability limits of the filling equipment, and to calculate the control limits. Step 3 followed immediately, and established sampling locations, frequencies, size, and methods of testing and reporting. The product under study was a one-pound can of granular food weighed and sealed at 120 cans per minute. Because of the volume of samples, the sample size was set at triplicates from each filling head on approximately half-hour intervals. A study of process outliers quickly led to the cause: overadjustment of the filling head scales.

Step 4 called for correcting these assignable causes and calculating the improved capability. On a test basis, this assignable cause of variation was temporarily removed by forbidding scale adjustment unless the product weight exceeded the process control limits. As expected, the weight distribution narrowed dramatically. This led to recalculating the new and tighter process capability with corresponding lowering of the target weight. The last of Step 4 required reporting to management the dollars saved, along with other accomplishments. By repeated iteration, the overfill on the test line was reduced from the $\frac{1}{4}$-ounce overweight to 0.001-pound overweight, with a   saving of close to \$30,000 of product per month.

Step 5 repeated the preceding two steps until no further improvements were apparent. It was at Step 5 that the first real break in the management's prior quality control philosophy was evidenced. Bear in mind that all of the above work was accomplished in a near-hostile environment by a single engineer. However, the success evidenced at Step 4 convinced management that statistical process control really does work (even though the study took nearly four months), and the engineer was rewarded with two new employees. In realistic terms, this meant that the final improvements expected from Step 5 were completed in just a few weeks on the remaining production lines.

Although the quality program suggested states that design of experiments to modify the process and thus improve productivity should come next as Step 6, Pareto analysis may show that there are more fruitful areas to explore before wringing out all of the increased productivity from this first project. Also, from a practical viewpoint, management might well be tired of hearing the same success story. So there are situations where Step 6 may be skipped for awhile and Step 7 pursued.

Step 7 points the way to another line or another function. It also suggests another department, but it would be well to wait a long time before having the confidence to tackle this very difficult quality control challenge. The chances are that there are enough glitches in the production area which demand the attention of statistical process control techniques without stepping out into the minefield of other departments.

Finally, Step 8 installs quality attribute acceptance sampling plans where indicated as safeguards for quality in the process and the finished product. A natural expansion of this effort leads to a company-wide audit system. Then—as so many statistical quality control plans demand—it starts all over again at Step No. 1. It should be easier this time!

# TRAINING QUALITY CONTROL TECHNICIANS

Most of the discussions so far have been concerned with training nontechnical personnel with statistical control techniques without teaching heavy statistical procedures. The language used with management, we have shown, is dollars saved, profits, productivity improvement, reduced complaints, and less scrap and rework. The language emphasized with operations personnel has been largely related to fewer defects, job simplification, process improvements, as well as cost savings.

Now we are concerned with the language to be used to train technical personnel who are required to collect and analyze data and processes, and to understand a host of technical terms. Some may need to learn how to operate chromatographs, pH meters, HPLCs, mass spectrometers, infra-red spectrophotometers, bacteriological and other multisyllable pieces of expensive scientific equipment. They may be asked to explain to line personnel the meaning of control charts which are posted on the production floor.

Perhaps this is not as much of a problem as it appears to be on the surface. If we agree that each person has specific abilities and learning limitations, then the training should proceed in directions that are comprehended and, hopefully, enjoyed. In large companies, the quality control personnel may become specialists in one or two of the technical fields. In smaller companies, they should be taught "everything," and supervised closely in those areas which seem not to be well understood. What we are leading up to is the recognition that people are different, companies are different, and specialization needs differ in each company environment. As a logical result, the training of quality control personnel cannot be standardized.

An excellent starting place for training is the quality control manual. As we pointed out earlier, the manual should cover in detail all of the principles, test techniques, calculating, product specifications, reporting and recording, audit procedures, sampling locations, quantities and frequencies, production flow diagrams with HACCP precautions, sanitation, etc. An old Chinese proverb says: *I hear and I forget; I see and I remember; I do and I understand.* There is much wisdom in that quotation. There is one more phrase which might be added: *I teach and I become expert.*

An effective method for instructing a new employee is to "attach" him to a seasoned inspector, to make duplicate rounds, inspections and tests, while the older inspector teaches the newer one to actually perform the functions under close supervision. As the sayings above indicate, this adds to the knowledge of both persons. A safe technique for training a newcomer to run a complex analytical instrument (with supervision) is to allow him to run duplicate tests of those run by experienced employees until he reports identical results.

It is suggested that statistical techniques be taught outside of the plant and laboratory. Excellent courses are available at community colleges, and through local chapters of quality control and mathematical societies. Seminars are constantly being offered by experts in statistical quality control, and some of these may be applicable to the food industry. Teaching the statistics in-house is certainly a possibility as well, but the time and personnel required might be utilized more economically on day-to-day quality functions.

# SUMMARY

In this chapter we have explored methods of implementing a statistical quality control program, and we have discussed some of the training methods available:

1. Use of an outside consultant program with management support and direction can be successful if the program content is compatible with the company's needs. Beware of the "one program fits all" program mill.
2. In-house team development takes time, but can work well. Management support can be earned through profit-generating project resolutions.
3. A universal program started from scratch without the initial support of management has been developed for several food companies, and has subsequently been enthusiastically adopted, implemented and endorsed by management.
4. A few suggestions for training quality control technicians have been offered.

# 15 The Computer and Process Control

Yesterday the computer was an electronic slide rule with pictures and limitless file folders. Today it can also be programmed to listen to commands, compare inputs, make decisions, and send orders to machines or other computers. Tomorrow—there is no limit! The explosive growth of computer applications has taken at least three major directions, although there is considerable overlapping. One has been total company management through computer integration (CIM); a second has been the development of Artificial Intelligence and Expert Systems; and the third has been the growth of processing automation by computer control.

## COMPUTER INTEGRATED MANAGEMENT (CIM)

The goal of CIM is to optimize processing through the use of linked (integrated) computers covering every stage of a business: design, engineering, human resources, purchasing, production scheduling, quality control, processing, maintenance, materials handling, safety, inventory control, marketing, legal, distribution, finance and accounting. For some companies, the initial goal of CIM may have been reduction of personnel through automation, but this powerful business tool has turned its focus to improved product quality and process efficiency (cost savings) in every phase of the organization. By integrating the data bases of each segment of the company, instantaneous communication of shared information can optimize the entire operation. In addition to providing real time information, a CIM system is an excellent tool which can provide rapid answers to "what if" scenarios.

The application of computers to quality control and to processing is a specialized segment of CIM. In theory, quality control techniques can be applied to any function of the corporation. Initially, analysis of production quality by the use of attribute and variable control charts is greatly simplified in a CIM system, since data should be immediately available at every step of the production process. By including data from distribution, purchasing, sales, and accounting, the analysis of process improvement tests can be more realistically evaluated from the overall company viewpoint. In the past, cost information required for a test run, for example, would usually be calculated from estimates of standard costs at each step; with CIM, precise costs should be readily available, eliminating some of the inaccuracies of the older methods.

There are over 500 software companies offering quality control computer programs available off the shelf, making it difficult to select the "right one." Each company offers up to six different programs, making a staggering total of over 1500 programs from which to make a selection. Some of the features offered by these programs are listed in Table 15-1. The Seven basic tools are in bold type.

### Table 15-1.   Statistical Software Examples

| | |
|---|---|
| Acceptable quality limit | Fourier transformation |
| Accuracy graphing | Fractional factorial designs |
| Analysis of variance | GMP training and control |
| Analysis of means | Gage control |
| **Attribute charts** | Gage reliability and repeatability |
| Average outgoing quality | Gantt charting |
| Baldridge award criteria | **Histogram** |
| Bar code access | Hypothesis test |
| Bar code generator | Cost of quality |
| Benchmarking | Criticality analysis |
| Box plot | Matrices |
| Brainstorm format | Mean time between failure |
| Calibration | Median charts |
| **Cause and effect diagram** | Moving average/range charting |
| **Check sheet** | Non-normal distributions |
| Chromatographic analysis | Normality test |
| Complaint management | Operating characteristic curves |
| Continuous quality improvement | **Pareto** |
| Confidence limits | Pearson curve fit |
| Contour plotter | Probability |
| Correlation | Process capability |
| Customer survey analysis | Process flow diagram |
| Cusum | Process improvement |
| Data acquisition | Quality manual format |
| Defect code management | Quincunx demo |
| Defect cost calculation | Real time charting |
| Design of experiments | Regression analysis |
| Distribution analysis | Reliability functions |
| Document control | Run chart |
| Failure mode effect analysis | Sampling plans |
| Fault tree analysis | **Scatter diagram** |
| Fishbone charting | Scrap and defect tracking |
| **Flowcharting** | Short run control |
| Forms design  · | Signal/noise ratio |

| | |
|---|---|
| Ingredient tracking | *t*-Test of means |
| ISO 9000 criteria | Taguchi analysis |
| Simulation | Thermal analysis |
| Spectrographic analysis | Three-dimension plotter |
| Standard deviation | Traceability |
| Statistical methods | Training |
| Statistical process control | Trend analysis |
| Stratification analysis | **Variables charts** |
| Supplier quality assurance | Weibull curve |

It is suggested that quality departments planning to obtain their first program should give consideration to those which contain the seven basic quality control tools referred to previously. The finer points available in advanced programs can be obtained later, after the basic procedures have been established. There is little need at the outset to have the ability to run Analysis of Variance, or to have *t*-test of Means available on disk. There is also the danger of blindly selecting a fragment of a program which looks interesting, draws complicated graphs, and expresses results of calculations in complex Greek letters, but is an incorrect application for the purpose intended. In one unfortunate instance, an inexperienced quality manager selected a process capability graphing routine from the menu of a new program, since it "looked good." He then incorrectly used it as a process control chart for package fill volume control, causing excessive fluctuations of this measurement in finished product, with accompanying increased production costs.

## ARTIFICIAL INTELLIGENCE AND EXPERT SYSTEMS

Some of the early advances in Expert Systems were made in the food processing industry. One company, troubled by the advancing age of their most knowledgeable processing engineer, realized that when he retired, he was taking with him a lifetime of trouble-shooting techniques that had proven invaluable over the years. They teamed him with a computer expert, with the goal of transferring his expert production experience to some type of artificially intelligent computer program. After nearly a year of full-time concentration on this project, a series of disks was prepared by which operators could readily find expert assistance at solving production problems with the touch of a computer keyboard. The mind of the expert engineer had been transferred to the computer for ready access! Additionally, this information was expanded with the help of other experts to provide a training program for new employees. Later another set of expert system disks was prepared for use by the maintenance department machinists. As more information became available, and as process changes required modifications, this program was readily expanded and modified. Since most of the information contained in this type of program is proprietary, few detailed procedures for this technique are available in the literature.

There is a real need for this type of information. Most experts tend to bury several steps in their reasoning process deep in their subconscious. After the expert has left the company, these reasoning steps may be lost forever, unless recorded as part of an expert system program. A simple example: once the expert has learned that "left makes loose, right makes tight," he no longer needs to go through this thinking process when he instructs an operator to "turn the controller knob counterclockwise." In addition, the expert is not available for consultation on each shift every day the plant is operating. Sometimes there are situations where two or three experts may be required to offer a solution, and one or more of them might be on a business trip, ill, or on vacation. In their absence, this function is handled in many organizations by the use of operating manuals; but the need to thumb through pages of the "trouble-shooting" section to locate the problem description and its cure often takes much too long to be of practical use.

With expert systems available, productivity can be expected to increase by having the latest expertise available at each worker's command, avoiding the necessity to shut down a process while seeking an expert. In addition, by having the best advice on-line, improvement in decisions affecting quality control are achievable.

Following is a simplified example of how a production operator is able to call up an "expert" in seconds to solve a processing problem. Assume that the process requires a precise temperature range for control of the product quality, and the operator observes that the temperature has fallen below the lower limit. He slips his Expert System disk into his workstation computer and observes a message on the screen (Figure 15-1).

By checking "Temperature", the next window immediately appears (Figure 15-2).

Checking "Too Low" brings up the next window (Figure 15-3).

This window explains the procedure to raise the temperature. In the event that the instructions do not produce the desired result, checking the "no" square will bring up additional instructions provided by the expert. If the instructions correct

```
┌─────────────────────────────────────────────┐
│  PROCESS: COOKER, POTATO STARCH #6543        │
├─────────────────────────────────────────────┤
│  Select Problem                              │
│                                              │
│         Temperature          [X]             │
│                                              │
│         Pressure             [ ]             │
│                                              │
│         Flow                 [ ]             │
│                                              │
│         Color                [ ]             │
│                                              │
└─────────────────────────────────────────────┘
```

**Figure 15-1.** Problem selection.

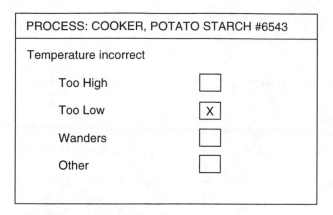

**Figure 15-2.** Problem definition.

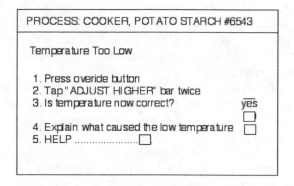

**Figure 15-3.** Problem solution.

the problem, the operator can ask for an explanation of the malfunction. In the event of an impending emergency, pressing "Help" will bring up the crisis procedure window (and perhaps sound an alarm, or notify the supervisor). In actual practice, instructions might be expected to be more complicated than the example shown. A more typical instruction might read: "If the filters are becoming saturated, and the pump speed is near maximum, increase the air flow, add 2% additional filter aid, and monitor the pH carefully. If it exceeds 7.3, slowly close the feed valve."

A useful refinement to the above system would be the addition of a series of windows showing the overall process flow, accompanied by photographs of each piece of equipment. Additionally, process values and characteristics can be included. A disk containing all of these elements would be ideal for training new operators, or product improvement teams.

By adding a data collection folder to the program, it would be a relatively uncomplicated step to include up-to-the-minute control charts for each of the parameters (temperature, pressure, flow, color) for quality control and process improvement studies. Periodic printed control charts could be made available to the production supervisor and the quality control department.

## COMPUTER-CONTROLLED PROCESSING

Before exploring the function of computers in quality control, a review of the principles of control might clarify this fairly complex subject. The three steps involved in quality control are measurement, evaluation, and response.

1. The measurement can be made manually, visually or by electrical, mechanical, biological, or chemical methods, or by combinations of these.
2. The evaluation compares the measurement to a setpoint (the optimum value), and generates an error signal. In statistical quality control terms, this signal may be expressed as "within 3-sigma, take no action," or "beyond the 3-sigma control limits, take action." This class of evaluations may be accomplished manually or automatically. Another type of evaluation, usually conducted by automatic instruments, provides for one of three modes:
   - *Proportional mode control.* Measures the magnitude of the error relative to an acceptable range.
   - *Integral or reset mode control.* Measures the magnitude of the error and relates it to the interval of time over which the error occurred.
   - *Derivative or rate mode of control.* Measures the rate at which the error is changing.
3. The response (or feedback signal) is generated by automatic instruments to reduce the error signal to zero. Older pneumatic and electronic controllers are classed as analog devices. Advances in digital computer technology have provided improved methods for control by the use of computer programs to evaluate the error. Commonly, the same modes of control are utilized: proportional, reset and rate.

A word of caution: unless the statistical principles of process variation are considered when installing instrumented process control, overcontrol might be programmed into a computer system. This may produce a progression of frantic searching by the computer to reduce error in a process which may be perfectly satisfactory if left alone. In the worst case, this can produce defects both above and below the control limits. A few "sea-going" examples should illustrate overcontrol.

To steer a canoe on a straight course, a correction is made with every stroke of the paddle. Occasionally, in a stiff breeze, an extra correction stroke might be required. Beyond this, an attempt to correct the oscillation of the canoe in mid-stroke would be overcontrol.

Once the course has been set and the rudder has been adjusted, a 100,000 ton tanker will continue in a straight line, requiring no correction for hours at a time.

A 10,000 ton freighter in a following sea (waves rolling in approximately the same direction of travel as the ship), will alternately swing slowly to the left and then to the right of the course if no rudder adjustment is made. Under this condition however, the overall direction of travel will be in a relatively straight line. If an attempt is made to correct the freighter's swing by steering in the opposite direction, the freighter will continue its swing for a few more minutes, and then will respond to both the rudder and the waves to overcorrect in the other direction. The net effect is to cause extremely wide oscillations, each larger than the preceding one. The overall direction of travel would be unpredictable.

Once a system is capable of operating successfully within satisfactory limits, using manual control or analog instrumentation, the next step many companies consider is the installation of computerized quality control. The ultimate goal is to reduce or eliminate the need for operator intervention. Continuous measurements of a process or product may result in large volumes of data which provide a guide to controlling quality. There are circumstances—perhaps a small company, or a short production run—where facilities and personnel may not be available to evaluate this data statistically. Here, too, the use of computer technology may simplify the control of quality.

Selection of the hardware and software to build a satisfactory control system can be accomplished in many ways. Perhaps the simplest method is to employ the services of a consultant who has successfully built effective computerized quality control systems for other companies in the food industry. Technical and production personnel should participate in the development and receive training in the program as it is created. If personnel in quality control and engineering are available, the company may elect to provide them with the specialized training to choose their own program from the hundreds available. The advantage of having an expert or two on the payroll is obvious; conversely, the disadvantage is the possible promotion or loss of these employees. The use of a team to assist the technical personnel may slow the process somewhat, but should provide the cooperation necessary to make any new system workable.

To construct a new or improved computer-controlled process, a number of lengthy observations and calculations must be prepared. Each stage of the process has both inputs and outputs from and to many associated process stages. These must be measured and arranged in tabular (or graphical) form so that the optimum level of each output can be calculated. Where there are many interrelated operations to the overall process, this exploratory step can be very lengthy. The next procedure used by many programmers is to connect a series of control instruments which simulate the process steps above, and which can be operated at various settings to validate the calculations.

For years, these two initial steps had to be performed manually, and usually at great expense. Software tools are now available to perform the analysis and simulations more quickly. These programs can then be combined to demonstrate how the operation would eventually be computer controlled, and how realistically the theoretical calculations were prepared. By running a series of simulations, these programs can accurately forecast the expected output of the process under a number of varying inputs.

Installing a computerized system of production or quality control requires major decisions in three areas:

- What hardware will do the job?
- What software will make it work?
- How do we prepare the workforce?

Successful installations are rarely those which attempt to cover every aspect of the company's operations with a single blanket. Once top management is convinced that computer control is the way to go, it is essential that a management team be created with the understanding that they will start small, growing as the payback is demonstrated. The team must also be prepared to solve many new production and quality problems, because the ease and speed of computerized data collection will most surely disclose several problems not previously recognized. Unless these are handled as they occur, the new system will probably fail. If problems are not solved promptly, the same "red lights" will continue to flash every day, and the effectiveness of the computer will be lost.

Some of the obvious advantages to computer-guided operations are their assistance in eliminating operator variability, avoiding use of the wrong or obsolete materials, instructing operators on performance of their jobs, and entering data directly from the measuring device output (calipers, pH meters, thermometers, balances, HPLC). For some applications, data can be color-coded as it is entered in order to inform the operator if the process is satisfactory (green), marginal (yellow), or in trouble (red). For simplicity, the data can be entered on a speedometer dial, a thermometer scale, or stair-step diagram to assist in interpreting the measurement before the numbers are entered into the statistical data bank for further processing and recording. For additional value, the data calculations may be shown on the computer screen as control charts in real-time to assist the operator, the supervisor, and the manager in evaluating the process while it is underway. Additionally, the program can continuously produce an updated histogram for daily, monthly or other comparisons. Of equal importance, a Pareto analysis of the problems encountered in the process can be simultaneously constructed and updated. This would permit anyone interested in the process to determine the most critical problems, at any time. Inventory control may continue to be a function of the warehouse, but the production floor computers may also be programmed to control inventory as production progresses. Where maintenance scheduling is critical, the process computer may also be used to keep track of machine hours, flashing a notice when maintenance performance is approaching. Even in the control laboratory, a dozen different quality tests can be available for display on the computer screen simultaneously, along with control or specification limits, so that out-of-limits data can be immediately signalled for a collection of analyses such as color, moisture, pH, flow, solubility, texture, alkalinity, etc. If desired, all of the information from warehouse, production floor, quality control laboratory, shipping and accounting can be "piped" to the CEO's office for real-time evaluation of each department as well as the entire operation. Finally, each analytical tool performed by the computer may be programmed to flash a warning light or screen message when an undesirable condition occurs.

## Data Input

In addition to the necessity of providing guidance in operating a process, there are many other reasons for taking data describing a product's quality characteristics. Some of the common ones:

1. A customer may require data as part of the purchase order. The requirement may be contractual to certify a lot's conformance to specifications, or it may be considered important to the customer's manufacturing or marketing systems. The data may or may not have any realistic value to the customer.
2. The CEO or the legal department may wish to have ample files as backup in the event of a product quality failure, resulting in a recall or a lawsuit.
3. A databank of purchased raw material performance may be a useful tool in negotiating price.
4. A confused or untrained management might ask for volumes of assorted data with the hope of finding a clue to areas for quality improvement.
5. Defect data might be accumulated by incoming shipments to assist in selection of a supplier.
6. A large file on competitive product quality might be a useful tool for benchmarking purposes.

It is likely that there are another half-dozen reasons for collecting data, but there are four general levels of information which seem to be commonly required by all companies:

1. Top management wants information on an $8\frac{1}{2} \times 11$ piece of paper which tells them "what's going on," an overview of what is happening.
2. On the shop floor, the workers need the type of information which will give them data regarding the variability of the process, allowing them to take action in real time.
3. Middle management (QC or Engineering) uses parcels of information to statistically analyze and compare—after the fact. They may be interested in comparing shifts, machines, products, processes, production days, operators, material supplier performance.

4. Product and process improvement teams or Research and Development usually require large masses of past data in their efforts to reach their goals.

Middle management needs considerable more information than is required for immediate use by workers on the production floor. The data may have to be tagged with shift, machine, product, serial number, lot size, etc. labels for further analysis.

The major advantages of electronic data collection and transmission are accuracy and speed. Collecting data by individual hand measurement, writing the data onto a form, calculating or summarizing, and copying it onto a chart presents opportunities for making an error (is that figure a 7 or a 5?), and requires far more time to perform than does an electronic system. It has been found in a detailed study that mistakes in writing down and reading numbers can run as high as 12% of the time. A major improvement over copying data by hand would be to enter the information through a computer keyboard, although the opportunity for errors still exists. Variable data can be collected from virtually any electronic output—gages, calipers, pH meters, chromatographs, scales, etc.—and be transmitted error-free through one of many relatively inexpensive "black boxes" to computers for display, storage, or further processing.

One effective method of collecting attribute data is the use of a bar code. Hand-held devices are available which can be programmed to read bar codes, store, calculate, or transmit the information to a master computer. They can be programmed to perform specific tasks, or they can be purchased pre-programmed for general use. In addition to the obvious functions of a bar code for identifying raw materials on the receiving dock and finished product in the warehouse, they are also used to classify blocks of data collected at specific stages on the production line.

Before investing in an electronic data collection system, one must be sure that there is no proprietary function in the black box which will make it impossible to use for a particular operation. There is a myriad of products available, but some of them are programmed to handle a specific function which might not be applicable to the process under consideration.

Some hand-held computers, such as personal organizer units, are available which can be programmed to contain the day's schedule, which accept pen input for notes or graphics, and which can be down-loaded to desktop computers for further processing or recording.

Modules are available which can accept voice data—in any language. Others can speak to the operator in phrases or whole pages which are highlighted on the screen. This capability is of value where the employees are not proficient in reading, or need translation.

Through the use of video cameras and video recorders, tapes of a process can be readily prepared while in operation, to be used as a training film for other employees at the same or different locations. If required, instantaneous transfer of such tapes can be made using a modem and telephone lines, thus permitting interchange of ideas between widely separated plants while operating or modifying a process.

## Sensing Devices

A sensor is a device which reacts to a physical or chemical stimulus to produce a corresponding electrical signal proportional to the physical parameters measured. As generally understood, a sensor is another word for a transducer. In the example above for an Expert System, four such sensors were used to control a process for cooking potato starch: temperature detector, pressure measuring device, flow rate instrument and color analyzer. Each of these sensors responded to specific stimuli, and converted the relative response to an electrical signal which in turn was sent ultimately to the computer at the workstation. The signal was then processed by a computer program and returned to an operating device (such as a solenoid valve) which would appropriately adjust the physical condition as required to maintain control. In the event of a physical condition beyond the control limits for which the program was defined, an out-of-limits notice was issued on screen for the operator's attention (the temperature, in the example shown). As an additional tool for the operator, an overall pictorial of the process under his control may be included in the program as shown in Figure 15-4. It shows the major equipment, and the relative location of the sensors and controllers.

Sensors are available for countless conditions and respond to stimuli using a wide variety of principles. Temperature sensors, for example, may be classified as thermocouples (TC), resistance temperature detectors (RTD), thermistors, and IC sensors.

Other sensors are listed in Table 15-2.

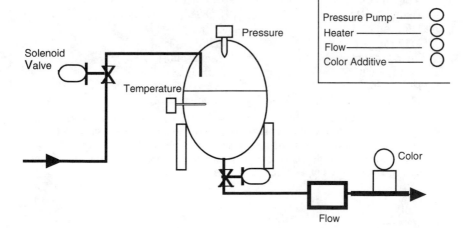

**Figure 15-4.** Process flow controls.

**Table 15-2.   Sensors**

| Sensors | Measurement |
| --- | --- |
| Infra-red, conductivity | Acid |
| Polymer resins | Aroma |
| Bioreactor | BOD (high speed) |
| Turbidimeters | Clarity |
| Colorimeter | Color |
| Infra-red | Composition (chemical) |
| Electrical resistance | Concentration |
| X-ray | Contamination, solid |
| Gamma ray, scales | Count |
| Gamma rays | Density |
| Reflectometer | Defect (visual) |
| Infra-red | Fat |
| Gamma ray | Fill level |
| Strain gage | Force |
| Differential pressure | Low rate |
| Polarized or IR light | Glass chips |
| Strain gage/Brinell | Hardness |
| Visible light | Labels (missing, faulty) |
| Float, strain gage | Level |
| Photocell | Light |
| Proximity switch | Motion |
| Strain gage | Mass (weight) |
| Infra-red | Moisture |
| pH cell | pH |
| Pressure transducer | Pressure |
| Photocell | Seal area contamination |
| Microphone | Sound |
| Viscometer | Shear |
| Refractometer | Solids |
| Infra-red | Sugar |
| Infra-red | Sulfur dioxide |
| Thermocouple | Temperature |
| Conductivity | Total acid |
| Rheometer | Viscosity |

Note that Flowmeters can be based on differential pressure, magnetic sensor, vortex, coriolis effect, thermal mass, positive displacement, turbine, ultrasonic, and other sensors. Level sensors may also be triggered by capacitance/RF, hydrostatic pressure, float switch, photocell, radar/microwave, ultrasonics, load cell, conductance instrument, point switches.

## Automatic Control

There are a number of precautions which should precede the discussion of automatic control of a process. First of all, equipment may unexpectedly fail to perform for a variety of reasons, some of which might not be at all apparent. Valves which have not been moved for long periods have been known to jam in their last position. Light-sensing devices, such as colorimeters or refractometers may develop cracked, etched, scaled or clouded lenses from product or environmental contact. Growth of plant life or insect colonies can clog flow measuring devices. Electrical power fluctuations may produce erroneous signals. Strain gages or spring-loaded devices may require recalibration due to age or wear. Vibration or temperature has been known to warp indicating arrows on meters. Barometric pressure changes may cause false reading in vacuum-activated measuring devices. Humidity can affect electrical circuits.

This is not to say that automatic controls cannot be trusted, but it does suggest that those responsible for quality control should be aware of the requirement that the ultimate success or failure of an automatic system depends on routinely maintained and calibrated equipment. In the final analysis, the success of a system is measured by the consistent, predictable uniform quality of the finished product.

Figure 15-5 contains the major elements of a computer-based process controlled system.

*Sensor.*   Senses a physical or chemical condition, converts this impulse to an electrical signal proportional to the condition measured.

*Signal Conditioner.*   Converts the sensor signal into a form acceptable to the Data Analyzer. In some instances, the signal may be amplified or filtered. The amplification of the signal received should be such that its maximum voltage range equals the maximum input of the Data Analyzer. In some cases, it may be required to reduce the electrical signal to protect the computer. Filtering may be desirable to remove unwanted signals from certain sensors. Some signals are logarithmic in nature, and their output should be reduced to a linear voltage for maximum sensitivity.

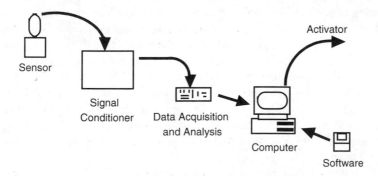

**Figure 15-5.**   Computer-based process control system.

*Data Acquisition Board (DAB).*   Accepts signals from a number of channels, at specified sampling rates and types of inputs. The sampling rates are often determined by the ability of a DAB to multiplex, that is, switch from channel to channel. These boards are available in a wide array of capabilities and outputs.

*Computer.*   Most computers have sufficient computational power to handle the output of available DABs. If necessary, however, special analysis hardware is available. Calculations can be performed in the millions per second, allowing a computer to handle many inputs from the DAB at extremely high speed.

*Software.*   Some companies might prefer to develop their own software to program the hardware. It is probably quicker (and certainly simpler) to purchase driver software from the manufacturer of the DAB. The use of application software will add presentation and analysis capabilities to the driver software.

*Activator.*   Outputs from the computer in automatic control systems are primarily concerned with operating control devices such as valves, motor drives, positioning arms, mixers, fans and other powered equipment. Expert System inputs/outputs can also be programmed into some systems. As discussed above, the computer quality control calculating output can take the form of data compilations, control charts, run charts, and many other graphs.

As mentioned above, the computer makes data analysis both easy and rapid. In addition to the control charts which can be instantly calculated and viewed by the line operator, the same data can be downloaded or connected through computer networks to quality control and process engineers for further analysis.

Consider the following simplified example. A small company manufactures a family of drink mixes, using a simple system as shown in the flow chart (Figure 15-6).

The raw ingredients are weighed manually, and dumped into a screw blendor which discharges through a feed bin into a four-scale filler head. The products are packaged in composite foil cannisters, and cased after labeling. Net weights are checked occasionally. The major reason for checkweighing is to reduce product giveaway and to avoid the possibility creating seam cutovers from over filling.

The principle customer has specified that the net weights should average 454 g, with an acceptable range of 445–463 g. Over the years, this customer has complained about underweights on several occasions, and has returned the offending shipments for reworking—that is, hand sorting and removing the underweight containers. Eventually, this customer tired of the delays resulting from the weight problem and threatened to take his business elsewhere. The customer suggested contacting the Smathers Scale Company, which manufactured automatic checkweighers.

The Smathers representative brought in an automatic checkweigher and data computer for a demonstration. While it was being assembled and wired into the

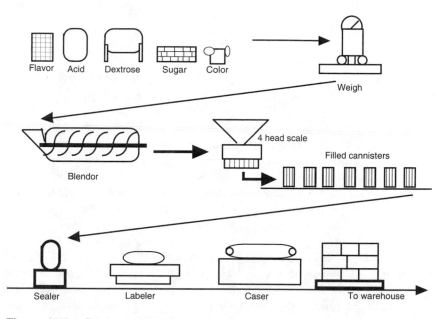

**Figure 15-6.**  Drink mix flow chart.

line, the representative took 50 consecutive samples of orange drink mix off the line as produced. Over the next several hours he weighed each can, subtracting the tare weight, then calculated the process capability based on these samples, using a hand-held calculator.

The calculations showed that over an extended run, the process would be expected to produce nearly 10% below the lower specification limit of 445 g, and about 6% above the upper specification limit of 463 g. This came as a shock to the owner, who had assumed that the four filler scales were always reliable.

The following day, the automatic Smathers equipment was ready to demonstrate. The checkweigher system gross weighed each cannister as it passed over a strain gage, adjusting for the tare. The electrical signal generated was picked up by a computer, which stored the data, and was programmed to perform a number of statistical functions. The line speed was running at the usual 45 cannisters per minute, and in less than two minutes, nearly 100 weighings were recorded in the computer. Smathers then pressed the PROCESS CAPABILITY button on the console, and selected the CALCULATE option. A number of statistical calculations appeared on the computer screen within seconds:

|                    |                   |
| ------------------ | ----------------- |
| Sample size        | 50                |
| Mean               | 453.206           |
| Standard deviation | 6.32257           |
| Three-Sigma limits | 434.238–472.174   |

| | |
|---|---|
| Lower specification limit | 445 |
| Upper specification limit | 463 |
| CPK | 0.43263 |
| CP Index | 0.474491 |
| % Under LSL of 445 | 9.7% |
| % Over USL of 463 | 6.1% |

To further demonstrate the effectiveness of the system, Smathers then selected the GRAPH option, and the following graph appeared on the computer screen within a few seconds. The data, calculations and graph all showed that the system was not capable of satisfying the customer's requirements.

The graph (Figure 15-7) and calculations convinced the owner that the scales were in need of maintenance. The counterweights were adjusted to the precise delivery weights, and the knife edge scale bearings were honed. With the assistance of the Smathers representative, another fifty samples were automatically weighed and recorded in the computer. The resulting chart (Figure 15-8) showed that the line was now capable of producing at least the minimum weight, although some overweights could still be expected. Changing the line speed would probably permit some control over the average weight value, but the process was marginal.

After much discussion, it was decided that better density control could narrow the weight range, and combined with the improved line maintenance, perhaps good control might be attained. The density could be adjusted by varying the blendor time for each batch (Figure 15-9). Once the ideal blendor duration was established, another set of data was fed into the computer and analyzed with the following results:

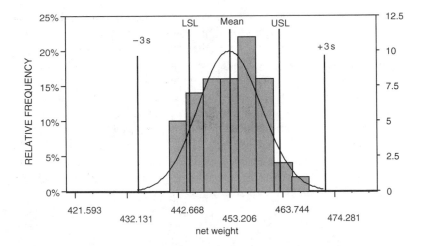

**Figure 15-7.**   Initial capability analysis.

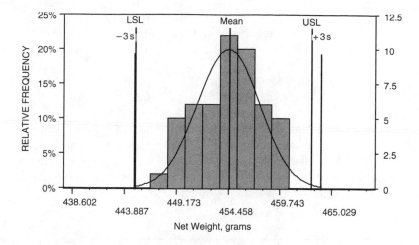

**Figure 15-8.**   Capability analysis after maintenance.

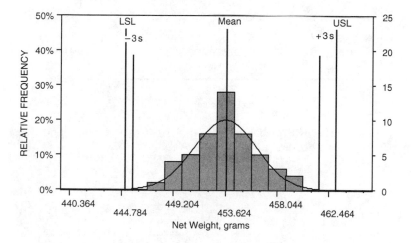

**Figure 15-9.**   Capability analysis with density control.

The PROCESS CAPABILITY–CALCULATE buttons were activated, producing the following data:

| | |
|---|---|
| Sample size | 50 |
| Mean | 453.624 |
| Standard deviation | 2.65196 |
| Three-Sigma limits | 445.668–461.58 |

| Lower specification limit | 445 |
| Upper specification limit | 463 |
| CPK | 1.08398 |
| CP Index | 1.13124 |
| % Under LSL of 445 | 0.1% |
| % Over USL of 463 | 0.0% |

To demonstrate the flexibility of the Smathers computer programs (and to further assist in closing the sale), some of the other features were demonstrated by the simple pressing of the appropriate computer panel buttons. Using the data from the capability study with density control (the final test conditions), the control charts were instantly constructed (Figures 15-10 and 15-11).

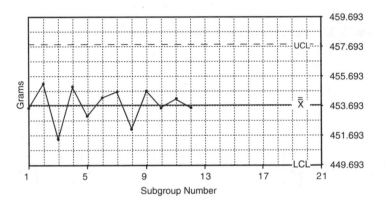

**Figure 15-10.** Net weight control chart (with maintenance, and density control).

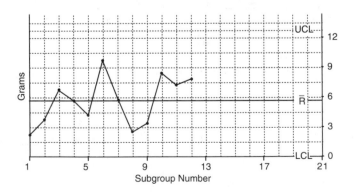

**Figure 15-11.** Net weight range control chart.

By depressing the DATA button, the fifty data points used to construct the preceding control charts were printed out in seconds. Two other charts were also demonstrated: trend analysis and X/Y chart.

This simple example has demonstrated the speed with which computers can solve production problems, calculate and graph results of changes, and provide high speed calculations in process improvement studies.

## SUMMARY

To summarize, a computer system can be installed on the factory floor to permit the operator to enter and observe data with reference to standards. The data may be combined with previous information to produce control charts for review by the operator, the supervisor, or the manager to evaluate the process at any time with respect to current or prior performance. The computer might serve as an expert with push-button information to guide the operator in performance of his work. Various signals or alarms may be included to notify the operator of the need for remedial action. The signals received by the computer may bypass the operator and automatically trigger control reactions to operate the process at optimum conditions. Finally, the data may be stored or transferred to other computer banks for further analysis or for historical purposes.

# 16    Six-Sigma

The six-sigma program for quality improvement and control outlined in Chapter 1 has provided many hardware companies with a method of greatly increasing profits by reducing costs associated with every department in the company, and by product reliability improvements. The program requires that every employee be accountable for understanding and implementing six-sigma. This conflicts with Dr. Deming's principle that 85% of the quality problems are faults of the system which can be corrected only by management.

"Quality" has been defined by Dr. Mikel Harry as a "state in which value entitlement is realized for the customer and provider in every aspect of the business relationship. When a process reaches six-sigma control, only 3.4 defects per million will be produced."

The major concerns with process improvement under this system are rework and repair costs. This has little relation to quality controls in the food industry. For example, one of the major steps in implementation of six-sigma is to arbitrarily increase the central tendency (average) quality value of a process by 1.5-sigma, since this is considered to be related to machine wear during processing. Again, this has little application in the food industry. In fact, for packaged food net weight quality control, arbitrarily moving the central tendency higher by 1.5-sigma could be a financial disaster.

The control limits of a hardware process using the six-sigma system are determined by counting the defects produced by a process and comparing this figure with a table. When the process average is shifted by 1.5-sigma, the "Defects per Million Opportunities" is shown below.

| Percent non-defective | Defects per million | Sigma |
|---|---|---|
| 93.3 | 66,807 | 3 |
| 99.4 | 6,210 | 4 |
| 99.97 | 233 | 5 |
| 99.999 | 3.4 | 6 |

Several major hardware and electronic equipment companies have adopted this system and are enthusiastic about the reduction of complaints and the savings achieved.

Unlike three-sigma statistical quality control procedures, the upper *specification* limit for a six-sigma system is at six-sigma, and the lower *specification* limit is set at zero. In the three-sigma statistical quality control procedure, the upper and lower specification limits are determined by customer requirements or by industry standards.

Furthermore, in the three-sigma system, if a product characteristic reaches the upper or lower *control limits*, the product is resampled immediately; and if still beyond the limits, the process is stopped until the problem is corrected. This does not suggest that any product has been produced beyond the specification limit. In fact, under the three-sigma system, it is possible that a defect will *rarely* be produced. As the process is improved, the three-sigma limits are reset closer together and further away from the specification limits, thus reducing the possibility of *ever* producing a defect.

It should be emphasized that in the food industry, critical defects which might be injurious to health are never permitted, whereas in some hardware industries, products can be returned under warranty for replacement or repair. As an example of this basic principle in the food industry, consider the need for absolute food safety in a fast-food hamburger restaurant chain.

- 500 hamburgers served each day by 1 restaurant
- 2,000,000 served each day by the 4,000 restaurants in the chain
- 730 million served by the chain in a year
- Assume that the chain can achieve five-sigma defect control
- Defect tables, at five-sigma, predict 233 defects per million
- If 233 hamburgers contained salmonella at five-sigma control, then 170,000 customers would become violently ill in a year.

Obviously this cannot be permitted. There was a case where a tainted shipment arrived at a fast-food restaurant, and several customers became ill. Worse still, one small child actually died. This event was so rare in the food industry, the story was on radio stations and in newspapers across the country. The company promptly initiated preventive measures, but came very close to becoming a failed business. Unsafe food products cannot leave the plant.

If, in spite of the cautions noted above, management of a food processing company insists on investigating a six-sigma system, following is a digest of the steps generally agreed upon by six-sigma enthusiasts.

## Step 1. Recognize

At the outset, the executive leading the process must understand that the major goal is determined not by the reduction of defects or improvement of profits, but rather it is meeting the requirements and satisfaction of the consumer. Defect reduction and profit improvement are by-products of this concept. This step requires studies to identify procedures which affect principle business systems, and how they affect customers. In the language of six-sigma, these systems are known as "processes." The areas investigated are identification of the core values, precise definition of products supplied to customers, and the interrelationships of all of the company's processes.

Identification of the company's processes is a vital beginning, but at the same time, "benchmarking" competitors' processes is obviously of value at the Recognize stage. Exploring noncompeting companies is suggested where possible, since they frequently have processes which can be adapted. These processes may not necessarily be proprietary, nor even directly applicable. However they assist in determining some of the processes to explore and improve. In addition to assisting in identification of the most important processes, benchmarking can frequently prioritize improvement candidates.

## Step 2. Define

The "Recognize" step may produce a large number of possibilities for improvement. The task during the "Define" stage is to select the three or four processes which are most likely to improve customer satisfaction. Emphasis should be placed on information gathered during customer surveys. From this information, establish standards and define which areas of the company require further study. This procedure should result in defining standards from the customer's point of view. Establish critical features of products required by the customer. Conclude how the company and the customer should interact. Expertise of management personnel who had originally prepared specifications based on their own experience may recognize that these standards require tailoring to customer needs.

Critical outputs from the processes selected are identified since they will subsequently be used to evaluate capability. In addition to defining key processes, it is also wise to identify key customers and determine the differing relationships of the processes selected. There are other considerations which may identify the most important processes: use of raw materials, potential cost savings, estimated time to solve, manpower requirements, effect on other processes, etc.

## Step 3. Measure

Based on the results of the customer survey, conduct statistical studies of those areas of the company which have a direct bearing on the customer satisfaction needs identified in Step 2. Determine from these studies those areas which require further attention. In addition to product and customer emphasis, these studies may provide the basis for improving process costs. An efficient system should be established to closely control critical customer requirements during processing. The sigma ratings for each process should be studied to determine which candidates are most critical. At this stage, statistical process control techniques should be evaluated: process capability, product failure analysis, control limits versus specifications. By introducing vigorous statistical techniques, changes in the process (both favorable and unfavorable) can be recognized for appropriate action.

## Step 4. Analyze

Prioritize the critical studies of Step 3, based on their standard deviations. Those with the lowest sigma will require the most improvement. Examples of critical studies are customer complaints, defective materials, marginal product performance, slow response time to complaints, price of the product, poor design when compared to competitors. Identify those processes which affect consumer requirements, and those which affect process costs and variability. Compare the project conclusions with organization goals.

## Step 5. Improve

By the use of teams headed by a six-sigma specialist, explore methods by which the low sigma operations can be improved. It is interesting to compare this part of the six-sigma process with the Deming concept of improvement. The six-sigma concept as practiced by some hardware industries includes line operators. Management recognizes that line operators know more about the details of their part of the manufacturing process than the heads of sales department, the engineers, the accountants, or the shipping department managers. Dr. Deming has stated that teams which include line operators may be disappointed with their abilities to suggest valid improvements to the operation. Line operators are probably unfamiliar with the technical requirements (biological, structural, legal, financial or safety) of their department when compared to other employees who are trained for these specialized responsibilities. There is also the probability that line operators may not have the same interest, education or drive necessary for successful team studies. There are other considerations. Many employees are content to twist the knobs, turn on the switches, and pick up their paycheck; some may fear that any change in the procedure might be more difficult, or might possibly lead to an end to their employment. In all other respects, the Deming approach emphasizes the use of management teams.

## Step 6. Control

Once the data confirms that the improvement goal of the project has been reached, written procedures are prepared to provide employees with the specific instructions that will produce the desired product or process. Procedures to ensure that the steps are followed and that the expected improvements are realized should be documented. During early operations of the new process, a leader should be assigned by management to ensure that the proper steps are being followed and that the desired results are achieved. Progress reports are to be prepared by the leader to include suggestions for solving problems which might occur. It is also suggested that at this stage, a report might be issued to explain the new process, giving credit to the team members who were instrumental in preparing the new process.

## Step 7. Standardize

One organization has organized this step into three phases: at the business level, at the operational level, and at the process level. If a particular process has gone through all of these steps and has proved successful, consideration should be given to applying the same principles to other processes in the company. There are companies where upper management prefers to operate based on experience and intuition. When six-sigma-designed process improvements are successful, and when the standardized success in one area can be applied to others, these managers are more likely to appreciate and utilize the six-sigma process.

## Step 8. Integrate

Having successfully passed through all of the above steps, one of the most difficult steps is the integration of the new philosophy into the company policy. This is partially because of inherent resistance to change. Many companies find that they can sometimes solve process problems through simple reasoning instead of the complex six-sigma technique. In some instances, rewards such as extra compensation for six-sigma accomplishments may convince management (and lower level employees as well) that this technique is worth supporting.

By standardizing the process of problem solving and integrating this process into company policy, future problems may be explored more efficiently using existing techniques.

# SUMMARY

The six-sigma process of quality control and problem solving is a carefully constructed and relatively complicated procedure which has been successfully used by several hardware manufacturers. It is a complicated extension of the Deming technique: Plan/Do/Check/Act (see Chapter 14). Food processors may find the technique useful, with the single exception of determining the sigma character of their processes by counting defects and setting the goal of six-sigma at the process control limits. The dangers of this technique have been explained previously.

# Appendix

**Table A-1. Summation of Terms of the Poisson Exponential Binomial Distribution Limit**

Entries in the table, when multiplied by $10^{-3}$, represent the probabilities of $c$ or fewer defects, defectives, or occurrences when the expected or average number is $np'$.

| $c$ / $c'$or $np'$ | 0 | 1 | 2 | 3 | 4 | 5 | 6 | 7 | 8 | 9 |
|---|---|---|---|---|---|---|---|---|---|---|
| 0.02 | 980 | 1,000 | | | | | | | | |
| 0.04 | 961 | 999 | 1,000 | | | | | | | |
| 0.06 | 942 | 998 | 1,000 | | | | | | | |
| 0.08 | 923 | 997 | 1,000 | | | | | | | |
| 0.10 | 905 | 995 | 1,000 | | | | | | | |
| 0.15 | 861 | 990 | 999 | 1,000 | | | | | | |
| 0.20 | 819 | 982 | 999 | 1,000 | | | | | | |
| 0.25 | 779 | 974 | 998 | 1,000 | | | | | | |
| 0.30 | 741 | 963 | 996 | 1,000 | | | | | | |
| 0.35 | 705 | 951 | 994 | 1,000 | | | | | | |
| 0.40 | 670 | 938 | 992 | 999 | 1,000 | | | | | |
| 0.45 | 638 | 925 | 989 | 999 | 1,000 | | | | | |
| 0.50 | 607 | 910 | 986 | 998 | 1,000 | | | | | |
| 0.55 | 577 | 894 | 982 | 998 | 1,000 | | | | | |
| 0.60 | 549 | 878 | 977 | 997 | 1,000 | | | | | |
| 0.65 | 522 | 861 | 972 | 996 | 999 | 1,000 | | | | |
| 0.70 | 497 | 844 | 966 | 994 | 999 | 1,000 | | | | |
| 0.75 | 472 | 827 | 959 | 993 | 999 | 1,000 | | | | |
| 0.80 | 449 | 809 | 953 | 991 | 999 | 1,000 | | | | |
| 0.85 | 427 | 791 | 945 | 989 | 998 | 1,000 | | | | |
| 0.90 | 407 | 772 | 937 | 987 | 998 | 1,000 | | | | |
| 0.95 | 387 | 754 | 929 | 984 | 997 | 1,000 | | | | |
| 1.00 | 368 | 736 | 920 | 981 | 996 | 999 | 1,000 | | | |

**Table A-1.** (*continued*)

| c'or np' \ c | 0 | 1 | 2 | 3 | 4 | 5 | 6 | 7 | 8 | 9 |
|---|---|---|---|---|---|---|---|---|---|---|
| 1.1 | 333 | 699 | 900 | 974 | 995 | 999 | 1,000 | | | |
| 1.2 | 301 | 663 | 879 | 966 | 992 | 998 | 1,000 | | | |
| 1.3 | 273 | 627 | 857 | 957 | 989 | 998 | 1,000 | | | |
| 1.4 | 247 | 592 | 833 | 946 | 986 | 997 | 999 | 1,000 | | |
| 1.5 | 223 | 558 | 809 | 934 | 981 | 996 | 999 | 1,000 | | |
| 1.6 | 202 | 525 | 783 | 921 | 976 | 994 | 999 | 1,000 | | |
| 1.7 | 183 | 493 | 757 | 907 | 970 | 992 | 998 | 1,000 | | |
| 1.8 | 165 | 463 | 731 | 891 | 964 | 990 | 997 | 999 | 1,000 | |
| 1.9 | 150 | 434 | 704 | 875 | 956 | 987 | 997 | 999 | 1,000 | |
| 2.0 | 135 | 406 | 677 | 857 | 947 | 983 | 995 | 999 | 1,000 | |
| 2.2 | 111 | 355 | 623 | 819 | 928 | 975 | 993 | 998 | 1,000 | |
| 2.4 | 091 | 308 | 570 | 779 | 904 | 964 | 988 | 997 | 999 | 1,000 |
| 2.6 | 074 | 267 | 518 | 736 | 877 | 951 | 983 | 995 | 999 | 1,000 |
| 2.8 | 061 | 231 | 469 | 692 | 848 | 935 | 976 | 992 | 998 | 999 |
| 3.0 | 050 | 199 | 423 | 647 | 815 | 916 | 966 | 988 | 996 | 999 |
| 3.2 | 041 | 171 | 380 | 603 | 781 | 895 | 955 | 983 | 994 | 998 |
| 3.4 | 033 | 147 | 340 | 558 | 744 | 871 | 942 | 977 | 992 | 997 |
| 3.6 | 027 | 126 | 303 | 515 | 706 | 844 | 927 | 969 | 988 | 996 |
| 3.8 | 022 | 107 | 269 | 473 | 668 | 816 | 909 | 960 | 984 | 994 |
| 4.0 | 018 | 092 | 238 | 433 | 629 | 785 | 889 | 949 | 979 | 992 |
| 4.2 | 015 | 078 | 210 | 395 | 590 | 753 | 867 | 936 | 972 | 989 |
| 4.4 | 012 | 066 | 185 | 359 | 551 | 720 | 844 | 921 | 964 | 985 |
| 4.6 | 010 | 056 | 163 | 326 | 513 | 686 | 818 | 905 | 955 | 980 |
| 4.8 | 008 | 048 | 143 | 294 | 476 | 651 | 791 | 887 | 944 | 975 |
| 5.0 | 007 | 040 | 125 | 265 | 440 | 616 | 762 | 867 | 932 | 968 |
| 5.2 | 006 | 034 | 109 | 238 | 406 | 581 | 732 | 845 | 918 | 960 |
| 5.4 | 005 | 029 | 095 | 213 | 373 | 546 | 702 | 822 | 903 | 951 |
| 5.6 | 004 | 024 | 082 | 191 | 342 | 512 | 670 | 797 | 886 | 941 |
| 5.8 | 003 | 021 | 072 | 170 | 313 | 478 | 638 | 771 | 867 | 929 |
| 6.0 | 002 | 017 | 062 | 151 | 285 | 446 | 606 | 744 | 847 | 916 |

| | 10 | 11 | 12 | 13 | 14 | 15 | 16 |
|---|---|---|---|---|---|---|---|
| 2.8 | 1,000 | | | | | | |
| 3.0 | 1,000 | | | | | | |
| 3.2 | 1,000 | | | | | | |
| 3.4 | 999 | 1,000 | | | | | |
| 3.6 | 999 | 1,000 | | | | | |
| 3.8 | 998 | 999 | 1,000 | | | | |
| 4.0 | 997 | 999 | 1,000 | | | | |
| 4.2 | 996 | 999 | 1,000 | | | | |
| 4.4 | 994 | 998 | 999 | 1,000 | | | |

**Table A-1.** *(continued)*

| c'or np' \ c | 10 | 11 | 12 | 13 | 14 | 15 | 16 | | | |
|---|---|---|---|---|---|---|---|---|---|---|
| 4.6 | 992 | 997 | 999 | 1,000 | | | | | | |
| 4.8 | 990 | 996 | 999 | 1,000 | | | | | | |
| 5.0 | 986 | 995 | 998 | 999 | 1,000 | | | | | |
| 5.2 | 982 | 993 | 997 | 999 | 1,000 | | | | | |
| 5.4 | 977 | 990 | 996 | 999 | 1,000 | | | | | |
| 5.6 | 972 | 988 | 995 | 998 | 999 | 1,000 | | | | |
| 5.8 | 965 | 984 | 993 | 997 | 999 | 1,000 | | | | |
| 6.0 | 957 | 980 | 991 | 996 | 999 | 999 | 1,000 | | | |

| | 0 | 1 | 2 | 3 | 4 | 5 | 6 | 7 | 8 | 9 |
|---|---|---|---|---|---|---|---|---|---|---|
| 6.2 | 002 | 015 | 054 | 134 | 259 | 414 | 574 | 716 | 826 | 902 |
| 6.4 | 002 | 012 | 046 | 119 | 235 | 384 | 542 | 687 | 803 | 886 |
| 6.6 | 001 | 010 | 040 | 105 | 213 | 355 | 511 | 658 | 780 | 869 |
| 6.8 | 001 | 009 | 034 | 093 | 192 | 327 | 480 | 628 | 755 | 850 |
| 7.0 | 001 | 007 | 030 | 082 | 173 | 301 | 450 | 599 | 729 | 830 |
| 7.2 | 001 | 006 | 025 | 072 | 156 | 276 | 420 | 569 | 703 | 810 |
| 7.4 | 001 | 005 | 022 | 063 | 140 | 253 | 392 | 539 | 676 | 788 |
| 7.6 | 001 | 004 | 019 | 055 | 125 | 231 | 365 | 510 | 648 | 765 |
| 7.8 | 0C0 | 004 | 016 | 048 | 112 | 210 | 338 | 481 | 620 | 741 |
| 8.0 | 000 | 003 | 014 | 042 | 100 | 191 | 313 | 453 | 593 | 717 |
| 8.5 | 000 | 002 | 009 | 030 | 074 | 150 | 256 | 386 | 523 | 653 |
| 9.0 | 000 | 001 | 006 | 021 | 055 | 116 | 207 | 324 | 456 | 587 |
| 9.5 | 000 | 001 | 004 | 015 | 040 | 089 | 165 | 269 | 392 | 522 |
| 10.0 | 000 | 000 | 003 | 010 | 029 | 067 | 130 | 220 | 333 | 458 |

| | 10 | 11 | 12 | 13 | 14 | 15 | 16 | 17 | 18 | 19 |
|---|---|---|---|---|---|---|---|---|---|---|
| 6.2 | 949 | 975 | 989 | 995 | 998 | 999 | 1,000 | | | |
| 6.4 | 939 | 969 | 986 | 994 | 997 | 999 | 1,000 | | | |
| 6.6 | 927 | 963 | 982 | 992 | 997 | 999 | 999 | 1,000 | | |
| 6.8 | 915 | 955 | 978 | 990 | 996 | 998 | 999 | 1,000 | | |
| 7.0 | 901 | 947 | 973 | 987 | 994 | 998 | 999 | 1,000 | | |
| 7.2 | 887 | 937 | 967 | 984 | 993 | 997 | 999 | 999 | 1,000 | |
| 7.4 | 871 | 926 | 961 | 980 | 991 | 996 | 998 | 999 | 1,000 | |
| 7.6 | 854 | 915 | 954 | 976 | 989 | 995 | 998 | 999 | 1,000 | |
| 7.8 | 835 | 902 | 945 | 971 | 986 | 993 | 997 | 999 | 1,000 | |
| 8.0 | 816 | 888 | 936 | 966 | 983 | 992 | 996 | 998 | 999 | 1,000 |
| 8.5 | 763 | 849 | 909 | 949 | 973 | 986 | 993 | 997 | 999 | 999 |
| 9.0 | 706 | 803 | 876 | 926 | 959 | 978 | 989 | 995 | 998 | 999 |
| 9.5 | 645 | 752 | 836 | 898 | 940 | 967 | 982 | 991 | 996 | 998 |
| 10.0 | 583 | 697 | 792 | 864 | 917 | 951 | 973 | 986 | 993 | 997 |

**Table A-1.**   (*continued*)

| c / c' or np' | 20 | 21 | 22 |
|---|---|---|---|
| 8.5 | 1,000 | | |
| 9.0 | 1,000 | | |
| 9.5 | 999 | 1,000 | |
| 10.0 | 998 | 999 | 1,000 |

| c' or np' c | 0 | 1 | 2 | 3 | 4 | 5 | 6 | 7 | 8 | 9 |
|---|---|---|---|---|---|---|---|---|---|---|
| 10.5 | 000 | 000 | 002 | 007 | 021 | 050 | 102 | 179 | 279 | 397 |
| 11.0 | 000 | 000 | 001 | 005 | 015 | 038 | 079 | 143 | 232 | 341 |
| 11.5 | 000 | 000 | 001 | 003 | 011 | 028 | 060 | 114 | 191 | 289 |
| 12.0 | 000 | 000 | 001 | 002 | 008 | 020 | 046 | 090 | 155 | 242 |
| 12.5 | 000 | 000 | 000 | 002 | 005 | 015 | 035 | 070 | 125 | 201 |
| 13.0 | 000 | 000 | 000 | 001 | 004 | 011 | 026 | 054 | 100 | 166 |
| 13.5 | 000 | 000 | 000 | 001 | 003 | 008 | 019 | 041 | 079 | 135 |
| 14.0 | 000 | 000 | 000 | 000 | 002 | 006 | 014 | 032 | 062 | 109 |
| 14.5 | 000 | 000 | 000 | 000 | 001 | 004 | 010 | 024 | 048 | 088 |
| 15.0 | 000 | 000 | 000 | 000 | 001 | 003 | 008 | 018 | 037 | 070 |

| | 10 | 11 | 12 | 13 | 14 | 15 | 16 | 17 | 18 | 19 |
|---|---|---|---|---|---|---|---|---|---|---|
| 10.5 | 521 | 639 | 742 | 825 | 888 | 932 | 960 | 978 | 988 | 994 |
| 11.0 | 460 | 579 | 689 | 781 | 854 | 907 | 944 | 968 | 982 | 991 |
| 11.5 | 402 | 520 | 633 | 733 | 815 | 878 | 924 | 954 | 974 | 986 |
| 12.0 | 347 | 462 | 576 | 682 | 772 | 844 | 899 | 937 | 963 | 979 |
| 12.5 | 297 | 406 | 519 | 628 | 725 | 806 | 869 | 916 | 948 | 969 |
| 13.0 | 252 | 353 | 463 | 573 | 675 | 764 | 835 | 890 | 930 | 957 |
| 13.5 | 211 | 304 | 409 | 518 | 623 | 718 | 798 | 861 | 908 | 942 |
| 14.0 | 176 | 260 | 358 | 464 | 570 | 669 | 756 | 827 | 883 | 923 |
| 14.5 | 145 | 220 | 311 | 413 | 518 | 619 | 711 | 790 | 853 | 901 |
| 15.0 | 118 | 185 | 268 | 363 | 466 | 568 | 664 | 749 | 819 | 875 |

| | 20 | 21 | 22 | 23 | 24 | 25 | 26 | 27 | 28 | 29 |
|---|---|---|---|---|---|---|---|---|---|---|
| 10.5 | 997 | 999 | 999 | 1,000 | | | | | | |
| 11.0 | 995 | 998 | 999 | 1,000 | | | | | | |
| 11.5 | 992 | 996 | 998 | 999 | 1,000 | | | | | |
| 12.0 | 988 | 994 | 997 | 999 | 999 | 1,000 | | | | |
| 12.5 | 983 | 991 | 995 | 998 | 999 | 999 | 1,000 | | | |
| 13.0 | 975 | 986 | 992 | 996 | 998 | 999 | 1,000 | | | |
| 13.5 | 965 | 980 | 989 | 994 | 997 | 998 | 999 | 1,000 | | |
| 14.0 | 952 | 971 | 983 | 991 | 995 | 997 | 999 | 999 | 1,000 | |
| 14.5 | 936 | 960 | 976 | 986 | 992 | 996 | 998 | 999 | 999 | 1,000 |
| 15.0 | 917 | 947 | 967 | 981 | 989 | 994 | 997 | 998 | 999 | 1,000 |

**Table A-1.** (*continued*)

| c' or np' \ c | 4 | 5 | 6 | 7 | 8 | 9 | 10 | 11 | 12 | 13 |
|---|---|---|---|---|---|---|---|---|---|---|
| 16 | 000 | 001 | 004 | 010 | 022 | 043 | 077 | 127 | 193 | 275 |
| 17 | 000 | 001 | 002 | 005 | 013 | 026 | 049 | 085 | 135 | 201 |
| 18 | 000 | 000 | 001 | 003 | 007 | 015 | 030 | 055 | 092 | 143 |
| 19 | 000 | 000 | 001 | 002 | 004 | 009 | 018 | 035 | 061 | 098 |
| 20 | 000 | 000 | 000 | 001 | 002 | 005 | 011 | 021 | 039 | 066 |
| 21 | 000 | 000 | 000 | 000 | 001 | 003 | 006 | 013 | 025 | 043 |
| 22 | 000 | 000 | 000 | 000 | 001 | 002 | 004 | 008 | 015 | 028 |
| 23 | 000 | 000 | 000 | 000 | 000 | 001 | 002 | 004 | 009 | 017 |
| 24 | 000 | 000 | 000 | 000 | 000 | 000 | 001 | 003 | 005 | 011 |
| 25 | 000 | 000 | 000 | 000 | 000 | 000 | 001 | 001 | 003 | 006 |

| | 14 | 15 | 16 | 17 | 18 | 19 | 20 | 21 | 22 | 23 |
|---|---|---|---|---|---|---|---|---|---|---|
| 16 | 368 | 467 | 566 | 659 | 742 | 812 | 868 | 911 | 942 | 963 |
| 17 | 281 | 371 | 468 | 564 | 655 | 736 | 805 | 861 | 905 | 937 |
| 18 | 208 | 287 | 375 | 469 | 562 | 651 | 731 | 799 | 855 | 899 |
| 19 | 150 | 215 | 292 | 378 | 469 | 561 | 647 | 725 | 793 | 849 |
| 20 | 105 | 157 | 221 | 297 | 381 | 470 | 559 | 644 | 721 | 787 |
| 21 | 072 | 111 | 163 | 227 | 302 | 384 | 471 | 558 | 640 | 716 |
| 22 | 048 | 077 | 117 | 169 | 232 | 306 | 387 | 472 | 556 | 637 |
| 23 | 031 | 052 | 082 | 123 | 175 | 238 | 310 | 389 | 472 | 555 |
| 24 | 020 | 034 | 056 | 087 | 128 | 180 | 243 | 314 | 392 | 473 |
| 25 | 012 | 022 | 038 | 060 | 092 | 134 | 185 | 247 | 318 | 394 |

| | 24 | 25 | 26 | 27 | 28 | 29 | 30 | 31 | 32 | 33 |
|---|---|---|---|---|---|---|---|---|---|---|
| 16 | 978 | 987 | 993 | 996 | 998 | 999 | 999 | 1,000 | | |
| 17 | 959 | 975 | 985 | 991 | 995 | 997 | 999 | 999 | 1,000 | |
| 18 | 932 | 955 | 972 | 983 | 990 | 994 | 997 | 998 | 999 | 1,000 |
| 19 | 893 | 927 | 951 | 969 | 980 | 988 | 993 | 996 | 998 | 999 |
| 20 | 843 | 888 | 922 | 948 | 966 | 978 | 987 | 992 | 995 | 997 |
| 21 | 782 | 838 | 883 | 917 | 944 | 963 | 976 | 985 | 991 | 994 |
| 22 | 712 | 777 | 832 | 877 | 913 | 940 | 959 | 973 | 983 | 989 |
| 23 | 635 | 708 | 772 | 827 | 873 | 908 | 936 | 956 | 971 | 981 |
| 24 | 554 | 632 | 704 | 768 | 823 | 868 | 904 | 932 | 953 | 969 |
| 25 | 473 | 553 | 629 | 700 | 763 | 818 | 863 | 900 | 929 | 950 |

| | 34 | 35 | 36 | 37 | 38 | 39 | 40 | 41 | 42 | 43 |
|---|---|---|---|---|---|---|---|---|---|---|
| 19 | 999 | 1,000 | | | | | | | | |
| 20 | 990 | 999 | 1,000 | | | | | | | |
| 21 | 997 | 998 | 999 | 999 | 1,000 | | | | | |
| 22 | 994 | 996 | 998 | 999 | 999 | 1,000 | | | | |
| 23 | 988 | 993 | 996 | 997 | 999 | 999 | 1,000 | | | |

**Table A-1.** (*continued*)

| c \ c' or np' | 34 | 35 | 36 | 37 | 38 | 39 | 40 | 41 | 42 | 43 |
|---|---|---|---|---|---|---|---|---|---|---|
| 24 | 979 | 987 | 992 | 995 | 997 | 998 | 999 | 999 | 1,000 | |
| 25 | 966 | 978 | 985 | 991 | 994 | 997 | 998 | 999 | 999 | 1,000 |

*Credit:* Eugene L. Grant, *Statistical quality control*, 3rd ed. (New York: McGraw-Hill Book Company, 1964).

**Table A-2.   Areas under the Normal Probability Curve**

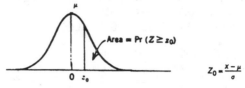

Area to the right of $z$ (or to the left of $-z$), or the probability of a random value of $z$ exceeding the marginal value.

| $z_0$ | .00 | .01 | .02 | .03 | .04 | .05 | .06 | .07 | .08 | .09 |
|---|---|---|---|---|---|---|---|---|---|---|
| 0.0 | .500 | .496 | .492 | .488 | .484 | .480 | .476 | .472 | .468 | .464 |
| 0.1 | .460 | .456 | .452 | .448 | .444 | .440 | .436 | .433 | .429 | .425 |
| 0.2 | .421 | .417 | .413 | .409 | .405 | .401 | .397 | .394 | .390 | .386 |
| 0.3 | .382 | .378 | .374 | .371 | .367 | .363 | .359 | .356 | .352 | .348 |
| 0.4 | .345 | .341 | .337 | .334 | .330 | .326 | .323 | .319 | .316 | .312 |
| 0.5 | .309 | .305 | .302 | .298 | .295 | .291 | .288 | .284 | .281 | .278 |
| 0.6 | .274 | .271 | .268 | .264 | .261 | .258 | .255 | .251 | .248 | .245 |
| 0.7 | .242 | .239 | .236 | .233 | .230 | .227 | .224 | .221 | .218 | .215 |
| 0.8 | .212 | .209 | .206 | .203 | .200 | .198 | .195 | .192 | .189 | .187 |
| 0.9 | .184 | .181 | .179 | .176 | .174 | .171 | .169 | .166 | .164 | .161 |
| 1.0 | .159 | .156 | .154 | .152 | .149 | .147 | .145 | .142 | .140 | .138 |
| 1.1 | .136 | .133 | .131 | .129 | .127 | .125 | .123 | .121 | .119 | .117 |
| 1.2 | .115 | .113 | .111 | .109 | .107 | .106 | .104 | .102 | .100 | .099 |
| 1.3 | .097 | .095 | .093 | .092 | .090 | .089 | .087 | .085 | .084 | .082 |
| 1.4 | .081 | .079 | .078 | .076 | .075 | .074 | .072 | .071 | .069 | .068 |
| 1.5 | .067 | .066 | .064 | .063 | .062 | .061 | .059 | .058 | .057 | .056 |
| 1.6 | .055 | .054 | .053 | .052 | .051 | .049 | .048 | .047 | .046 | .046 |
| 1.7 | .045 | .044 | .043 | .042 | .041 | .040 | .039 | .038 | .038 | .037 |
| 1.8 | .036 | .035 | .034 | .034 | .033 | .032 | .031 | .031 | .030 | .029 |
| 1.9 | .029 | .028 | .027 | .027 | .026 | .026 | .025 | .024 | .024 | .023 |
| 2.0 | .023 | .022 | .022 | .021 | .021 | .020 | .020 | .019 | .019 | .018 |
| 2.1 | .018 | .017 | .017 | .017 | .016 | .016 | .015 | .015 | .015 | .014 |
| 2.2 | .014 | .014 | .013 | .013 | .013 | .012 | .012 | .012 | .011 | .011 |
| 2.3 | .011 | .010 | .010 | .010 | .010 | .009 | .009 | .009 | .009 | .008 |
| 2.4 | .008 | .008 | .008 | .008 | .007 | .007 | .007 | .007 | .007 | .006 |

**Table A-2.**   *(continued)*

| $z_0$ | Detail of tail ($._2135$, for example, means .00135) | | | | | | | | | |
|------|------|------|------|------|------|------|------|------|------|------|
| 2.5 | .006 | .006 | .006 | .006 | .006 | .005 | .005 | .005 | .005 | .005 |
| 2.6 | .005 | .005 | .004 | .004 | .004 | .004 | .004 | .004 | .004 | .004 |
| 2.7 | .003 | .003 | .003 | .003 | .003 | .003 | .003 | .003 | .003 | .003 |
| 2.8 | .003 | .002 | .002 | .002 | .002 | .002 | .002 | .002 | .002 | .002 |
| 2.9 | .002 | .002 | .002 | .002 | .002 | .002 | .002 | .001 | .001 | .001 |
| 2. | $._1228$ | $._1179$ | $._1139$ | $._1107$ | $._2820$ | $._2621$ | $._2466$ | $._2347$ | $._2256$ | $._2187$ |
| 3. | $._2135$ | $._3968$ | $._3687$ | $._3483$ | $._3337$ | $._3233$ | $._3159$ | $._3108$ | $._4723$ | $._4481$ |
| 4. | $._4317$ | $._4207$ | $._4133$ | $._5854$ | $._5541$ | $._5340$ | $._5211$ | $._5130$ | $._6793$ | $._6479$ |
| 5. | $._6287$ | $._6170$ | $._7996$ | $._7579$ | $._7333$ | $._7190$ | $._7107$ | $._8599$ | $._8332$ | $._8182$ |
|      | 0.0 | 0.1 | 0.2 | 0.3 | 0.4 | 0.5 | 0.6 | 0.7 | 0.8 | 0.9 |

*Credit:* Ronald J. Wonnacott and Thomas H. Wonnacott, *Introductory statistics*, 4th ed. (New York: John Wiley & Sons, 1985).

**Table A-3.    Values of *t***

| $\bar{\alpha}$ $\nu$ | Two-tail critical values | | | | | | |
|------|---------|--------|--------|--------|--------|--------|--------|
|      | 0.50 | 0.25 | 0.10 | 0.05 | 0.025 | 0.01 | 0.005 |
| 1 | 1.00000 | 2.4142 | 6.3138 | 12.706 | 25.452 | 63.657 | 127.32 |
| 2 | 0.81650 | 1.6036 | 2.9200 | 4.3027 | 6.2053 | 9.9248 | 14.089 |
| 3 | 0.76489 | 1.4226 | 2.3534 | 3.1825 | 4.1765 | 5.8409 | 7.4533 |
| 4 | 0.74070 | 1.3444 | 2.1318 | 2.7764 | 3.4954 | 4.6041 | 5.5976 |
| 5 | 0.72669 | 1.3009 | 2.0150 | 2.5706 | 3.1634 | 4.0321 | 4.7733 |
| 6 | 0.71756 | 1.2733 | 1.9432 | 2.4469 | 2.9687 | 3.7074 | 4.3168 |
| 7 | 0.71114 | 1.2543 | 1.8946 | 2.3646 | 2.8412 | 3.4995 | 4.0293 |
| 8 | 0.70639 | 1.2403 | 1.8595 | 2.3060 | 2.7515 | 3.3554 | 3.8325 |
| 9 | 0.70272 | 1.2297 | 1.8331 | 2.2622 | 2.6850 | 3.2498 | 3.6897 |
| 10 | 0.69981 | 1.2213 | 1.8125 | 2.2281 | 2.6338 | 3.1693 | 3.5814 |
| 11 | 0.69745 | 1.2145 | 1.7959 | 2.2010 | 2.5931 | 3.1058 | 3.4966 |
| 12 | 0.69548 | 1.2089 | 1.7823 | 2.1788 | 2.5600 | 3.0545 | 3.4284 |
| 13 | 0.69384 | 1.2041 | 1.7709 | 2.1604 | 2.5326 | 3.0123 | 3.3725 |
| 14 | 0.69242 | 1.2001 | 1.7613 | 2.1448 | 2.5096 | 2.9768 | 3.3257 |
| 15 | 0.69120 | 1.1967 | 1.7530 | 2.1315 | 2.4899 | 2.9467 | 3.2860 |
| 16 | 0.69013 | 1.1937 | 1.7459 | 2.1199 | 2.4729 | 2.9208 | 3.2520 |
| 17 | 0.68919 | 1.1910 | 1.7396 | 2.1098 | 2.4581 | 2.8982 | 3.2225 |
| 18 | 0.68837 | 1.1887 | 1.7341 | 2.1009 | 2.4450 | 2.8784 | 3.1966 |
| 19 | 0.68763 | 1.1866 | 1.7291 | 2.0930 | 2.4334 | 2.8609 | 3.1737 |
| 20 | 0.68696 | 1.1848 | 1.7247 | 2.0860 | 2.4231 | 2.8453 | 3.1534 |
| 21 | 0.68635 | 1.1831 | 1.7207 | 2.0796 | 2.4138 | 2.8314 | 3.1352 |
| 22 | 0.68580 | 1.1816 | 1.7171 | 2.0739 | 2.4055 | 2.8188 | 3.1188 |

**Table A-3.**   (*continued*)

| ᾱ | Two-tail critical values | | | | | | |
|---|---|---|---|---|---|---|---|
| ν | 0.50 | 0.25 | 0.10 | 0.05 | 0.025 | 0.01 | 0.005 |
| 23 | 0.68531 | 1.1802 | 1.7139 | 2.0687 | 2.3979 | 2.8073 | 3.1040 |
| 24 | 0.68485 | 1.1789 | 1.7109 | 2.0639 | 2.3910 | 2.7969 | 3.0905 |
| 25 | 0.68443 | 1.1777 | 1.7081 | 2.0595 | 2.3846 | 2.7874 | 3.0782 |
| 26 | 0.68405 | 1.1766 | 1.7056 | 2.0555 | 2.3788 | 2.7787 | 3.0669 |
| 27 | 0.68370 | 1.1757 | 1.7033 | 2.0518 | 2.3734 | 2.7707 | 3.0565 |
| 28 | 0.68335 | 1.1748 | 1.7011 | 2.0484 | 2.3685 | 2.7633 | 3.0469 |
| 29 | 0.68304 | 1.1739 | 1.6991 | 2.0452 | 2.3638 | 2.7564 | 3.0380 |
| 30 | 0.68276 | 1.1731 | 1.6973 | 2.0423 | 2.3596 | 2.7500 | 3.0298 |
| 40 | 0.68066 | 1.1673 | 1.6839 | 2.0211 | 2.3289 | 2.7045 | 2.9712 |
| 60 | 0.67862 | 1.1616 | 1.6707 | 2.0003 | 2.2991 | 2.6603 | 2.9146 |
| 120 | 0.67656 | 1.1559 | 1.6577 | 1.9799 | 2.2699 | 2.6174 | 2.8599 |
| ∞ | 0.67449 | 1.1503 | 1.6449 | 1.9600 | 2.2414 | 2.5758 | 2.8070 |
| ν | 0.25 | 0.125 | 0.05 | 0.025 | 0.0125 | 0.005 | 0.0025 |
| α | One-tail critical values | | | | | | |

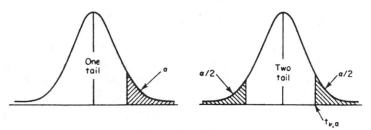

*Credit:* E. S. Pearson, Critical values of Student's *t* distribution. *Biometrika* 32 (1941): 168–181.

**Table A-4.   Binomial Coefficients**

| $n$ | $\binom{n}{0}$ | $\binom{n}{1}$ | $\binom{n}{2}$ | $\binom{n}{3}$ | $\binom{n}{4}$ | $\binom{n}{5}$ | $\binom{n}{6}$ | $\binom{n}{7}$ | $\binom{n}{8}$ | $\binom{n}{9}$ | $\binom{n}{10}$ |
|---|---|---|---|---|---|---|---|---|---|---|---|
| 0 | 1 | | | | | | | | | | |
| 1 | 1 | 1 | | | | | | | | | |
| 2 | 1 | 2 | 1 | | | | | | | | |
| 3 | 1 | 3 | 3 | 1 | | | | | | | |
| 4 | 1 | 4 | 6 | 4 | 1 | | | | | | |
| 5 | 1 | 5 | 10 | 10 | 5 | 1 | | | | | |
| 6 | 1 | 6 | 15 | 20 | 15 | 6 | 1 | | | | |

## Table A-4. (continued)

| $n$ | $\binom{n}{0}$ | $\binom{n}{1}$ | $\binom{n}{2}$ | $\binom{n}{3}$ | $\binom{n}{4}$ | $\binom{n}{5}$ | $\binom{n}{6}$ | $\binom{n}{7}$ | $\binom{n}{8}$ | $\binom{n}{9}$ | $\binom{n}{10}$ |
|---|---|---|---|---|---|---|---|---|---|---|---|
| 7 | 1 | 7 | 21 | 35 | 35 | 21 | 7 | 1 | | | |
| 8 | 1 | 8 | 28 | 56 | 70 | 56 | 28 | 8 | 1 | | |
| 9 | 1 | 9 | 36 | 84 | 126 | 126 | 84 | 36 | 9 | 1 | |
| 10 | 1 | 10 | 45 | 120 | 210 | 252 | 210 | 120 | 45 | 10 | 1 |
| 11 | 1 | 11 | 55 | 165 | 330 | 462 | 462 | 330 | 165 | 55 | 11 |
| 12 | 1 | 12 | 66 | 220 | 495 | 792 | 924 | 792 | 495 | 220 | 66 |
| 13 | 1 | 13 | 78 | 286 | 715 | 1287 | 1716 | 1716 | 1287 | 715 | 286 |
| 14 | 1 | 14 | 91 | 364 | 1001 | 2002 | 3003 | 3432 | 3003 | 2002 | 1001 |
| 15 | 1 | 15 | 105 | 455 | 1365 | 3003 | 5005 | 6435 | 6435 | 5005 | 3003 |
| 16 | 1 | 16 | 120 | 560 | 1820 | 4368 | 8008 | 11440 | 12870 | 11440 | 8008 |
| 17 | 1 | 17 | 136 | 680 | 2380 | 6188 | 12376 | 19448 | 24310 | 24310 | 19448 |
| 18 | 1 | 18 | 153 | 816 | 3060 | 8568 | 18564 | 31824 | 43758 | 48620 | 43758 |
| 19 | 1 | 19 | 171 | 969 | 3876 | 11628 | 27132 | 50388 | 75582 | 92378 | 92378 |
| 20 | 1 | 20 | 190 | 1140 | 4845 | 15504 | 38760 | 77520 | 125970 | 167960 | 184756 |

$$C_r^n = \binom{n}{r} = \frac{n!}{n(n-r)!} = \binom{n}{n-r}, \quad \binom{n}{0} = 1.$$

$$(q+p)^n = q^n + C_1^n pq^{n-1} + \cdots + C_r^n p^r q^{n-r} + \cdots + p^n.$$

Example: $(q+p)^4 = q^4 + 4q^3p + 6q^2p^2 + 4qp^3 + p^4$

## Table A-5. Random Numbers

| | | | | |
|---|---|---|---|---|
| 10 27 53 96 23 | 71 50 54 36 23 | 54 31 04 82 98 | 04 14 12 15 09 | 26 78 25 47 47 |
| 28 41 50 61 88 | 64 85 27 20 18 | 83 36 36 05 56 | 39 71 65 09 62 | 94 76 62 11 89 |
| 34 21 42 57 02 | 59 19 18 97 48 | 80 30 03 30 98 | 05 24 67 70 07 | 84 97 50 87 46 |
| 61 81 77 23 23 | 82 82 11 54 08 | 53 28 70 58 96 | 44 07 39 55 43 | 42 34 43 39 28 |
| 61 15 18 13 54 | 16 86 20 26 88 | 90 74 80 55 09 | 14 53 90 51 17 | 52 01 63 01 59 |
| 91 76 21 64 64 | 44 91 13 32 97 | 75 31 62 66 54 | 84 80 32 75 77 | 56 08 25 70 29 |
| 00 97 79 08 06 | 37 30 28 59 85 | 53 56 68 53 40 | 01 74 39 59 73 | 30 19 99 85 48 |
| 36 46 18 34 94 | 75 20 80 27 77 | 78 91 69 16 00 | 08 43 18 73 68 | 67 69 61 34 25 |
| 88 98 99 60 50 | 65 95 79 42 94 | 93 62 40 89 96 | 43 56 47 71 66 | 46 76 29 67 02 |
| 04 37 59 87 21 | 05 02 03 24 17 | 47 97 81 56 51 | 92 34 86 01 82 | 55 51 33 12 91 |
| 63 62 06 34 41 | 94 21 78 55 09 | 72 76 45 16 94 | 29 95 81 83 83 | 79 88 01 97 30 |
| 78 47 23 53 90 | 34 41 92 45 71 | 09 23 70 70 07 | 12 38 92 79 43 | 14 85 11 47 23 |
| 87 68 62 15 43 | 53 14 36 59 25 | 54 47 33 70 15 | 59 24 48 40 35 | 50 03 42 99 36 |
| 47 60 92 10 77 | 88 59 53 11 52 | 66 25 69 07 04 | 48 68 64 71 06 | 61 65 70 22 12 |
| 56 88 87 59 41 | 65 28 04 67 53 | 95 79 88 37 31 | 50 41 06 94 76 | 81 83 17 16 33 |

## Table A-5.  (*continued*)

| | | | | |
|---|---|---|---|---|
| 02 57 45 86 67 | 73 43 07 34 48 | 44 26 87 93 29 | 77 09 61 67 84 | 06 69 44 77 75 |
| 31 54 14 13 17 | 48 62 11 90 60 | 68 12 93 64 28 | 46 24 79 16 76 | 14 60 25 51 01 |
| 28 50 16 43 36 | 28 97 85 58 99 | 67 22 52 76 23 | 24 70 36 54 54 | 59 28 61 71 96 |
| 63 29 62 66 50 | 02 63 45 52 38 | 67 63 47 54 75 | 83 24 78 43 20 | 92 63 13 47 48 |
| | | | | |
| 45 65 58 26 51 | 76 96 59 38 72 | 86 57 45 71 46 | 44 67 76 14 55 | 44 88 01 62 12 |
| 39 65 36 63 70 | 77 45 85 50 51 | 74 13 39 35 22 | 30 53 36 02 95 | 49 34 88 73 61 |
| 73 71 98 16 04 | 29 18 94 51 23 | 76 51 94 84 86 | 79 93 96 38 63 | 08 58 25 58 94 |
| 72 20 56 20 11 | 72 65 71 08 86 | 79 57 95 13 91 | 97 48 72 66 48 | 09 71 17 24 89 |
| 75 17 26 99 76 | 89 37 20 70 01 | 77 31 61 95 46 | 26 97 05 73 51 | 53 33 18 72 87 |
| 37 48 60 82 29 | 81 30 15 39 14 | 48 38 75 93 29 | 06 87 37 78 48 | 45 56 00 84 47 |
| | | | | |
| 68 08 02 80 72 | 83 71 46 30 49 | 89 17 95 88 29 | 02 39 56 03 46 | 97 74 06 56 17 |
| 14 23 98 61 67 | 70 52 85 01 50 | 01 84 02 78 43 | 10 62 98 19 41 | 18 83 99 47 99 |
| 49 08 96 21 44 | 25 27 99 41 28 | 07 41 08 34 66 | 19 42 74 39 91 | 41 96 53 78 72 |
| 78 37 06 08 43 | 63 61 62 42 29 | 39 68 95 10 96 | 09 24 23 00 62 | 56 12 80 73 16 |
| 37 21 34 17 68 | 68 96 83 23 56 | 32 84 60 15 31 | 44 73 67 34 77 | 91 15 79 74 58 |
| 14 29 09 34 04 | 87 83 07 55 07 | 76 58 30 83 64 | 87 29 25 58 84 | 86 50 60 00 25 |
| 58 43 28 06 36 | 49 52 83 51 14 | 47 56 91 29 34 | 05 87 31 06 95 | 12 45 57 09 09 |
| 10 43 67 29 70 | 80 62 80 03 42 | 10 80 21 38 84 | 90 56 35 03 09 | 43 12 74 49 14 |
| 44 38 88 39 54 | 86 97 37 44 22 | 00 95 01 31 76 | 17 16 29 56 63 | 38 78 94 49 81 |
| 90 69 59 19 51 | 85 39 52 85 13 | 07 28 37 07 61 | 11 16 36 27 03 | 78 86 72 04 95 |
| 41 47 10 25 62 | 97 05 31 03 61 | 20 26 36 31 62 | 68 69 86 95 44 | 84 95 48 46 45 |
| 91 94 14 63 19 | 75 89 11 47 11 | 31 56 34 19 09 | 79 57 92 36 59 | 14 93 87 81 40 |
| 80 06 54 18 66 | 09 18 94 06 19 | 98 40 07 17 81 | 22 45 44 84 11 | 24 62 20 42 31 |
| 67 72 77 63 48 | 84 08 31 55 58 | 24 33 45 77 58 | 80 45 67 93 82 | 75 70 16 08 24 |
| 59 40 24 13 27 | 79 26 88 86 30 | 01 31 60 10 39 | 53 58 47 70 93 | 85 81 56 39 38 |
| 05 90 35 89 95 | 01 61 16 96 94 | 50 78 13 69 36 | 37 68 53 37 31 | 71 26 35 03 71 |
| 44 43 80 69 98 | 46 68 05 14 82 | 90 78 50 05 62 | 77 79 13 57 44 | 59 60 10 39 66 |
| 61 81 31 96 82 | 00 57 25 60 59 | 46 72 60 18 77 | 55 66 12 62 11 | 08 99 55 64 57 |
| 42 88 07 10 05 | 24 98 65 63 21 | 47 21 61 88 32 | 27 80 30 21 60 | 10 92 35 36 12 |
| 77 94 30 05 39 | 28 10 99 00 27 | 12 73 73 99 12 | 49 99 57 94 82 | 96 88 57 17 91 |
| 78 83 19 76 16 | 94 11 68 84 26 | 23 54 20 86 85 | 23 86 66 99 07 | 36 37 34 92 09 |
| 87 76 59 61 81 | 43 63 64 61 61 | 65 76 36 95 90 | 18 48 27 45 68 | 27 23 65 30 72 |
| 91 43 05 96 47 | 55 78 99 95 24 | 37 55 85 78 78 | 01 48 41 19 10 | 35 19 54 07 73 |
| 84 97 77 72 73 | 09 62 06 65 72 | 87 12 49 03 60 | 41 15 20 76 27 | 50 47 02 29 16 |
| 87 41 60 76 83 | 44 88 96 07 80 | 83 05 83 38 96 | 73 70 66 81 90 | 30 56 10 48 59 |

Table is taken from Table XXXIII of Fisher & Yates' *Statistical tables for biological agricultural and medical research* published by Longman Group UK Ltd., London (previously published by Oliver and Boyd Ltd., Edinburgh) and by permission of the authors and publishers.

# Table A-6.  F Distribution

$\alpha = 0.01$

| $v_2$ \ $v_1$ | 1 | 2 | 3 | 4 | 5 | 6 | 7 | 8 | 9 | 10 | 12 | 15 | 20 | 24 | 30 | 40 | 60 | 120 | ∞ |
|---|---|---|---|---|---|---|---|---|---|---|---|---|---|---|---|---|---|---|---|
| 1 | 4052.2 | 4999.5 | 5403.3 | 5624.6 | 5763.7 | 5859.0 | 5928.3 | 5981.6 | 6022.5 | 6055.8 | 6106.3 | 6157.3 | 6208.7 | 6234.6 | 6260.7 | 6286.8 | 6313.0 | 6339.4 | 6366.0 |
| 2 | 98.503 | 99.000 | 99.166 | 99.249 | 99.299 | 99.332 | 99.356 | 99.374 | 99.388 | 99.399 | 99.416 | 99.432 | 99.449 | 99.458 | 99.466 | 99.474 | 99.483 | 99.491 | 99.501 |
| 3 | 34.116 | 30.817 | 29.457 | 28.710 | 28.237 | 27.911 | 27.672 | 27.489 | 27.345 | 27.229 | 27.052 | 26.872 | 26.690 | 26.598 | 26.505 | 26.411 | 26.316 | 26.221 | 26.125 |
| 4 | 21.198 | 18.000 | 16.694 | 15.977 | 15.522 | 15.207 | 14.976 | 14.799 | 14.659 | 14.546 | 14.374 | 14.198 | 14.020 | 13.929 | 13.838 | 13.745 | 13.652 | 13.558 | 13.463 |
| 5 | 16.258 | 13.274 | 12.060 | 11.392 | 10.967 | 10.672 | 10.456 | 10.289 | 10.158 | 10.051 | 9.8883 | 9.7222 | 9.5527 | 9.4665 | 9.3793 | 9.2912 | 9.2020 | 9.1118 | 9.0204 |
| 6 | 13.745 | 10.925 | 9.7795 | 9.1483 | 8.7459 | 8.4661 | 8.2600 | 8.1016 | 7.9761 | 7.8741 | 7.7183 | 7.5590 | 7.3958 | 7.3127 | 7.2285 | 7.1432 | 7.0568 | 6.9690 | 6.8801 |
| 7 | 12.246 | 9.5466 | 8.4513 | 7.8467 | 7.4604 | 7.1914 | 6.9928 | 6.8401 | 6.7188 | 6.6201 | 6.4691 | 6.3143 | 6.1554 | 6.0743 | 5.9921 | 5.9084 | 5.8236 | 5.7372 | 5.6495 |
| 8 | 11.259 | 8.6491 | 7.5910 | 7.0060 | 6.6318 | 6.3707 | 6.1776 | 6.0289 | 5.9106 | 5.8143 | 5.6668 | 5.5151 | 5.3591 | 5.2793 | 5.1981 | 5.1156 | 5.0316 | 4.9460 | 4.8588 |
| 9 | 10.561 | 8.0215 | 6.9919 | 6.4221 | 6.0569 | 5.8018 | 5.6129 | 5.4671 | 5.3511 | 5.2565 | 5.1114 | 4.9621 | 4.8080 | 4.7290 | 4.6486 | 4.5667 | 4.4831 | 4.3978 | 4.3105 |
| 10 | 10.044 | 7.5594 | 6.5523 | 5.9943 | 5.6363 | 5.3858 | 5.2001 | 5.0567 | 4.9424 | 4.8492 | 4.7059 | 4.5582 | 4.4054 | 4.3269 | 4.2469 | 4.1653 | 4.0819 | 3.9965 | 3.9090 |
| 11 | 9.6460 | 7.2057 | 6.2167 | 5.6683 | 5.3160 | 5.0692 | 4.8861 | 4.7445 | 4.6315 | 4.5393 | 4.3974 | 4.2509 | 4.0990 | 4.0209 | 3.9411 | 3.8596 | 3.7761 | 3.6904 | 3.6025 |
| 12 | 9.3302 | 6.9266 | 5.9526 | 5.4119 | 5.0643 | 4.8206 | 4.6395 | 4.4994 | 4.3875 | 4.2961 | 4.1553 | 4.0096 | 3.8584 | 3.7805 | 3.7008 | 3.6192 | 3.5355 | 3.4494 | 3.3608 |
| 13 | 9.0738 | 6.7010 | 5.7394 | 5.2053 | 4.8616 | 4.6204 | 4.4410 | 4.3021 | 4.1911 | 4.1003 | 3.9603 | 3.8154 | 3.6646 | 3.5868 | 3.5070 | 3.4253 | 3.3413 | 3.2548 | 3.1654 |
| 14 | 8.8616 | 6.5149 | 5.5639 | 5.0354 | 4.6950 | 4.4558 | 4.2779 | 4.1399 | 4.0297 | 3.9394 | 3.8001 | 3.6557 | 3.5052 | 3.4274 | 3.3476 | 3.2656 | 3.1813 | 3.0942 | 3.0040 |
| 15 | 8.6831 | 6.3589 | 5.4170 | 4.8932 | 4.5556 | 4.3183 | 4.1415 | 4.0045 | 3.8948 | 3.8049 | 3.6662 | 3.5222 | 3.3719 | 3.2940 | 3.2141 | 3.1319 | 3.0471 | 2.9595 | 2.8684 |
| 16 | 8.5310 | 6.2262 | 5.2922 | 4.7726 | 4.4374 | 4.2016 | 4.0259 | 3.8896 | 3.7804 | 3.6909 | 3.5527 | 3.4089 | 3.2588 | 3.1808 | 3.1007 | 3.0182 | 2.9330 | 2.8447 | 2.7528 |
| 17 | 8.3997 | 6.1121 | 5.1850 | 4.6690 | 4.3359 | 4.1015 | 3.9267 | 3.7910 | 3.6822 | 3.5931 | 3.4552 | 3.3117 | 3.1615 | 3.0835 | 3.0032 | 2.9205 | 2.8348 | 2.7459 | 2.6530 |
| 18 | 8.2854 | 6.0129 | 5.0919 | 4.5790 | 4.2479 | 4.0146 | 3.8406 | 3.7054 | 3.5971 | 3.5082 | 3.3706 | 3.2273 | 3.0771 | 2.9990 | 2.9185 | 2.8354 | 2.7493 | 2.6597 | 2.5660 |
| 19 | 8.1850 | 5.9259 | 5.0103 | 4.5003 | 4.1708 | 3.9386 | 3.7653 | 3.6305 | 3.5225 | 3.4338 | 3.2965 | 3.1533 | 3.0031 | 2.9249 | 2.8442 | 2.7608 | 2.6742 | 2.5839 | 2.4893 |
| 20 | 8.0960 | 5.8489 | 4.9382 | 4.4307 | 4.1027 | 3.8714 | 3.6987 | 3.5644 | 3.4567 | 3.3682 | 3.2311 | 3.0880 | 2.9377 | 2.8594 | 2.7785 | 2.6947 | 2.6077 | 2.5168 | 2.4212 |
| 21 | 8.0166 | 5.7804 | 4.8740 | 4.3688 | 4.0421 | 3.8117 | 3.6396 | 3.5056 | 3.3981 | 3.3098 | 3.1729 | 3.0299 | 2.8796 | 2.8011 | 2.7200 | 2.6359 | 2.5484 | 2.4568 | 2.3603 |
| 22 | 7.9454 | 5.7190 | 4.8166 | 4.3134 | 3.9880 | 3.7583 | 3.5867 | 3.4530 | 3.3458 | 3.2576 | 3.1209 | 2.9780 | 2.8274 | 2.7488 | 2.6675 | 2.5831 | 2.4951 | 2.4029 | 2.3055 |
| 23 | 7.8811 | 5.6637 | 4.7649 | 4.2635 | 3.9392 | 3.7102 | 3.5390 | 3.4057 | 3.2986 | 3.2106 | 3.0740 | 2.9311 | 2.7805 | 2.7017 | 2.6202 | 2.5355 | 2.4471 | 2.3542 | 2.2559 |
| 24 | 7.8229 | 5.6136 | 4.7181 | 4.2184 | 3.8951 | 3.6667 | 3.4959 | 3.3629 | 3.2560 | 3.1681 | 3.0316 | 2.8887 | 2.7380 | 2.6591 | 2.5773 | 2.4923 | 2.4035 | 2.3099 | 2.2107 |

**Table A-6.** (continued)

α = 0.01

| $\nu_2$\\$\nu_1$ | 1 | 2 | 3 | 4 | 5 | 6 | 7 | 8 | 9 | 10 | 12 | 15 | 20 | 24 | 30 | 40 | 60 | 120 | ∞ |
|---|---|---|---|---|---|---|---|---|---|---|---|---|---|---|---|---|---|---|---|
| 25 | 7.7698 | 5.5680 | 4.6755 | 4.1774 | 3.8550 | 3.6272 | 3.4568 | 3.3239 | 3.2172 | 3.1294 | 2.9931 | 2.8502 | 2.6993 | 2.6203 | 2.5383 | 2.4530 | 2.3637 | 2.2695 | 2.1694 |
| 26 | 7.7213 | 5.5263 | 4.6366 | 4.1400 | 3.8183 | 3.5911 | 3.4210 | 3.2884 | 3.1818 | 3.0941 | 2.9579 | 2.8150 | 2.6640 | 2.5848 | 2.5026 | 2.4170 | 2.3273 | 2.2325 | 2.1315 |
| 27 | 7.6767 | 5.4881 | 4.6009 | 4.1056 | 3.7848 | 3.5580 | 3.3882 | 3.2558 | 3.1494 | 3.0618 | 2.9256 | 2.7827 | 2.6316 | 2.5522 | 2.4699 | 2.3840 | 2.2938 | 2.1984 | 2.0965 |
| 28 | 7.6356 | 5.4529 | 4.5681 | 4.0740 | 3.7539 | 3.5276 | 3.3581 | 3.2259 | 3.1195 | 3.0320 | 2.8959 | 2.7530 | 2.6017 | 2.5223 | 2.4397 | 2.3535 | 2.2629 | 2.1670 | 2.0642 |
| 29 | 7.5976 | 5.4205 | 4.5378 | 4.0449 | 3.7254 | 3.4995 | 3.3302 | 3.1982 | 3.0920 | 3.0045 | 2.8685 | 2.7256 | 2.5742 | 2.4946 | 2.4118 | 2.3253 | 2.2344 | 2.1378 | 2.0342 |
| 30 | 7.5625 | 5.3904 | 4.5097 | 4.0179 | 3.6990 | 3.4735 | 3.3045 | 3.1726 | 3.0665 | 2.9791 | 2.8431 | 2.7002 | 2.5487 | 2.4689 | 2.3860 | 2.2992 | 2.2079 | 2.1107 | 2.0062 |
| 40 | 7.3141 | 5.1785 | 4.3126 | 3.8283 | 3.5138 | 3.2910 | 3.1238 | 2.9930 | 2.8876 | 2.8005 | 2.6648 | 2.5216 | 2.3689 | 2.2880 | 2.2034 | 2.1142 | 2.0194 | 1.9172 | 1.8047 |
| 60 | 7.0771 | 4.9774 | 4.1259 | 3.6491 | 3.3389 | 3.1187 | 2.9530 | 2.8233 | 2.7185 | 2.6318 | 2.4961 | 2.3523 | 2.1978 | 2.1154 | 2.0285 | 1.9360 | 1.8363 | 1.7263 | 1.6006 |
| 120 | 6.8510 | 4.7865 | 3.9493 | 3.4796 | 3.1735 | 2.9559 | 2.7918 | 2.6629 | 2.5586 | 2.4721 | 2.3363 | 2.1915 | 2.0346 | 1.9500 | 1.8600 | 1.7628 | 1.6557 | 1.5330 | 1.3805 |
| ∞ | 6.6349 | 4.6052 | 3.7816 | 3.3192 | 3.0173 | 2.8020 | 2.6393 | 2.5113 | 2.4073 | 2.3209 | 2.1848 | 2.0385 | 1.8783 | 1.7908 | 1.6964 | 1.5923 | 1.4730 | 1.3146 | 1.0000 |
| 1 | 161.45 | 199.50 | 215.71 | 224.58 | 230.16 | 233.99 | 236.77 | 238.88 | 240.54 | 241.88 | 243.91 | 245.95 | 248.01 | 249.05 | 250.09 | 251.14 | 252.20 | 253.25 | 254.32 |
| 2 | 18.513 | 19.000 | 19.164 | 19.247 | 19.296 | 19.330 | 19.353 | 19.371 | 19.385 | 19.396 | 19.413 | 19.429 | 19.446 | 19.454 | 19.462 | 19.471 | 19.479 | 19.487 | 19.496 |
| 3 | 10.128 | 9.5521 | 9.2766 | 9.1172 | 9.0135 | 8.9406 | 8.8868 | 8.8452 | 8.8123 | 8.7855 | 8.7446 | 8.7029 | 8.6602 | 8.6385 | 8.6166 | 8.5944 | 8.5720 | 8.5494 | 8.5265 |
| 4 | 7.7086 | 6.9443 | 6.5914 | 6.3883 | 6.2560 | 6.1631 | 6.0942 | 6.0410 | 5.9988 | 5.9644 | 5.9117 | 5.8578 | 5.8025 | 5.7744 | 5.7459 | 5.7170 | 5.6878 | 5.6581 | 5.6281 |
| 5 | 6.6079 | 5.7861 | 5.4095 | 5.1922 | 5.0503 | 4.9503 | 4.8759 | 4.8183 | 4.7725 | 4.7351 | 4.6777 | 4.6188 | 4.5581 | 4.5272 | 4.4957 | 4.4638 | 4.4314 | 4.3984 | 4.3650 |
| 6 | 5.9874 | 5.1433 | 4.7571 | 4.5337 | 4.3874 | 4.2839 | 4.2066 | 4.1468 | 4.0990 | 4.0600 | 3.9999 | 3.9381 | 3.8742 | 3.8415 | 3.8082 | 3.7743 | 3.7398 | 3.7047 | 3.6688 |
| 7 | 5.5914 | 4.7374 | 4.3468 | 4.1203 | 3.9715 | 3.8660 | 3.7870 | 3.7257 | 3.6767 | 3.6365 | 3.5747 | 3.5108 | 3.4445 | 3.4105 | 3.3758 | 3.3404 | 3.3043 | 3.2674 | 3.2298 |
| 8 | 5.3177 | 4.4590 | 4.0662 | 3.8378 | 3.6875 | 3.5806 | 3.5005 | 3.4381 | 3.3881 | 3.3472 | 3.2840 | 3.2184 | 3.1503 | 3.1152 | 3.0794 | 3.0428 | 3.0053 | 2.9669 | 2.9276 |
| 9 | 5.1174 | 4.2565 | 3.8626 | 3.6331 | 3.4817 | 3.3738 | 3.2927 | 3.2296 | 3.1789 | 3.1373 | 3.0729 | 3.0061 | 2.9365 | 2.9005 | 2.8637 | 2.8259 | 2.7872 | 2.7475 | 2.7067 |

| | | | | | | | | | | | | | | | | | | | |
|---|---|---|---|---|---|---|---|---|---|---|---|---|---|---|---|---|---|---|---|
| 10 | 4.9646 | 4.1028 | 3.7083 | 3.4780 | 3.3258 | 3.2172 | 3.1355 | 3.0717 | 3.0204 | 2.9782 | 2.9130 | 2.8450 | 2.7740 | 2.7372 | 2.6996 | 2.6609 | 2.6211 | 2.5801 | 2.5379 |
| 11 | 4.8443 | 3.9823 | 3.5874 | 3.3567 | 3.2039 | 3.0946 | 3.0123 | 2.9480 | 2.8962 | 2.8536 | 2.7876 | 2.7186 | 2.6464 | 2.6090 | 2.5705 | 2.5309 | 2.4901 | 2.4480 | 2.4045 |
| 12 | 4.7472 | 3.8853 | 3.4903 | 3.2592 | 3.1059 | 2.9961 | 2.9134 | 2.8486 | 2.7964 | 2.7534 | 2.6866 | 2.6169 | 2.5436 | 2.5055 | 2.4663 | 2.4259 | 2.3842 | 2.3410 | 2.2962 |
| 13 | 4.6672 | 3.8056 | 3.4105 | 3.1791 | 3.0254 | 2.9153 | 2.8321 | 2.7669 | 2.7144 | 2.6710 | 2.6037 | 2.5331 | 2.4589 | 2.4202 | 2.3803 | 2.3392 | 2.2966 | 2.2524 | 2.2064 |
| 14 | 4.6001 | 3.7389 | 3.3439 | 3.1122 | 2.9582 | 2.8477 | 2.7642 | 2.6987 | 2.6458 | 2.6021 | 2.5342 | 2.4630 | 2.3879 | 2.3487 | 2.3082 | 2.2664 | 2.2230 | 2.1778 | 2.1307 |
| 15 | 4.5431 | 3.6823 | 3.2874 | 3.0556 | 2.9013 | 2.7905 | 2.7066 | 2.6408 | 2.5876 | 2.5437 | 2.4753 | 2.4035 | 2.3275 | 2.2878 | 2.2468 | 2.2043 | 2.1601 | 2.1141 | 2.0658 |
| 16 | 4.4940 | 3.6337 | 3.2389 | 3.0069 | 2.8524 | 2.7413 | 2.6572 | 2.5911 | 2.5377 | 2.4935 | 2.4247 | 2.3522 | 2.2756 | 2.2354 | 2.1938 | 2.1507 | 2.1058 | 2.0589 | 2.0096 |
| 17 | 4.4513 | 3.5915 | 3.1968 | 2.9647 | 2.8100 | 2.6987 | 2.6143 | 2.5480 | 2.4943 | 2.4499 | 2.3807 | 2.3077 | 2.2304 | 2.1898 | 2.1477 | 2.1040 | 2.0584 | 2.0107 | 1.9604 |
| 18 | 4.4139 | 3.5546 | 3.1599 | 2.9277 | 2.7729 | 2.6613 | 2.5767 | 2.5102 | 2.4563 | 2.4117 | 2.3421 | 2.2686 | 2.1906 | 2.1497 | 2.1071 | 2.0629 | 2.0166 | 1.9681 | 1.9168 |
| 19 | 4.3808 | 3.5219 | 3.1274 | 2.8951 | 2.7401 | 2.6283 | 2.5435 | 2.4768 | 2.4227 | 2.3779 | 2.3080 | 2.2341 | 2.1555 | 2.1141 | 2.0712 | 2.0264 | 1.9796 | 1.9302 | 1.8780 |
| 20 | 4.3513 | 3.4928 | 3.0984 | 2.8661 | 2.7109 | 2.5990 | 2.5140 | 2.4471 | 2.3928 | 2.3479 | 2.2776 | 2.2033 | 2.1242 | 2.0825 | 2.0391 | 1.9938 | 1.9464 | 1.8963 | 1.8432 |
| 21 | 4.3248 | 3.4668 | 3.0725 | 2.8401 | 2.6848 | 2.5727 | 2.4876 | 2.4205 | 2.3661 | 2.3210 | 2.2504 | 2.1757 | 2.0960 | 2.0540 | 2.0102 | 1.9645 | 1.9165 | 1.8657 | 1.8117 |
| 22 | 4.3009 | 3.4434 | 3.0491 | 2.8167 | 2.6613 | 2.5491 | 2.4638 | 2.3965 | 2.3419 | 2.2967 | 2.2258 | 2.1508 | 2.0707 | 2.0283 | 1.9842 | 1.9380 | 1.8895 | 1.8380 | 1.7831 |
| 23 | 4.2793 | 3.4221 | 3.0280 | 2.7955 | 2.6400 | 2.5277 | 2.4422 | 2.3748 | 2.3201 | 2.2747 | 2.2036 | 2.1282 | 2.0476 | 2.0050 | 1.9605 | 1.9139 | 1.8649 | 1.8128 | 1.7570 |
| 24 | 4.2597 | 3.4028 | 3.0088 | 2.7763 | 2.6207 | 2.5082 | 2.4226 | 2.3551 | 2.3002 | 2.2547 | 2.1834 | 2.1077 | 2.0267 | 1.9838 | 1.9390 | 1.8920 | 1.8424 | 1.7897 | 1.7331 |
| 25 | 4.2417 | 3.3852 | 2.9912 | 2.7587 | 2.6030 | 2.4904 | 2.4047 | 2.3371 | 2.2821 | 2.2365 | 2.1649 | 2.0889 | 2.0075 | 1.9643 | 1.9192 | 1.8718 | 1.8217 | 1.7684 | 1.7110 |
| 26 | 4.2252 | 3.3690 | 2.9751 | 2.7426 | 2.5868 | 2.4741 | 2.3883 | 2.3205 | 2.2655 | 2.2197 | 2.1479 | 2.0716 | 1.9898 | 1.9464 | 1.9010 | 1.8533 | 1.8027 | 1.7488 | 1.6906 |
| 27 | 4.2100 | 3.3541 | 2.9604 | 2.7278 | 2.5719 | 2.4591 | 2.3732 | 2.3053 | 2.2501 | 2.2043 | 2.1323 | 2.0558 | 1.9736 | 1.9299 | 1.8842 | 1.8361 | 1.7851 | 1.7307 | 1.6717 |
| 28 | 4.1960 | 3.3404 | 2.9467 | 2.7141 | 2.5581 | 2.4453 | 2.3593 | 2.2913 | 2.2360 | 2.1900 | 2.1179 | 2.0411 | 1.9586 | 1.9147 | 1.8687 | 1.8203 | 1.7689 | 1.7138 | 1.6541 |
| 29 | 4.1830 | 3.3277 | 2.9340 | 2.7014 | 2.5454 | 2.4324 | 2.3463 | 2.2782 | 2.2229 | 2.1768 | 2.1045 | 2.0275 | 1.9446 | 1.9005 | 1.8543 | 1.8055 | 1.7537 | 1.6981 | 1.6377 |
| 30 | 4.1709 | 3.3158 | 2.9223 | 2.6896 | 2.5336 | 2.4205 | 2.3343 | 2.2662 | 2.2107 | 2.1646 | 2.0921 | 2.0148 | 1.9317 | 1.8874 | 1.8409 | 1.7918 | 1.7396 | 1.6835 | 1.6223 |
| 40 | 4.0848 | 3.2317 | 2.8387 | 2.6060 | 2.4495 | 2.3359 | 2.2490 | 2.1802 | 2.1240 | 2.0772 | 2.0035 | 1.9245 | 1.8389 | 1.7929 | 1.7444 | 1.6928 | 1.6373 | 1.5766 | 1.5089 |
| 60 | 4.0012 | 3.1504 | 2.7581 | 2.5252 | 2.3683 | 2.2540 | 2.1665 | 2.0970 | 2.0401 | 1.9926 | 1.9174 | 1.8364 | 1.7480 | 1.7001 | 1.6491 | 1.5943 | 1.5343 | 1.4673 | 1.3893 |
| 120 | 3.9201 | 3.0718 | 2.6802 | 2.4472 | 2.2900 | 2.1750 | 2.0867 | 2.0164 | 1.9588 | 1.9105 | 1.8337 | 1.7505 | 1.6587 | 1.6084 | 1.5543 | 1.4952 | 1.4290 | 1.3519 | 1.2539 |
| ∞ | 3.8415 | 2.9957 | 2.6049 | 2.3719 | 2.2141 | 2.0986 | 2.0096 | 1.9384 | 1.8799 | 1.8307 | 1.7522 | 1.6664 | 1.5705 | 1.5173 | 1.4591 | 1.3940 | 1.3180 | 1.2214 | 1.0000 |

Credit: E. S. Pearson, Tables of percentage points of the inverted beta (F) distribution. Biometrika 32 (1943): 73–88.

## Table A-7.   Duncan's Multiple Ranges

The numbers given in this table are the values of $Q_p$ used to find $R_p = Q_p s_{\bar{x}}$. The value of $R_p$, then, is the shortest significant range for comparing the largest and smallest of $p$ means arranged in order of magnitude.

### (5% Level)

| Degrees of freedom | $p$ = number of means within range being tested | | | | | | | | | | | |
|---|---|---|---|---|---|---|---|---|---|---|---|---|
| | 2 | 3 | 4 | 5 | 6 | 7 | 8 | 9 | 10 | 20 | 50 | 100 |
| 1 | 18.00 | 18.00 | 18.00 | 18.00 | 18.00 | 18.00 | 18.00 | 18.00 | 18.00 | 18.00 | 18.00 | 18.00 |
| 2 | 6.08 | 6.08 | 6.08 | 6.08 | 6.08 | 6.08 | 6.08 | 6.08 | 6.08 | 6.08 | 6.08 | 6.08 |
| 3 | 4.50 | 4.52 | 4.52 | 4.52 | 4.52 | 4.52 | 4.52 | 4.52 | 4.52 | 4.52 | 4.52 | 4.52 |
| 4 | 3.93 | 4.01 | 4.03 | 4.03 | 4.03 | 4.03 | 4.03 | 4.03 | 4.03 | 4.03 | 4.03 | 4.03 |
| 5 | 3.64 | 3.75 | 3.80 | 3.81 | 3.81 | 3.81 | 3.81 | 3.81 | 3.81 | 3.81 | 3.81 | 3.81 |
| 6 | 3.46 | 3.59 | 3.65 | 3.68 | 3.69 | 3.70 | 3.70 | 3.70 | 3.70 | 3.70 | 3.70 | 3.70 |
| 7 | 3.34 | 3.48 | 3.55 | 3.59 | 3.61 | 3.62 | 3.63 | 3.63 | 3.63 | 3.63 | 3.63 | 3.63 |
| 8 | 3.26 | 3.40 | 3.48 | 3.52 | 3.55 | 3.57 | 3.58 | 3.58 | 3.58 | 3.58 | 3.58 | 3.58 |
| 9 | 3.20 | 3.34 | 3.42 | 3.47 | 3.50 | 3.52 | 3.54 | 3.54 | 3.55 | 3.55 | 3.55 | 3.55 |
| 10 | 3.15 | 3.29 | 3.38 | 3.43 | 3.46 | 3.49 | 3.50 | 3.52 | 3.52 | 3.53 | 3.53 | 3.53 |
| 11 | 3.11 | 3.26 | 3.34 | 3.40 | 3.44 | 3.46 | 3.48 | 3.49 | 3.50 | 3.51 | 3.51 | 3.51 |
| 12 | 3.08 | 3.22 | 3.31 | 3.37 | 3.41 | 3.44 | 3.46 | 3.47 | 3.48 | 3.50 | 3.50 | 3.50 |
| 13 | 3.06 | 3.20 | 3.29 | 3.35 | 3.39 | 3.42 | 3.44 | 3.46 | 3.47 | 3.49 | 3.49 | 3.49 |
| 14 | 3.03 | 3.18 | 3.27 | 3.33 | 3.37 | 3.40 | 3.43 | 3.44 | 3.46 | 3.48 | 3.48 | 3.48 |
| 15 | 3.01 | 3.16 | 3.25 | 3.31 | 3.36 | 3.39 | 3.41 | 3.43 | 3.45 | 3.48 | 3.48 | 3.48 |
| 16 | 3.00 | 3.14 | 3.24 | 3.30 | 3.34 | 3.38 | 3.40 | 3.42 | 3.44 | 3.48 | 3.48 | 3.48 |
| 17 | 2.98 | 3.13 | 3.22 | 3.28 | 3.33 | 3.37 | 3.39 | 3.41 | 3.43 | 3.48 | 3.48 | 3.48 |
| 18 | 2.97 | 3.12 | 3.21 | 3.27 | 3.32 | 3.36 | 3.38 | 3.40 | 3.42 | 3.47 | 3.47 | 3.47 |
| 19 | 2.96 | 3.11 | 3.20 | 3.26 | 3.31 | 3.35 | 3.38 | 3.40 | 3.42 | 3.47 | 3.47 | 3.47 |
| 20 | 2.95 | 3.10 | 3.19 | 3.26 | 3.30 | 3.34 | 3.37 | 3.39 | 3.41 | 3.47 | 3.47 | 3.47 |
| 30 | 2.89 | 3.04 | 3.13 | 3.20 | 3.25 | 3.29 | 3.32 | 3.35 | 3.37 | 3.47 | 3.49 | 3.49 |
| 40 | 2.86 | 3.01 | 3.10 | 3.17 | 3.22 | 3.27 | 3.30 | 3.33 | 3.35 | 3.47 | 3.50 | 3.50 |
| 60 | 2.83 | 2.98 | 3.07 | 3.14 | 3.20 | 3.24 | 3.28 | 3.31 | 3.33 | 3.47 | 3.54 | 3.54 |
| 120 | 2.80 | 2.95 | 3.04 | 3.12 | 3.17 | 3.22 | 3.25 | 3.29 | 3.31 | 3.47 | 3.58 | 3.60 |
| ∞ | 2.77 | 2.92 | 3.02 | 3.09 | 3.15 | 3.19 | 3.23 | 3.26 | 3.29 | 3.47 | 3.64 | 3.74 |

Credit: D. B. Duncan, *Biometrics* 11 (1955): 1–42 and modified by H. L. Harter, *Biometrics* 16 (1960): 671–685 and *Biometrics* 17 (1961): 321–324.

**Table A-7.**  (*continued*)

(1% Level)

| Degrees of freedom | $p$ = number of means within range being tested | | | | | | | | | | | |
|---|---|---|---|---|---|---|---|---|---|---|---|---|
| | 2 | 3 | 4 | 5 | 6 | 7 | 8 | 9 | 10 | 20 | 50 | 100 |
| 1 | 90.00 | 90.00 | 90.00 | 90.00 | 90.00 | 90.00 | 90.00 | 90.00 | 90.00 | 90.00 | 90.00 | 90.00 |
| 2 | 14.00 | 14.00 | 14.00 | 14.00 | 14.00 | 14.00 | 14.00 | 14.00 | 14.00 | 14.00 | 14.00 | 14.00 |
| 3 | 8.26 | 8.32 | 8.32 | 8.32 | 8.32 | 8.32 | 8.32 | 8.32 | 8.32 | 8.32 | 8.32 | 8.32 |
| 4 | 6.51 | 6.68 | 6.74 | 6.76 | 6.76 | 6.76 | 6.76 | 6.76 | 6.76 | 6.76 | 6.76 | 6.76 |
| 5 | 5.70 | 5.89 | 6.00 | 6.04 | 6.06 | 6.07 | 6.07 | 6.07 | 6.07 | 6.07 | 6.07 | 6.07 |
| 6 | 5.25 | 5.44 | 5.55 | 5.61 | 5.66 | 5.68 | 5.69 | 5.70 | 5.70 | 5.70 | 5.70 | 5.70 |
| 7 | 4.95 | 5.14 | 5.26 | 5.33 | 5.38 | 5.42 | 5.44 | 5.45 | 5.46 | 5.47 | 5.47 | 5.47 |
| 8 | 4.75 | 4.94 | 5.06 | 5.14 | 5.19 | 5.23 | 5.26 | 5.28 | 5.29 | 5.32 | 5.32 | 5.32 |
| 9 | 4.60 | 4.79 | 4.91 | 4.99 | 5.04 | 5.09 | 5.12 | 5.14 | 5.16 | 5.21 | 5.21 | 5.21 |
| 10 | 4.48 | 4.67 | 4.79 | 4.87 | 4.93 | 4.98 | 5.01 | 5.04 | 5.06 | 5.12 | 5.12 | 5.12 |
| 11 | 4.39 | 4.58 | 4.70 | 4.78 | 4.84 | 4.89 | 4.92 | 4.95 | 4.98 | 5.06 | 5.06 | 5.06 |
| 12 | 4.32 | 4.50 | 4.62 | 4.71 | 4.77 | 4.82 | 4.85 | 4.88 | 4.91 | 5.01 | 5.01 | 5.01 |
| 13 | 4.26 | 4.44 | 4.56 | 4.64 | 4.71 | 4.76 | 4.79 | 4.82 | 4.85 | 4.96 | 4.97 | 4.97 |
| 14 | 4.21 | 4.39 | 4.51 | 4.59 | 4.65 | 4.70 | 4.74 | 4.78 | 4.80 | 4.92 | 4.94 | 4.94 |
| 15 | 4.17 | 4.35 | 4.46 | 4.55 | 4.61 | 4.66 | 4.70 | 4.73 | 4.76 | 4.89 | 4.91 | 4.91 |
| 16 | 4.13 | 4.31 | 4.42 | 4.51 | 4.57 | 4.62 | 4.66 | 4.70 | 4.72 | 4.86 | 4.89 | 4.89 |
| 17 | 4.10 | 4.28 | 4.39 | 4.48 | 4.54 | 4.59 | 4.63 | 4.66 | 4.69 | 4.83 | 4.87 | 4.87 |
| 18 | 4.07 | 4.25 | 4.36 | 4.44 | 4.51 | 4.56 | 4.60 | 4.64 | 4.66 | 4.81 | 4.86 | 4.86 |
| 19 | 4.05 | 4.22 | 4.34 | 4.42 | 4.48 | 4.53 | 4.58 | 4.61 | 4.64 | 4.79 | 4.84 | 4.84 |
| 20 | 4.02 | 4.20 | 4.31 | 4.40 | 4.46 | 4.51 | 4.55 | 4.59 | 4.62 | 4.77 | 4.83 | 4.83 |
| 30 | 3.89 | 4.06 | 4.17 | 4.25 | 4.31 | 4.37 | 4.41 | 4.44 | 4.48 | 4.65 | 4.77 | 4.78 |
| 40 | 3.82 | 3.99 | 4.10 | 4.18 | 4.24 | 4.30 | 4.34 | 4.38 | 4.41 | 4.59 | 4.74 | 4.76 |
| 60 | 3.76 | 3.92 | 4.03 | 4.11 | 4.17 | 4.23 | 4.27 | 4.31 | 4.34 | 4.53 | 4.71 | 4.76 |
| 120 | 3.70 | 3.86 | 3.96 | 4.04 | 4.11 | 4.16 | 4.20 | 4.24 | 4.27 | 4.47 | 4.67 | 4.77 |
| $\infty$ | 3.64 | 3.80 | 3.90 | 3.98 | 4.04 | 4.09 | 4.14 | 4.17 | 4.20 | 4.41 | 4.64 | 4.78 |

## Table A-8.   Factors for Computing Control Chart Limits

| Observations in sample, $n$ | For averages $A_2$ | For $R$ chart | |
|---|---|---|---|
| | | $D_3$ | $D_4$ |
| 2 ........ | 1.880 | 0 | 3.267 |
| 3 ........ | 1.023 | 0 | 2.575 |
| 4 ........ | 0.729 | 0 | 2.282 |
| 5 ........ | 0.577 | 0 | 2.115 |
| 6 ........ | 0.483 | 0 | 2.004 |
| 7 ........ | 0.419 | 0.076 | 1.924 |
| 8 ........ | 0.373 | 0.136 | 1.864 |
| 9 ........ | 0.337 | 0.184 | 1.816 |
| 10 ........ | 0.308 | 0.223 | 1.777 |
| 11 ........ | 0.285 | 0.256 | 1.744 |
| 12 ........ | 0.266 | 0.284 | 1.716 |
| 13 ........ | 0.249 | 0.308 | 1.692 |
| 14 ........ | 0.235 | 0.329 | 1.671 |
| 15 ........ | 0.223 | 0.348 | 1.652 |
| 16 ........ | 0.212 | 0.364 | 1.636 |
| 17 ........ | 0.203 | 0.379 | 1.621 |
| 18 ........ | 0.194 | 0.392 | 1.608 |
| 19 ........ | 0.187 | 0.404 | 1.596 |
| 20 ........ | 0.180 | 0.414 | 1.586 |

**Table A-9.   Military Standard 105E Sample Size Code Letters**

| Lot or batch size | | Special inspection levels—Used for destructive tests or where sample sizes must be small | | | | General inspection levels—Confidence levels | | |
| --- | --- | --- | --- | --- | --- | --- | --- | --- |
| | | | | | | 90% | 95% | 99% |
| | | S-1 | S-2 | S-3 | S-4 | I | II | III |
| 2 to | 8 | A | A | A | A | A | A | B |
| 9 to | 15 | A | A | A | A | A | B | C |
| 16 to | 25 | A | A | B | B | B | C | D |
| 26 to | 50 | A | B | B | C | C | D | E |
| 51 to | 90 | B | B | C | C | C | E | F |
| 91 to | 150 | B | B | C | D | D | F | G |
| 151 to | 280 | B | C | D | E | E | G | H |
| 281 to | 500 | B | C | D | E | F | H | J |
| 501 to | 1200 | C | C | E | F | G | J | K |
| 1201 to | 3200 | C | D | E | G | H | K | L |
| 3201 to | 10000 | C | D | F | G | J | L | M |
| 10001 to | 35000 | C | D | F | H | K | M | N |
| 35001 to | 150000 | D | E | G | J | L | N | P |
| 150001 to | 500000 | D | E | G | J | M | P | Q |
| 500001 and over | | D | E | H | K | N | Q | R |

## Table A-10. Master Table for Normal Inspection (Single Sampling—MIL-STD-105E)

Acceptable quality levels (normal inspection). Each entry is given as "Ac Re" (acceptance number / rejection number).

| Sample size code letter | Sample size | 0.010 | 0.015 | 0.025 | 0.040 | 0.065 | 0.10 | 0.15 | 0.25 | 0.40 | 0.65 | 1.0 | 1.5 | 2.5 | 4.0 | 6.5 | 10 | 15 | 25 | 40 | 65 | 100 | 150 | 250 | 400 | 650 | 1000 |
|---|---|---|---|---|---|---|---|---|---|---|---|---|---|---|---|---|---|---|---|---|---|---|---|---|---|---|---|
| A | 2 | | | | | | | | | | | | | | | | ↓ | 0 1 | 1 2 | 2 3 | 3 4 | 5 6 | 7 8 | 10 11 | 14 15 | 21 22 | 30 31 |
| B | 3 | | | | | | | | | | | | | | | ↓ | 0 1 | 1 2 | 2 3 | 3 4 | 5 6 | 7 8 | 10 11 | 14 15 | 21 22 | 30 31 | 44 45 |
| C | 5 | | | | | | | | | | | | | | ↓ | 0 1 | 1 2 | 2 3 | 3 4 | 5 6 | 7 8 | 10 11 | 14 15 | 21 22 | 30 31 | 44 45 | ↑ |
| D | 8 | | | | | | | | | | | | | ↓ | 0 1 | 1 2 | 2 3 | 3 4 | 5 6 | 7 8 | 10 11 | 14 15 | 21 22 | 30 31 | 44 45 | ↑ | |
| E | 13 | | | | | | | | | | | | ↓ | 0 1 | 1 2 | 2 3 | 3 4 | 5 6 | 7 8 | 10 11 | 14 15 | 21 22 | 30 31 | 44 45 | ↑ | | |
| F | 20 | | | | | | | | | | | ↓ | 0 1 | 1 2 | 2 3 | 3 4 | 5 6 | 7 8 | 10 11 | 14 15 | 21 22 | 30 31 | 44 45 | ↑ | | | |
| G | 32 | | | | | | | | | | ↓ | 0 1 | 1 2 | 2 3 | 3 4 | 5 6 | 7 8 | 10 11 | 14 15 | 21 22 | 30 31 | 44 45 | ↑ | | | | |
| H | 50 | | | | | | | | | ↓ | 0 1 | 1 2 | 2 3 | 3 4 | 5 6 | 7 8 | 10 11 | 14 15 | 21 22 | 30 31 | 44 45 | ↑ | | | | | |
| J | 80 | | | | | | | | ↓ | 0 1 | 1 2 | 2 3 | 3 4 | 5 6 | 7 8 | 10 11 | 14 15 | 21 22 | 30 31 | 44 45 | ↑ | | | | | | |
| K | 125 | | | | | | | ↓ | 0 1 | 1 2 | 2 3 | 3 4 | 5 6 | 7 8 | 10 11 | 14 15 | 21 22 | 30 31 | 44 45 | ↑ | | | | | | | |
| L | 200 | | | | | | ↓ | 0 1 | 1 2 | 2 3 | 3 4 | 5 6 | 7 8 | 10 11 | 14 15 | 21 22 | 30 31 | 44 45 | ↑ | | | | | | | | |
| M | 315 | | | | | ↓ | 0 1 | 1 2 | 2 3 | 3 4 | 5 6 | 7 8 | 10 11 | 14 15 | 21 22 | 30 31 | 44 45 | ↑ | | | | | | | | | |
| N | 500 | | | | ↓ | 0 1 | 1 2 | 2 3 | 3 4 | 5 6 | 7 8 | 10 11 | 14 15 | 21 22 | 30 31 | 44 45 | ↑ | | | | | | | | | | |
| P | 800 | | | ↓ | 0 1 | 1 2 | 2 3 | 3 4 | 5 6 | 7 8 | 10 11 | 14 15 | 21 22 | 30 31 | 44 45 | ↑ | | | | | | | | | | | |
| Q | 1,250 | | ↓ | 0 1 | 1 2 | 2 3 | 3 4 | 5 6 | 7 8 | 10 11 | 14 15 | 21 22 | 30 31 | 44 45 | ↑ | | | | | | | | | | | | |
| R | 2,000 | ↓ | 0 1 | 1 2 | 2 3 | 3 4 | 5 6 | 7 8 | 10 11 | 14 15 | 21 22 | 30 31 | 44 45 | ↑ | | | | | | | | | | | | | |

↓ = use first sampling plan below arrow. If sample size equals, or exceeds, lot or batch size, do 100% inspection.
↑ = use first sampling plan above arrow.
Ac = acceptance number.
Re = rejection number.

**Table A-11. Percentage Points of the Studentized Range, $q = (x_n - x_1)/S_v$**

Lower 5% points

| $v$ \ $n$ | 2 | 3 | 4 | 5 | 6 | 7 | 8 | 9 | 10 |
|---|---|---|---|---|---|---|---|---|---|
| 1 | 18.00 | 27.00 | 32.80 | 37.10 | 40.40 | 43.10 | 45.40 | 47.10 | 49.10 |
| 2 | 6.09 | 8.30 | 9.80 | 10.90 | 11.70 | 12.40 | 13.00 | 13.50 | 14.00 |
| 3 | 4.50 | 5.91 | 6.82 | 7.50 | 8.04 | 8.48 | 8.85 | 9.18 | 9.46 |
| 4 | 3.93 | 5.04 | 5.76 | 6.29 | 6.71 | 7.05 | 7.35 | 7.60 | 7.83 |
| 5 | 3.64 | 4.60 | 5.22 | 5.67 | 6.03 | 6.33 | 6.58 | 6.80 | 6.99 |
| 6 | 3.46 | 4.34 | 4.90 | 5.31 | 5.83 | 5.89 | 6.12 | 6.32 | 6.49 |
| 7 | 3.34 | 4.16 | 4.68 | 5.06 | 5.36 | 5.61 | 5.82 | 6.00 | 6.16 |
| 8 | 3.26 | 4.04 | 4.53 | 4.89 | 5.17 | 5.40 | 5.60 | 5.77 | 5.92 |
| 9 | 3.20 | 3.95 | 4.42 | 4.76 | 5.02 | 5.24 | 5.43 | 5.60 | 5.74 |
| 10 | 3.15 | 3.88 | 4.33 | 4.65 | 4.91 | 5.12 | 5.30 | 5.46 | 5.60 |
| 11 | 3.11 | 3.82 | 4.26 | 4.57 | 4.82 | 5.03 | 5.20 | 5.35 | 5.49 |
| 12 | 3.08 | 3.77 | 4.20 | 4.51 | 4.75 | 4.95 | 5.12 | 5.27 | 5.40 |
| 13 | 3.06 | 3.73 | 4.15 | 4.45 | 4.69 | 4.88 | 5.05 | 5.19 | 5.32 |
| 14 | 3.03 | 3.70 | 4.11 | 4.41 | 4.64 | 4.83 | 4.99 | 5.13 | 5.25 |
| 15 | 3.01 | 3.67 | 4.08 | 4.37 | 4.60 | 4.78 | 4.94 | 5.08 | 5.20 |
| 20 | 2.95 | 3.58 | 3.96 | 4.23 | 4.45 | 4.62 | 4.77 | 4.90 | 5.01 |
| 40 | 2.86 | 3.44 | 3.79 | 4.04 | 4.23 | 4.39 | 4.52 | 4.63 | 4.74 |
| 60 | 2.83 | 3.40 | 3.74 | 3.98 | 4.16 | 4.31 | 4.44 | 4.55 | 4.65 |
| 120 | 2.80 | 3.36 | 3.69 | 3.92 | 4.10 | 4.24 | 4.36 | 4.48 | 4.56 |
| $\infty$ | 2.77 | 3.31 | 3.68 | 3.86 | 4.03 | 4.17 | 4.29 | 4.39 | 4.47 |

Upper 5% points

| | 2 | 3 | 4 | 5 | 6 | 7 | 8 | 9 | 10 |
|---|---|---|---|---|---|---|---|---|---|
| 10 | 0.09 | 0.75 | 1.20 | 1.52 | 1.74 | 1.91 | 2.05 | 2.17 | 2.26 |
| 11 | 0.09 | 0.75 | 1.21 | 1.52 | 1.75 | 1.92 | 2.07 | 2.18 | 2.28 |
| 12 | 0.09 | 0.75 | 1.21 | 1.53 | 1.76 | 1.93 | 2.08 | 2.20 | 2.30 |
| 13 | 0.09 | 0.75 | 1.22 | 1.53 | 1.76 | 1.94 | 2.09 | 2.21 | 2.31 |
| 14 | 0.09 | 0.75 | 1.22 | 1.54 | 1.77 | 1.95 | 2.10 | 2.22 | 2.33 |
| 15 | 0.09 | 0.75 | 1.22 | 1.54 | 1.77 | 1.95 | 2.11 | 2.23 | 2.34 |
| 16 | 0.09 | 0.75 | 1.22 | 1.54 | 1.78 | 1.96 | 2.11 | 2.24 | 2.34 |
| 17 | 0.09 | 0.75 | 1.22 | 1.55 | 1.78 | 1.97 | 2.12 | 2.25 | 2.35 |
| 18 | 0.09 | 0.75 | 1.22 | 1.55 | 1.79 | 1.97 | 2.12 | 2.25 | 2.36 |
| 19 | 0.09 | 0.75 | 1.23 | 1.55 | 1.79 | 1.98 | 2.13 | 2.26 | 2.37 |
| 20 | 0.09 | 0.75 | 1.23 | 1.55 | 1.79 | 1.98 | 2.13 | 2.27 | 2.37 |
| 40 | 0.09 | 0.76 | 1.24 | 1.57 | 1.82 | 2.02 | 2.18 | 2.32 | 2.43 |
| $\infty$ | 0.09 | 0.76 | 1.25 | 1.60 | 1.86 | 2.07 | 2.24 | 2.39 | 2.52 |

**Table A-12.   Table of $\alpha$ (Producer's Risk at 5%)**

| df \ k= | 2 | 3 | 4 | 5 | 6 | 8 | 10 | 15 | 20 | 30 |
|---|---|---|---|---|---|---|---|---|---|---|
| 6 | 1.73 | 2.59 | 2.94 | 3.19 | 3.37 | | | | | |
| 8 | 1.63 | 2.39 | 2.71 | 2.92 | 3.09 | 3.33 | | | | |
| 10 | 1.58 | 2.29 | 2.58 | 2.78 | 2.93 | 3.15 | 3.31 | | | |
| 15 | 1.51 | 2.16 | 2.42 | 2.60 | 2.74 | 2.93 | 3.07 | 3.32 | | |
| 20 | 1.48 | 2.10 | 2.35 | 2.52 | 2.64 | 2.83 | 2.96 | 3.18 | 3.33 | |
| 30 | 1.44 | 2.04 | 2.28 | 2.44 | 2.56 | 2.73 | 2.86 | 3.06 | 3.19 | 3.37 |
| 40 | 1.43 | 2.01 | 2.25 | 2.40 | 2.52 | 2.69 | 2.80 | 3.00 | 3.13 | 3.29 |
| 60 | 1.41 | 1.98 | 2.21 | 2.36 | 2.48 | 2.64 | 2.76 | 2.94 | 3.06 | 3.22 |
| 120 | 1.40 | 1.95 | 2.18 | 2.33 | 2.44 | 2.60 | 2.71 | 2.88 | 3.00 | 3.15 |
| $\infty$ | 1.39 | 1.93 | 2.15 | 2.29 | 2.40 | 2.55 | 2.65 | 2.82 | 2.94 | 3.08 |

$k$ = number of means being compared.
$df$ = degrees of freedom in estimate of error variance, $s_{\bar{x}}^2 = \hat{\sigma}_{\bar{x}}^2$.

# References

Alvarez, R. J. 1994. Managing total quality in foodservice. *Food Technology* 48(9): 140–143.

American Institute of Baking. 1988. *Quality assurance manual for food processors.* Marbottam. KS: AIB.

American National Standards Institute. 1981. *ANSLASQC ZI.4, Military Standard 105D. Sampling procedures and tables for inspection by attributes.* New York: American Standards Association.

American Society for Quality Control. 1986. *Food processing industry quality system guidelines.* Milwaukee, WI: ASQC.

American Society for Testing and Materials. 1968. *Basic Principles of sensory evaluation.* ASTM Special Technical Publication # 433. Philadelphia, PA: ASTM.

———, 1977. *Manual on sensory testing methods.* Philadelphia, PA: ASTM.

American Society for Quality Control. T15 1, 1986. *Food processing industry quality system guidelines.*

American Supplier Institute. 1990. *Total quality management* Dearborn, MI: American Supplier Institute.

ASM International (formerly American Society for Metals). *Metals handbook.* 8th ed. Vol. 11. Metals Park. OH: ASM International.

Boclens, and Haring 1982. Paper presented at the annual meeting of the Institute of Food Technologists.

Box. G. J. S. Hunter, and W. G. Hunter. 1978. *Statistics for experiments.* New York: Wiley.

Box. G., and N. R. Draper. 1969. *Evolutionary operation.* New York: Wiley.

Braverman, D. 1981. *Fundamentals of statistical quality control.*

Centers for Disease Control. 1981. *Foodborne disease outbreaks* Atlanta, GAL CDC.

Cianfrani, C., J. Tsiakals, and J. West. 2000. *ISO 9001:2000 Explained.* ISBN 0-87389-508-8. Quality Press.

Crosby, P. 1979. *Quality is free.* New York: McGraw-Hill.

Dambolenn, I., and A. Rao, 1994. What is six sigma anyway? *Quality.* 33(1): 10.

DataMyte Corporation. 1987. *DataMyte handbook.* Minretonka, MN: DataMyte Corporation.

Deming, S. N., and S. L. Morgan. 1988. *Experimental design: A cherometric approach* New York: Elsevier.

Deming, W. E. 1975. On some statistical aids toward economic production. *Interfaces* 5 (The Institute of Management Science) 4

———, 1982, *Quality, productivity, and competitive position*. Cambridge, MA: MIT Center for Advanced Engineering Study.

———, 1986, *Out of the crisis*, Cambridge, MA: MIT Press.

Dodge, H. F. 1943. Administration of a sampling inspection plan. *Industrial Quality Control* 5(3):12–19.

Feigenbaum, A. V. 1983. *Total quality control, engineering and management*. 3rd ed. New York: McGraw-Hill.

Food and Drug Administration. 1989. *Code of Federal Regulations 21, Section 101, 105*. Washington, DC: Government Printing Office.

Food Processors Institute. 1982. *Practical net weight control*. Washington, DC: FPI.

Ford Motor Company, 1985. *Continuing process control and process capability improvement*. Plymouth, MI: Ford Motor Company.

Freedman, W. 1984. *Products liability*, New York: Van Nostrand Reinhold.

Grant, E. L., and R. S. Leavenworth. 1988. *Statistical quality control*. 6th ed. New York: McGraw-Hill.

Grocery Manufacturers of America. 1974. *Guidelines for product recall*. Washington, DC: GMA.

Hansen, B. 1952. *Quality control theory and applications*. New York: Wiley.

Heckel, J. 1988. Built in quality at Greely. *Quality* 27:20.

Hertz, P. 1987. The Deming method, *Quality*, pp. Q-32-34.

Hradesky, J. L. 1988. *Productivity quality improvement*. New York: McGraw-Hill.

Hubbard, M. R. 1987. New quality control techniques for the food industry. *Hornblower* 39(5): 8–10.

Institute of Food Technologists, Sensory Evaluation Division, 1981 Sensory evaluation guide for testing food *Food Technology November*: 50.

Ishikawa, K. 1982. *Guide to quality control*. Tokyo: Asian Productivity Center.

———, 1985. *What is total quality control?* Englewood Cliffs, NJ: Prentice-Hall.

Juran, J. M. 1988. *Quality control handbook*. New York: McGraw-Hill.

Juran, J. M., and F. M. Gryna. 1970. *Quality planning and analysis*. 2nd ed. New York: McGraw-Hill.

Kennedy, C. W., and D. E. Andrews. 1987. *Inspection and gaging*. 5th ed. New York: Industrial Press.

McGonnagle, W. J. 1971. *Nondestructive testing*. New York: Gordon and Breech.

Merrill Lynch. 1995. *Global Investment Strategy*. March 21.

Mikel, H., and R. Schroeder. 2000. *Six sigma*. ISBN 0-385-49437-8. Doubleday.

Miller, J. C., and J. N. Miller. 1988. *Statistics for analytical chemistry*. New York: Wiley (Halsted).

National Institute of Standards and Technology (formerly National Bureau of Standards). 1988. *Handbook 133*. 3rd ed., with Supplements 1,2, and 3, dated 1990, 1991, and 1992. *Checking the net contents of packaged goods*. Washington, DC: Government Printing Office.

———, 1995. *Handbook 130, Uniform laws and regulations*. Washington, DC: Government Printing Office.

Ott, E. 1975. *Process and control: Trouble-shooting and interpretation of data*. New York: McGraw-Hill.

Pande, P., R. Neuman, and R. Cavanagh. 2000. *The six sigma way*, ISVN 0-07-135806-4. New York, McGraw-Hill.

Peters, T., and R. H. Waterman, Jr. 1982. *In search of excellence*. New York: Harper & Row.

Pierson, M. D., and D. A. Corlett. 1992. *HACCP principles and applications*. New York: Chapman & Hall.

*Quality Progress*. 1990. The tools of quality. June through December.

Rickmers, A. D., and H. N. Todd. 1967. *Statistics: An introduction*. New York: McGraw-Hill.

Schilling, E. G. *Statistical quality control*. 6th ed. New York: McGraw-Hill.

Tadikamalla, P. R. 1994. The confusion over six-sigma quality. *Quality Progress* 27(11): 83–91.

U.S. Department of Agriculture, Food Safety and Quality Service, 1988. *Code of Federal Regulations 9, Section 317.2*. Washington, DC: Government Printing Office.

U.S. General Accounting Office. 1991. *Management Practices*. GAO/NSLAD-91-190.

University of Iowa ASQC Section 1303. 1965. *Quality control training manual*. 2nd ed. Iowa City. IA: University of Iowa.

——. 1986. *Statistical quality control handbook*. Iowa City, IA: University of Iowa.

# Index

Acceptable Quality Level, 135, 150, 276
Acceptance testing, 39
Accuracy, 12, 26, 38, 43, 120, 121, 178, 192, 321
Administration, 12, 134, 152, 154, 259
Alpha risk, 154
Alternate hypothesis, 95–98
American Society for Quality, 28
Analysis of variance, 3, 105–113, 190, 191, 232–234, 248–291
  difference testing, 108, 109
  one-way classification, 193
ANOVA (see analysis of variance)
AQL, 135
Areas under the curve, 84–87
Artificial intelligence (see computer)
ASQC, 131, 149, 335
ASTM 335
Attribute
  charts, 52, 57–69, 178, 290
  sampling plans, 115–150
Audits, 17, 20, 26–35, 121, 143, 264, 274

Baldridge award, 27, 28, 290
Benchmarking, 168, 27, 173–175, 280, 287, 311, 354
Beta risk, 151
Binomial distribution, 72, 78–91, 315

C, (see Charts)
Calibration, 26, 33–38, 45, 169, 178, 264, 271, 290, 301
Capability
  defined, 157
  process, 36, 163–178, 286, 305

Capability *contd.*
  six-sigma, 7, 8, 16, 45, 170, 173, 309–313
  study, 158, 160
Cause
  Assignable, 3, 5, 164, 286
  common, 3, 163, 164
  special, 57, 67, 163
Cause and effect diagram, 9, 11, 290
Central tendency, 71, 309
Certification (see Vendor)
Charts
  as management tool, 184, 185
  attribute, 52–69, 179, 290
  bar, 13, 49, 180
  c, 57, 67–69, 178
  control, 9, 30–44, 49–70, 77–85, 120, 167–185, 248–302
  defectives, 59, 63, 65
  defects, 67–69
  m, 9, 57, 65
  mu, 52
  np, 52, 57, 59, 65, 67
  percent defective, 52, 57–65, 211, 215
  pie, 51
  R, 52–55
  three-dimensional, 49
  X-bar, 3, 10, 51–56, 178–183
Chart patterns
  abnormal uniform, 181
  excessive variability, 182
  high/low, 183, 184
  interpreting, 64, 205–211, 296
  normal, 58
  random cycles, 180
  recurring cycles, 181, 182

Chart patterns *contd.*
  trend, 9, 181
  uniform, 181
  wild outliers, 183
Check sheet, 9–13, 143, 290
Checkweigher, 120, 184, 208, 302, 303
Chi square distribution, 78
CIM. See computer
Compliance, 19, 20, 36, 169, 201, 204–218
Computer, 289–307
  activator, 301, 302
  advantages, 246, 288, 290, 296, 298
  artificial intelligence, 289, 293
  automatic control, 301, 302
  bar codes, 298
  controlled processing 289, 299–308
  dangers, 266, 313
  data acquisition board, 302
  data input, 297
  expert systems, 289–293
  feedback, 294
  integrated management, 6, 289
  sensing devices, 299, 301
  signal conditioner, 301
  software, 290, 295, 296, 301, 302
Consumer, 1, 16, 34–52, 101, 102, 132–136,
    141–149, 187–201, 242–247, 257–263
  Complaint, 10, 43, 44, 119, 134, 143, 152,
    174, 285
  risk, 97
Control, 177–186
  material, 5, 34, 187, 260
  test and measuring equipment, 3, 21, 25, 37,
    164, 271, 276
Control charts, 3, 49–70
  analysis (see design of experiments)
  as management tool, 184
  limits, 55
  interpreting, 64, 205–212, 296
Corrective action, 5, 6, 17, 18, 26, 27, 32, 38,
    42, 65, 139–141, 147, 179, 185, 283
Critical control points, 6, 10, 62, 138–146, 272
Crosby: 5, 255
Customer driven excellence, 44

Data control, 260
  input (see computer)
Defectives, 22, 52, 58–67, 77–81
Defects
  classes 132, 133, 259
  classification, 132–134
  critical few, 134
  trivial many, 134

Delivery, 21, 24–26, 126, 137, 159, 256, 263,
    266, 271, 272, 280, 304
Deming, 3, 4, 46, 164, 174, 278, 282, 309, 312,
    313
  cycle, (see Plan/Do/Check/Act)
Design
  assurance, 17
Design of experiments: 9, 219, 220, 286, 290
  central composite, 236, 237, 242, 243
  control chart analysis, 138, 248–250
  experimental variables, 222, 252
  external variables, 221
  extraneous variables, 221–223, 225
  fractional factorial, 227, 232–236, 290
  full factorial, 227–232
  goal, 4–8, 28–35, 40–42, 130–135, 173–179,
    183, 219–223, 272–277, 282–289
  incomplete block, 224, 225, 244
  interaction effect, 228–231
  Latin square, 220, 224
  mixture, 189, 193, 227, 239–247
  response surface, 9, 227, 236–239, 242,
    246, 247
  review, 17, 24
  star points, 237, 243
Difference testing, 108, 109, 193
Dispersion, 71–73, 92, 106, 107, 221
Distribution
  binomial, 77–83, 90, 91, 316
  difference between means, 100, 101, 109, 251
  F, 78, 104, 105, 325
  normal, 77, 81–85, 92, 97, 283, 290
  Poisson, 80, 81, 90
  sample means, 80–89, 105–107
  t, 16, 17, 92, 93, 100, 101, 117, 206,
    208, 322
Document control, 38, 200
Documentation, 18, 25, 27, 33, 34, 139, 145,
    147, 269, 270
Duncan's Multiple Range Test, 108, 109,
    251, 252

Education, 4, 18, 23, 264, 278, 312
Equipment, 3, 18–26, 30–67, 135–182,
    213–222, 259–268, 278–310
Evolutionary operations (see EVOP)
EVOP, 11, 222
Experiment. See Design of experiments.
Experimental error, 107, 223, 236
Expert systems (see computer)

F-distribution, 77, 78, 104, 105, 325, 327
Factorial design (see Design of experiments)

FDA, 6, 19, 20, 34, 36, 46, 134, 138–151, 159, 259, 271
Federal agencies, 15
Fiegenbaum, 4, 163, 226
Finished product: testing, 38, 39
Flow chart, 8–11, 22, 144–147, 283, 302, 303
Food processing industry quality system guide- lines, 15–18
Food quality systems, 15
Frequency, 5, 9, 13, 32, 36, 74, 75, 77, 82, 83, 96, 115, 120, 137, 138, 141, 142, 145, 149 189, 192, 267, 283–285, 290, 305

Goals, 4, 7, 16, 22, 28, 34, 35, 40–42, 57, 118, 135, 174, 201, 213, 221, 260, 262, 272, 277, 282, 283, 298, 312
Good manufacturing practice, 18, 20, 30, 36, 141, 148, 150, 201, 271, 272
Gross domestic product, 1

HAACP, 16, 138, 145–148, 283, 287, 337
Handbook
  packaged goods, 202, 204, 209–218
Handling and production functions, 17, 19, 25, 39, 62, 204, 226, 259, 268, 269, 290
Health and Human Services, 6, 201
Histogram, 9–13, 75–83, 90, 283, 290, 296
Hypothesis, 95–98, 102, 105, 290

Implementing quality program, 275–288
  hardware industry, 25–32
  in-house, 152, 192–197, 277, 287, 288
  management commitment, 5, 20, 22, 275
  PDCA cycle, 282
  professional training, 290
  teams, 20, 30, 266–284, 293, 298, 312
  training technicians, 287, 288
Imports, 1
Incomplete block, 224, 225, 244
Inspection, 4
  and test status, 24
  and testing, 24, 257
  equipment, 24
Internal audits, 26
Ishikawa, 11, 256–262
ISO 9000, 6, 17, 20, 26, 32, 272, 291

JIT, 265
Juran, 4, 163, 278, 337
Just in time (see JIT)

Latin square, 223–225
Leadership through quality, 28

Least significant difference, 124
Lot size, 58, 62, 136, 149

m charts, 57, 65
Malcolm Baldridge, 7, 27
Management:
  commitment, 5, 275–277
  control, 45
  responsibility, 20–22
Manual, 21, 22, 31, 32, 47, 287
  HACCP, 146–147
Manufacturing operations, 19
MAV, 202, 203, 214.
  (see also Net content control)
Mean, 71–74, 84–94, 100–111
Median, 71, 73
Methods and standards, 33
MilStd 105E, 132, 149, 332
Motivation, 47

Net content control, 201–218
  categories, 202
  evaluation, 205, 208, 217
  regulations, 201
  risk analysis 210, 212
  setting fill targets, 213–218
  (see also Handbook)
Non-conforming product, 20
Noncomformity, 20, 267
Normal distribution, 77, 61–86, 92, 97, 283
np, 57–59, 80–82, 315
Null hypothesis, 95, 96, 98, 101

Odor, 188, 189
Operations: proprietary, 37, 160
Operator control, 45

p charts, 52–65, 180, 184
Packaging supplier quality assurance, 30–35, 265–273
Paired observations, 100–103
Panels
  selection, 195–200
  training, 197–200
  types, 194–196
Parameter, 84
Pareto chart, 9, 12, 13
Partitioning sum of squares, 106–109
PDCA, 282
  (see Plan/Do/Check/Act)
Pearce, S, 17
Percent defective, 52, 57–65, 211, 215

Personnel, 15, 23–27, 36–47, 136–147, 191–193, 275–279, 293–295
Peters, 5
Pie charts, 51
Plan/Do/Check/Act, 28, 282, 313
Plant, 16, 117
Poisson distribution, 80–82
Policy, 30–32, 41, 45, 217, 218
Precision, 55, 221
Preventive action, 27
Probability, 26–29, 76, 77, 80–83, 95, 96
Process
  capability, 5, 157–176
    study, 36, 167, 168, 178
  control, 18, 177–186, 299–308
    (also see computer)
  drift, 116, 168, 169
  flow diagram, 139, 140, 282, 283
Processes, 18
Product
  Coding 38, 259, 271
  performance, 42
  research and development, 33
  safety and liability, 43
  verification, 20, 25, 37, 38
Production quality control, 36–38, 117
Purchased goods (see Vendor)
Purchasing
  materials, 17, 18, 62, 140–158, 183–185, 219–221, 257–273
  supplies, 10, 57–258, 263, 266

Quality
  authorities, 34
  control programs, 3–7
  control tools, 8–14
  documentation and records 40–43
  goal of R&D, 33
  goals, 4, 34, 40–42, 201, 221, 260, 262, 282
  improvement, 4–6, 26–31, 280–292
  improvement process, 33
  in marketing, 33
  in production, 17, 32, 36–39, 289
  leadership, 4–7, 21, 27, 28
  personnel, 36, 40, 46, 47, 184, 185, 222, 277
  policies, 4, 47
  programs, 5, 8
  proud program, 28
  records, 39, 40, 264
  strategies, 28
  system guidelines, 16–30, 45, 335
Quality: system, 15–48, 121, 279–281
Quality control
  manual, 20, 31, 32, 287
  production, 17, 28, 32, 36–41

R charts, 53–56, 183, 218
Randomized block, 222–224
Range
  control, 177–182
  Duncan multiple, 108, 109
  exaggerated, 72, 73
  procedure, 54–57
Raw materials, 257, 258
Recall, 39, 45, 46, 336
Records retention, 40
Records: quality, 39, 40
Relationships, 39, 40
Reports: progress, 34
Response surface (see Design of experiments)
Reviews, 22, 34, 45, 169
Risk
  alpha, 135, 136
  beta, 135
  consumer, 135
  minimum, 214
  producer, 135
  sampling, 135
  Type I, 135
  Type II, 135

Sampling, 115–150
  attribute sampling 149, 150
  cluster, 127, 128
  constant percentage, 121
  different distributions, 117
  double, 131, 150
  frequency, 120
  location, 5, 270, 287
  multiple, 131, 149, 150
  100 percent, 119, 261
  parameters, 115
  plans, 6, 9, 77, 80, 115, 128, 131, 135, 149, 150, 202–205, 272
  procedures, 116, 123, 131, 149, 259, 335
  risks, 135, 149
  selection, 128, 137, 138, 184
  simple random, 123–128
  single, 68, 131, 139, 177, 190, 191
  size, 32, 55, 64–68, 78, 84–86, 90–92, 106, 115, 118–124, 128–131, 136, 149, 150, 203, 248, 331
  spot checking, 121
  square root, 69, 73, 121–123, 172
  statistical 123
  stratified, 126, 127
  systematic 124, 125
  types, 128–131
Sanitary facilities, 18, 19
Sanitary operations, 17
Scatter diagram, 9–13

Sensors, 299–301
Sensory testing
  cold, 189
  general, 188–199
  heat, 19, 189
  odor, 188, 189
  pain, 190
  sight, 189
  sound, 189
  taste, 189
  touch, 189
  type, 191, 192
  (see also Panels)
Servicing, 7, 28
Shewhart, 3, 58
Software (see computer)
Specifications
  ingredients, 157–161
  process, 31, 160, 162, 177
  product, 157–161
SQC techniques, 10, 138
Standard deviation, 45, 72–75, 78, 83, 89, 92,
    100, 101, 104–106, 109, 167, 172,
    213–216, 252
Standards, 6, 16, 20, 33, 131, 132, 134, 149,
    154, 183, 194, 202, 258, 259, 311, 331,
    336
Star points (see design of experiments)
Statistics, 85, 336, 337
  methods, 3, 26, 153
  software (see computer)
  techniques, 9, 10, 287
Storage, 18, 19, 25, 39, 187, 193, 194
Supplier certification (see Vendor)
Systems: food quality, 15–48
  guidelines, 15–48

t-distribution, 78, 92, 93, 100, 101, 322
Taste, 188, 189, 192, 193, 198
Teams: 30, 277–284, 312
  quality systems, 28, 30
Test equipment: calibration, 37, 38
Testing
  basic library, 152
  equipment, 3, 25, 27, 38, 271
  methods, 33, 189–193, 335
  packaging, 266–270
  rapid analyses, 155
  references, 151–154, 335–337
  sensory 153–256, 187–199, 335, 336

Tools
  basic, 9–13
  selection, 8
Total quality
  commitment, 21, 22, 28
  management, 275
  service, 28
  systems, 28
TQM: definition, 28
Training, 4–6, 16–18, 21–37, 45–47, 168–177,
    195–199, 275–288, 291–298, 337
Two criteria classification, 111–114
Type I risk 135, 136
Type II risk, 136

Universal way, 28
Unpaired observations, 116
USDA, 34, 134, 138–145, 203, 271, 272
User contact, 16, 18
Utensils, 18

Variability, 2, 92, 134, 169, 182, 221
Variables
  design, 223
  examples, 129–131
  outcome, 220, 223, 233, 236, 240
  response, 221, 223, 227, 233, 239,
    241, 247
Variance 72, 73, 78, 105–113, 190, 191,
    232–234, 248–252, 291
Vendor
  certification, 256, 266
  Ishikawa, 11, 256, 262, 336
  JIT, 265
  materials specifications, 17, 34, 254,
    257–259
  quality, 253–274
  price, 263
  records, 270
  selection, 35, 263, 264
  single source, 265–273
  vendor-vendee relations, 255–266

Warehousing, 19, 146, 267
Weibull distribution, 78
Weight control (see Net content control)

X-Bar charts, 3, 52–56, 180–183, 218

Z, 83